REVIEW OF THE 21ST CENTURY TRUCK PARTNERSHIP, THIRD REPORT

Committee to Review the 21st Century Truck Partnership, Phase 3

Board on Energy and Environmental Systems

Division on Engineering and Physical Sciences

The National Academies of
SCIENCES · ENGINEERING · MEDICINE

THE NATIONAL ACADEMIES PRESS
Washington, DC
www.nap.edu

THE NATIONAL ACADEMIES PRESS • 500 Fifth Street, NW • Washington, DC 20001

NOTICE: The project that is the subject of this report was approved by the Governing Board of the National Research Council, whose members are drawn from the councils of the National Academy of Sciences, the National Academy of Engineering, and the Institute of Medicine. The members of the committee responsible for the report were chosen for their special competencies and with regard for appropriate balance.

This study was supported by Contract No. DE-EE0002931 between the National Academy of Sciences and the U.S. Department of Energy. Any opinions, findings, conclusions, or recommendations expressed in this publication do not necessarily reflect the views of any organization or agency that provided support for the project.

International Standard Book Number-13: 978-0-309-37710-2
International Standard Book Number-10: 0-309-37710-2

Additional copies of this report are available for sale from the National Academies Press, 500 Fifth Street, NW, Keck 360, Washington, DC 20001; (800) 624-6242 or (202) 334-3313; http://www.nap.edu.

Printed in the United States of America

Suggested citation: National Academies of Sciences, Engineering, and Medicine. 2015. *Review of the 21st Century Truck Partnership, Third Report*. Washington, DC: The National Academies Press.

The National Academies of

SCIENCES • ENGINEERING • MEDICINE

COMMITTEE TO REVIEW THE 21st CENTURY TRUCK PARTNERSHIP, PHASE 3

JOHN H. JOHNSON, Michigan Technological University, Houghton, Michigan, *Chair*
JULIE CHEN, University of Massachusetts-Lowell
DAVID FOSTER, University of Wisconsin Engine Research Center, Madison
THOMAS M. JAHNS, NAE,[1] University of Wisconsin-Madison
TIMOTHY V. JOHNSON, Corning Incorporated, Corning, New York
PAUL MENIG, Tech-I.M., Sherwood, Oregon
JAMES W. MORRIS, Volvo Powertrain NA (retired), Hagerstown, Maryland
THOMAS REINHART, Southwest Research Institute, San Antonio, Texas
BERNARD ROBERTSON, NAE, Chrysler Corporation (retired), Bloomfield Hills, Michigan
SUBHASH C. SINGHAL, NAE, Pacific Northwest National Laboratory (retired), Richland, Washington
JAMES A. SPEAROT, Mountain Ridgeline Consulting, LLC, Breckenridge, Colorado
KATHLEEN TAYLOR, NAE, General Motors Corporation (retired), Fort Myers, Florida
JOHN WOODROOFFE, University of Michigan Transportation Institute, Ann Arbor, Michigan

Staff

JAMES J. ZUCCHETTO, Senior Program Officer/Director, Board on Energy and Environmental Systems, Study Director
DANA CAINES, Financial Manager
LINDA CASOLA, Senior Program Assistant
ELIZABETH EULLER, Senior Program Assistant
LaNITA JONES, Administrative Coordinator
E. JONATHAN YANGER, Research Associate

[1] NAE, National Academy of Engineering.

BOARD ON ENERGY AND ENVIRONMENTAL SYSTEMS

ANDREW BROWN, NAE,[1] Delphi Corporation (retired), Troy, Michigan, *Chair*
DAVID ALLEN, University of Texas at Austin
W. TERRY BOSTON, NAE, PJM Interconnection, LLC, Audobon, Pennsylvania
WILLIAM BRINKMAN, NAS,[2] Princeton University, Princeton, New Jersey
EMILY CARTER, NAS, Princeton University, Princeton, New Jersey
CHRISTINE EHLIG-ECONOMIDES, NAE, Texas A&M University, College Station
DEBBIE NIEMEIER, University of California, Davis
MARGO OGE, Environmental Protection Agency (retired), McLean, Virginia
MICHAEL OPPENHEIMER, Princeton University, Princeton, New Jersey
JACKALYNE PFANNENSTIEL, Independent Consultant, Piedmont, California
DAN REICHER, Stanford University, Stanford, California
BERNARD ROBERTSON, NAE, DaimlerChrysler (retired), Bloomfield Hills, Michigan
DOROTHY ROBYN, Consultant, Washington, D.C.
GARY ROGERS, Roush Industries, Livonia, Michigan
ALISON SILVERSTEIN, Consultant, Pflugerville, Texas
MARK THIEMENS, NAS, University of California, San Diego
ADRIAN ZACCARIA, NAE, Bechtel Group, Inc. (retired), Frederick, Maryland
MARY LOU ZOBACK, NAS, Stanford University, Stanford, California

Staff

JAMES J. ZUCCHETTO, Director, Board on Energy and Environmental Systems
DANA CAINES, Financial Manager
LINDA CASOLA, Senior Program Assistant
ALAN CRANE, Senior Scientist
ELIZABETH EULLER, Senior Program Assistant
K. JOHN HOLMES, Associate Board Director
LaNITA JONES, Administrative Coordinator
MARTIN OFFUTT, Senior Program Officer
E. JONATHAN YANGER, Research Associate
ELIZABETH ZEITLER, Associate Program Officer

[1] NAE, National Academy of Engineering.
[2] NAS, National Academies of Sciences, Engineering, and Medicine.

Preface

This report is the third in a series of three by the National Academies of Sciences, Engineering, and Medicine[1] that have reviewed the research and development (R&D) projects carried out by the 21st Century Truck Partnership (21CTP), which was formed in 2001 to reduce fuel usage and emissions in trucks of Classes 3 through 8. The 21CTP has made significant progress since the Academies issued its first report in 2008. The early R&D was largely component-based, but, as a result of U.S. Department of Energy (DOE) American Recovery and Reinvestment Act (ARRA) funds in 2009, 21CTP was able to fund four SuperTruck projects, which combined all the component technology and aerodynamic improvements of the tractor and trailer into a Class 8 tractor-trailer to demonstrate and achieve the goal of 50 percent brake thermal efficiency (BTE) for the diesel engine in a cruise condition, while meeting the 2010 heavy-duty diesel emissions standards. One truck has achieved a freight efficiency of over 175 ton-miles per gallon, compared to a 2009 model baseline efficiency of 99 ton-miles per gallon. In terms of fuel economy, the truck achieved 10.7 miles per gallon (mpg), compared to the baseline truck at 6.45 mpg. As for load-specific fuel consumption (LSFC), the truck achieved 5.7 gallons/1,000 ton-miles, down 43 percent from the baseline LSFC of 10.0 gallons/1,000 ton-miles. A portion of the improvement on a ton-mile basis came from weight reduction, which allows extra freight to be carried.

A second truck has doubled fuel economy from a 2009 baseline of 6.1 mpg to 12.2 mpg over one long-haul route, with a 120 percent increase in freight efficiency in ton-miles per gallon from a 2009 baseline of 94 ton-miles per gallon to 206 ton-miles per gallon. LSFC was reduced by 55 percent on one route and by 49 percent on a second, lower speed route. On the route that produced the 12.2 mpg result, the LSFC was 4.85 gallons per 1,000 ton-miles, compared to 10.6 gallons per 1,000 ton-miles for the 2009 baseline.

The 50 percent BTE and the 4.85 and 5.7 gallons per 1,000 ton-miles LSFC values are significant accomplishments and could not have been achieved without the ARRA funds since the overall DOE budget in normal years was not sufficient to take on a project like this. The results of this R&D program will have an impact on reducing the demand for diesel fuel used in heavy-duty vehicles, which is projected to increase each year. The report makes a number of recommendations to further the R&D goals of the 21CTP in the next 5 years.

The committee appreciates the effort by the personnel from DOE, the Department of Transportation (DOT), the Environmental Protection Agency (EPA), the Department of Defense (DOD)-Army, and all the companies and national laboratories that prepared presentations and hosted our visits. The help of these members of the Partnership enabled us to get the latest data and information, which was very important for the committee's preparation of this report.

John H. Johnson, *Chair*
Committee to Review the 21st Century
Truck Partnership, Phase 3

[1] Effective July 1, 2015, the institution is called the National Academies of Sciences, Engineering, and Medicine. References in this report to the National Research Council are used in a historic context identifying programs prior to July 1.

Acknowledgments

The Committee to Review the 21st Century Truck Partnership, Phase 3, is grateful to the representatives of the 21st Century Truck Partnership, including the four government agencies—the U.S. Department of Energy, the U.S. Environmental Protection Agency, the U.S. Department of Transportation, and the U.S. Department of Defense-Army—and to the representatives of companies and national laboratories who contributed significantly of their time and effort to this National Academies of Sciences, Engineering, and Medicine[1] study by giving presentations at meetings or responding to committee requests for information, as well as hosting members of the committee at site visits. The committee also acknowledges the valuable contributions of other individuals who provided information and presentations at the committee's open meetings. Appendix B lists all of those presentations.

The committee offers its special appreciation to Ken Howden, director, 21st Century Truck Partnership, U.S. Department of Energy, Vehicle Technologies Office, for his significant contributions in coordinating responses to its questions and in making presentations to the committee, as well as Michael Laughlin, Energetics Incorporated, who assisted Ken in submitting data and information to the committee. Finally, the chairman wishes to recognize the committee members and the staff of the Board on Energy and Environmental Systems for organizing and planning the committee meetings and gathering information and drafting sections of the report. Jim Zucchetto in particular has done an outstanding job of facilitating the work of the committee and helping it to write a focused and timely report. Liz Euller provided efficient and very helpful support to its meetings, site visits, and report production.

This report has been reviewed in draft form by individuals chosen for their diverse perspectives and technical expertise, in accordance with procedures approved by the Report Review Committee. The purpose of this independent review is to provide candid and critical comments that will assist the institution in making its published report as sound as possible and to ensure that the report meets institutional standards for objectivity, evidence, and responsiveness to the study charge. The review comments and draft manuscript remain confidential to protect the integrity of the deliberative process. We wish to thank the following individuals for their review of this report:

Paul Blumberg (NAE), Ford Motor Company (retired),
William Brinkman (NAS), Princeton University,
Andrew Brown, Jr. (NAE), Delphi Corporation (retired),
Joseph Colucci (NAE), General Motors Company (retired),
Duke Drinkard, Southeastern Freight Lines,
David Merrion, Merrion Expert Consulting,
Margo Oge, U.S. Environmental Protection Agency (retired),
Craig Philip (NAE), Vanderbilt University,
Gary Rogers, Roush Industries, Inc.,
Dale Stein (NAE), Michigan Technological University,
R. Rhoads Stephenson, Technology Consultant, and
Michael Tunnell, American Transportation Research Institute.

Although the reviewers listed above have provided many constructive comments and suggestions, they were not asked to endorse the conclusions or recommendations, nor did they see the final draft of the report before its release. The review of this report was overseen by Douglas Chapin (NAE), who was responsible for making certain that an independent examination of this report was carried out in accordance with institutional procedures and that all review comments were carefully considered. Responsibility for the final content of this report rests entirely with the authoring committee and the institution.

[1] The National Research Council is now referred to as the National Academies of Sciences, Engineering, and Medicine.

Contents

APPENDIXES

Figures, Tables, and Boxes

FIGURES

TABLES

BOXES

Summary

BACKGROUND AND INTRODUCTION

This third review by the National Academies of Sciences, Engineering, and Medicine[1] Committee on Review of the 21st Century Truck Partnership (21CTP), Phase 3, hereinafter called the committee, follows on the Phase 1 and Phase 2 reviews by the National Research Council (NRC, 2008; 2012). The 21st Century Truck Partnership (21CTP—or, sometimes, "the Partnership") is a cooperative research and development (R&D) partnership made up of four federal agencies and 15 industrial partners.[2] The Partnership aims to "accelerate the introduction of advanced truck and bus technologies that use less fuel, have greater fuel diversity, operate more safely, are more reliable, meet future emissions standards, and are cost effective" (21CTP, 2013). It supports research, development, and demonstration (RD&D) that can lead to commercially viable products and systems. Its strategic approach includes (1) develop and implement an integrated vehicle systems R&D approach that validates and deploys advanced technology; (2) promote research on engines, combustion, exhaust aftertreatment, fuels, and advanced materials; (3) promote research on advanced hybrid propulsion systems; (4) promote research to reduce vehicle power demands; (5) promote the development of technologies to improve truck safety, (6) promote the development and deployment of technologies that substantially reduce energy consumption and exhaust emissions during idling; (7) promote the validation, demonstration, and deployment of advanced truck and bus technologies, and improve their reliability to the point where they can be adopted in the commercial marketplace; and (8) research, validate, and deploy technologies and methods that save fuel through more efficient operations of trucks and transportation systems, with an overall goal of better freight efficiency (21CTP, 2013).

The majority of the federal funding for RD&D projects supporting the goals of 21CTP comes from the DOE's Vehicle Technologies Office (VTO), with funds from the other three agencies as well. The American Recovery and Reinvestment Act of 2009 (ARRA, also known as the "stimulus") injected additional funding during the past few years in the SuperTruck program, supporting part of the four teams conducting R&D and integrating a variety of technologies into Class 8 tractor-trailer demonstration vehicles.

ASSESSMENT OF PROGRESS

With the focus of the Partnership on reducing fuel consumption and following on the NRC Phase 2 report (NRC, 2012) and the National Highway Traffic Safety Administration/Environmental Protection Agency (NHTSA/EPA) regulations that are being promulgated to reduce fuel consumption and greenhouse gas (GHG) emissions from medium- and heavy-duty vehicles (MHDVs), the Partnership has an important role to play in bringing together the government agencies and the private sector companies.[3] The Partnership is a means for facilitating communication among four government agencies, the national laboratories, and the private sector. Through regular meetings and exchanges of information on the various projects that are being funded, it seeks to avoid duplication of R&D efforts and identify industry needs for R&D projects, which then create a higher

[1] Effective July 1, 2015, the institution is called the National Academies of Sciences, Engineering, and Medicine. References in this report to the National Research Council are used in a historic context identifying programs prior to July 1.

[2] The agencies are the Department of Energy (DOE), Department of Transportation (DOT), Department of Defense (DOD), and U.S. Environmental Protection Agency (EPA). The 15 industrial partners are Allison Transmission, ArvinMeritor, BAE Systems, Caterpillar, Cummins Engine, Daimler Trucks North America (Freightliner and Detroit Diesel), Eaton Corporation, Honeywell International, Navistar, Mack Trucks, NovaBUS, Oshkosh Truck, PACCAR, and Volvo Trucks North America.

[3] NHTSA/EPA issued a Phase 1 regulation for model years 2014 to 2017 and issued a Notice of Proposed Rulemaking in June 2015 for a Phase 2 regulation covering model year 2018 and beyond.

likelihood that they will help the private sector develop products that can be commercially deployed and reduce fuel consumption, GHG emissions, and meet criteria pollutant standards. Although this is not a centrally directed program with a single-point authority over budgets and priorities, the Partnership has made good progress since the NRC Phase 1 and 2 reviews in improving communications, coordination and collaboration among the partners, documenting most of the projects and budgets under the 21CTP umbrella, and making some impressive technical progress in selected areas under its umbrella.

21CTP has become increasingly important in carrying out fuel consumption R&D as the federal government issues MHDV fuel consumption regulations. In addition, anticipated emission standards for oxides of nitrogen (NO_x) in California and possibly at the federal level may affect engine and emission control technologies that will be deployed. These regulatory measures regarding fuel consumption as well as emissions imply that the federal government has a role, and perhaps an increasing role, in the development of technologies to help the private sector achieve these policy goals and also to help U.S. firms remain competitive in the face of international competition. The Partnership plays an important role in bringing together, to the extent possible, a wide variety of different groups in a fairly fragmented trucking industry that does not have an overall organization that coordinates long-term R&D. As pointed out in the NRC Phase 1 report,

> very few U.S. manufacturers of trucks and buses or heavy-duty vehicle components have the R&D resources to develop new technologies individually. The 21CTP is giving some of those companies access to extraordinary expertise and equipment of the federal laboratories, in addition to seed funding that draws financial commitment from the companies to push forward in new technology areas (NRC, 2008).

A number of important accomplishments, which are addressed in more detail in the remainder of this Summary and in the report, have occurred since the earlier NRC reviews:

- The engine systems Goal 1 of a 50 percent brake thermal efficiency (BTE) for an emissions compliant engine has been achieved. A pathway to achieve 55 percent is being developed (Chapter 3).
- The four SuperTruck projects jointly funded by DOE and the private sector are impressive projects that integrate a wide variety of engine and vehicle technologies to significantly reduce the fuel consumption of Class 8 tractor-trailer vehicles, which consume the greatest part of the fuel used in the United States for heavy vehicles. These efforts follow on the recommendations in the NRC Phase 1 report for the full system integration of technologies and away from component-only R&D. These projects have brought together a wide variety of companies, the national laboratories, and universities (Chapter 8).
- The SuperTruck projects incorporated a number of vehicle power demand technologies that accounted for about 56 to 74 percent of the total fuel consumption reductions, with 26 to 44 percent coming from engine efficiency improvements (Chapter 8).
- Following on previous NRC recommendations, the DOE has proposed the development of an annual dedicated report on 21CTP activities and gave the committee a first draft proposal (Chapter 2).
- Hybrid vehicle systems have demonstrated significant fuel consumption and emissions reductions in a number of MHDV applications, but their cost prohibits commercial deployment, especially at foreseeable fuel prices. In addition, the SuperTruck project results thus far show a limited potential benefit on long-haul duty cycles for hybrid systems using currently available technology. The 21CTP hybrid team is considering a proposal to restructure its mission and focus, which the committee supports (Chapter 4).

The committee notes, however, that there are still remaining issues that the Partnership should continue to endeavor to address, some of which have been of concern in the Phase 1 and 2 reviews as well:

- The Partnership has identified particular areas to address but in some areas research funding has not been commensurate with the goals for those areas. In some cases, e.g., efficient operations or hybrid vehicles, funding has been insufficient to meet the goals. In those areas that have not received funding, adjustments to the goals should be made.
- The Partnership needs to develop an ongoing and systematic approach to identify which projects fall under the 21CTP umbrella and how they contribute to the Partnership's goals, as well as monitoring the results of the projects relative to the goals on an ongoing basis.
- The Partnership has yet to develop a brief annual report but, as noted above, is in the process of developing one.
- Previous reviews have suggested that additional truck manufacturers and suppliers be recruited for membership to the Partnership but the members have remained the same. With the changes occurring in the industry, this should be revisited.
- Given the expected constraints on future budgets, it will probably be increasingly important to identify the federal government's role after assessing both domestic and overseas heavy-duty vehicle R&D. Assessing overseas R&D was recommended in previous reviews but it is not clear whether this was ever conducted.

The technology integration efforts of the SuperTruck projects led to significant efforts by the project teams to address component R&D of engine idle reduction (Chapter 6) and vehicle power demands (Chapter 5, e.g., aerodynamics of the tractor and trailer, tire rolling resistance, friction reduction, weight reduction, and other approaches to reducing fuel consumption). Consequently, because they have been addressed in the SuperTruck projects, these areas have received much reduced funding through DOE for individual projects in these areas. Furthermore, the relatively new area of efficient operations (Chapter 9) has not received much emphasis because of lack of funding. The discussion of these areas is left to the individual chapters and not included in the Summary.

MANAGEMENT STRATEGY AND PRIORITY SETTING

Since the previous Phase 1 and 2 reviews, the Partnership has evolved in the face of changing budgets and new initiatives. The main leadership resides with the DOE's VTO, which manages a number of DOE-funded RD&D programs directly related to MHDV technologies. The other agencies simply bring their own existing programs that are relevant to the goals of the 21CTP under the 21CTP umbrella. The other complicating factor is that the budgetary aspects of the different agencies are all controlled by different committees in Congress. Consequently, the Partnership is unlike a traditional R&D program with central control and responsibility for budgets and priorities. DOE staff organize meetings and conference calls, maintain the information-flow infrastructure (such as websites and e-mail lists), and have led the discussions for and preparation of the updated 21CTP roadmap and white papers laying out Partnership goals, which was issued in February 2013 (21CTP, 2013). The management of individual projects under the 21CTP umbrella rests with the individual federal agencies that have funded the work. These agencies use the 21CTP information-sharing infrastructure to coordinate efforts and ensure that valuable R&D results are communicated and that overlap of activities is reduced. As was noted in the NRC Phase 2 report, the NRC's review of the overall 21CTP has helped to communicate to the various stakeholders and Congress the ongoing R&D efforts in the agencies and on the various projects (NRC, 2012).

The NRC Phase 2 review called for the preparation of a specific list of projects within each agency deemed to fall under the 21CTP umbrella, the associated line-item funding, and the overall budget for 21CTP. While DOE was able to provide this information for its own projects, the previous reviews were not able to secure this information from DOT, DOD or EPA. The situation has improved in this Phase 3 review: Led by DOE, the Partnership provided an inventory of projects categorized as falling under 21CTP, with the associated funding levels (see Appendix D and Figure 1-1 in Chapter 1).

Finding 2-1. The 21CTP remains a virtual organization facilitating communication among four government agencies, the national laboratories, and industry, led by DOE but with no single-point authority over its activities, priorities, or budgets. While far from optimal, this structure is necessitated by the separate reporting and budgeting mechanisms for each agency. Led by DOE, the Partnership has made good progress in adapting to this reality by improving communications, coordination, and collaboration among the partners, and documenting most of the projects and budgets under the 21CTP umbrella.

Recommendation 2-1. The DOE is urged to continue this improvement by maintaining and publishing the inventory of projects and budgets across all four agencies, tying those projects into the specific 21CTP goals and promoting the use of a portfolio management approach or the DOE's Office of Energy Efficiency and Renewable Energy's Project Management Center (EERE PMC) equivalent within the other agencies. Furthermore, EPA, DOT, and DOD should appoint a dedicated counterpart to DOE's designated 21CTP leader, who in turn should report directly to the director of the Vehicle Technologies Office on 21CTP matters.

Recommendation 2-2. The Partnership should develop and adopt criteria for including projects under the 21CTP umbrella, such as "Does the project clearly address one of the specific goals of 21CTP?" and "Does the project fall within the R&D interests of the member Partners of the 21CTP?" The committee recognizes that there will be at least two levels of projects—those tightly connected to specific 21CTP goals and a supporting set of projects that have a longer term impact. Better definition of the criteria for including a project and at what level would assist in evaluating and increasing the effectiveness of the Partnership.

ENGINE SYSTEMS

"Engine systems" comprises the engine, the aftertreatment, and the fuel as an interlinked system. Two 21CTP engine goals are focused on a significant increase in energy efficiency: (1) develop and demonstrate an emissions compliant engine system for Classes 7 and 8 highway trucks that achieves 50 percent BTE in an over-the-road cruise condition and (2) achieve 55 percent BTE in prototype engine systems in the laboratory. R&D on engine systems includes the projects funded by DOE and, in some cases, cost-shared with industry; these range from fundamental experimental work, to kinetic mechanism development, to mechanism evaluation and simplification, to development of advanced numerical methods, and to further development of the computational codes. The advanced combustion engines program is well managed and there is good collaboration and synergy among the DOE 21CTP individual engine projects. The SuperTruck engine projects were instrumental in meeting Goal 1, 50 percent BTE, and in carrying out the research to define a path to meeting Goal 2, 55 percent BTE. Of the federal funding for the SuperTruck teams, the amounts spent on diesel engine systems R&D to achieve the 50 and 55 percent BTE goals are approximately these: Volvo, $7.6 million; Navistar, $12.7 million; Daimler, $15.8 million; Cummins, $15.5 million.

Finding 3-1. The 21CTP has successfully met Goal 1, to develop and demonstrate an emissions-compliant diesel engine system for Class 7 and 8 highway trucks that achieves 50 percent brake thermal efficiency in an over-the-road cruise condition. The engine uses a waste heat recovery system.

Finding 3-2. The projects in the engine systems portion of 21CTP represent a closely coordinated set of research activities that are pursuing a better fundamental understanding of processes critical to efficient engine operation. Fundamentals associated with fuel injection, sprays, gas exchange, in-cylinder flows, advanced combustion processes, plus comprehensive yet robust kinetic routines for realistic fuels are being investigated. The learning from these activities is being incorporated into models, both detailed and phenomenological, that serve as tools for advanced engine development. Integral to this effort is the continued advancement of the base computer program itself and the solvers that facilitate rapid computational turnaround time. The program is well managed and interfaces well with industry stakeholders.

Recommendation 3-1. With the increased importance of advanced computational fluid dynamics (CFD) for developing the engines and operating scenarios necessary for minimum fuel consumption and in light of DOE's role in the generation of new knowledge that gets incorporated into these CFD codes as submodels, a critical review of the Partnership's program to develop the next-generation code (KIVA 4) should be performed. Feedback from participants in the high-performance computing workshop should be matched against the current code development activities, and the adequacy of the current program should be assessed. If necessary, the next-generation code development should be adjusted.

Finding 3-5. Achieving Goal 2, 55 percent BTE in a laboratory engine, will be very challenging. This is a high-risk, high-reward fundamental research program. It is an important stretch goal because it will facilitate identifying the potential of different advanced engine, fuel, and combustion concepts for increased engine efficiency, even though these concepts may not be commercially viable in the near future.

Recommendation 3-2. The fundamental diesel engine research program pursuing advanced technologies and combustion processes and engine architectures to achieve 55 percent BTE should continue to be a focus of the 21CTP engine activities. However, the experiments and modeling should maintain a focus on dynamometer R&D, as opposed to attempting to build a demonstration vehicle. The achievement of this goal should be extended from 2015 to 2020 in order to have sufficient time to carry out R&D on this stretch goal. Also, this activity should not be at the expense of efforts to reduce load-specific fuel consumption via system integration and road load reductions.

Aftertreatment

The considerable effort and research funding focused on improving diesel emission control systems is important to the development of the system and the engine in the vehicle relative to the system cost, weight, and volume.

Finding 3-6. The research agenda for 21CTP is focused on a wide diversity of heavy-duty emissions control work. There are impressive fundamental studies on selective catalytic reduction (SCR) catalysts, diesel particulate filter (DPF) fundamentals, low-temperature SCR and oxidation catalysts, passive NO_x adsorbers, multifunctional components, emissions measurement and modeling, system models, fuel effects, aging, and sensor development. These programs are delivering valuable results, but there are no program goals to guide future directions.

Recommendation 3-3. The DOE should develop specific aftertreatment goals for the 21CTP. These goals will serve as a focal point for researchers to submit proposals and for the DOE to assess them.

Recommendation 3-4. The Partnership should continue to fund work on improved SCR NO_x efficiency (mainly low-temperature efficiency without compromising high-temperature efficiency) and aging and poisoning effects. California's and, potentially, EPA's move toward further heavy-duty NO_x reductions to meet National Ambient Air Quality Standards for ozone will be critical. These new targets need to be set for the research efforts.

Finding 3-8. To achieve 50 percent BTE in the SuperTruck Program (Chapter 8), the engine compartment has limited space for the cooling system, the waste heat recovery system, and the aftertreatment system. The aftertreatment system volume, weight, and cost are important for the design of the engine compartment for trucks that are developed for 50-55 percent BTE.

Recommendation 3-6. Technologies such as an SCR catalyst on a DPF or others that have the potential to reduce the volume, weight, and cost of the aftertreatment system should be a part of the program to develop a 55 percent BTE engine.

Fuels

Finding 3-10. A series of fuels for advanced combustion engines (FACE) and surrogates have been identified in cooperation between the DOE and the Coordinating Research Council (CRC). These fuels have specific physical and chemical properties and are being used in several advanced combustion research programs, including the evaluation of various low-temperature combustion concepts, the development of CFD models for in-nozzle flow, spray formation, and combustion, and the development of new analytical techniques.

Recommendation 3-8. The DOE should continue to explore how the United States might use its abundant petroleum, natural gas, and biofuel resources in the most efficient manner. Studies, some of which are under way that contribute to this objective, should strive to answer the following questions:

1. What fuel properties (e.g., ignition characteristics, volatility, composition) of diesel fuel and gasoline provide for maximum efficiency of various advanced combustion engines? FACE and a common set of surrogate fuels should be utilized by all DOE facilities involved in combustion research programs in order to provide consistent fuel characteristics when evaluating laboratory experiments and engine test results.
2. Based on well-to-tank analyses, what fuel properties and processing procedures result in the lowest GHG emissions for hydrocarbon-based and bio-based fuel components?

HYBRID VEHICLES

Hybrid systems have demonstrated significant fuel consumption and emissions reductions in a number of MHDV applications, but cost effectiveness has been a barrier faced by many of today's hybrid drive manufacturers. As fuel consumption and GHG emissions standards become more stringent, however, there is a need for 21CTP to support the development of advanced technologies such as battery-electric and hybrid drives that will help meet these goals. The cost of hybrid drive equipment is not likely to fall sufficiently fast to meet commercially acceptable cost/benefit ratios in the near future. As a result, there is a need for 21CTP to support R&D that will in time lead to commercially viable hybrid drive technologies.

Finding 4-1. The 21CTP is considering a proposal to restructure its hybrid team so that it can work on drivetrain efficiency improvements, including other types of system integration opportunities that incorporate hybrid drive equipment.

Recommendation 4-1. The 21CTP hybrid team is encouraged to use this opportunity to redefine its mission in a manner that will lead to vehicle efficiency and emissions reduction improvements via a range of technology options, including promising opportunities for electrification and other types of innovative drivetrain improvements. During the course of this restructuring, the six R&D stretch goals developed in 2011 for the MHDV hybridization program should be redefined as part of the development of strategic objectives of the restructured advanced drivetrain initiative. At the conclusion of this process, the 21CTP leadership, working together with DOE and the other 21CTP partner federal agencies, should make a serious effort to secure funding to pursue whatever goals emerge so that they have a realistic chance of being achieved.

Finding 4-2. Several manufacturers have commercialized medium-duty hybrid trucks during the past several years and successfully demonstrated their ability to significantly reduce fuel consumption and emissions, particularly in vocational[4] and delivery truck applications. Despite this progress, the high cost of the hybrid drive train equipment and batteries combined with dropping prices for natural gas and oil have significantly retarded their market penetration in the United States. This has caused economic hardships for many hybrid truck manufacturers, causing a widespread reevaluation of the current hybrid truck business viability, at least in North America. At the same time, there is evidence that business opportunities for MHDV hybrid equipment are growing in other parts of the world, particularly in China, where government mandates are having a major impact.

Recommendation 4-2. Recognizing the advantages that hybridization can offer in trucks, 21CTP should support the development of new technology that offers promise for significantly improving the performance and cost-effectiveness of hybrid truck technology in the longer term. Project opportunities should be pursued to evaluate cost-effective vehicle electrification configurations for trucks, including hybrid drives with optimized component ratings to minimize their payback periods in different vehicle classes and applications. This future work should take advantage of technology advances originally made and commercialized for light-duty vehicles, including new battery technologies as well as opportunities for integrated microelectrification of truck functions such as start/stop operation, idle reduction, waste heat recovery, engine starting, and accessory electrification.

Finding 4-3. Although EPA and NHTSA have made considerable progress toward specifying the certification proce-

[4] Vocational vehicles cover a wide range of vehicles, including delivery trucks, dump trucks, cement trucks, buses, cranes, bucket trucks, and others. They are typically sold as an incomplete chassis with multiple "outfitters," such as an engine manufacturer, a body manufacturer, and an equipment manufacturer.

dures for fuel consumption and emissions in hybrid MHDVs, these procedures are still incomplete and imprecise in some important areas, particularly with regard to chassis dynamometer testing of complete hybrid MHDVs, and dynamometer testing of hybrid drivetrain power packs to determine their emissions and fuel consumption performance.

Recommendation 4-3. 21CTP should make it a priority to encourage EPA and NHTSA to accelerate their efforts to strengthen and finalize procedures for certifying the fuel consumption and emissions of hybrid MHDVs, including procedures for chassis dynamometer testing of complete hybrid vehicles and dynamometer testing of hybrid propulsion drivetrains alone. The 21CTP leadership is encouraged to work together with EPA and NHTSA to inform and educate the 21CTP stakeholders and the broader MHDV manufacturing community about the details of these procedures when they become available.

SAFETY

The 21CTP includes goals to ensure that advancements in truck design and technology to improve fuel efficiency do not have negative impacts on safety, and ensure that efforts to improve safety do not reduce efficiency.

Finding 7-3. The current generation of commercially available Forward Collision Avoidance and Mitigation (F-CAM) systems should reduce fatalities in truck-striking rear-end collisions by 24 percent, injuries by 25 percent, and property damage only crashes by 9 percent. Second- and third-generation versions of the systems will bring substantially greater benefits.[5]

Recommendation 7-3. 21CTP should assess future generation Forward Collision Avoidance and Mitigation (F-CAM) system development in order to identify barriers to development and establish incentives to foster commercialization.

SUPERTRUCK PROGRAM

An important and major component of the 21CTP during the past 5 years has been the SuperTruck program that was designed to reduce the fuel consumption of Class 8 long-haul tractor-trailer freight trucks. These SuperTruck vehicles are employing and integrating a wide range of technologies, many of which have been developed at the component or subsystem level under various 21CTP projects. The SuperTruck program aligns with the findings and recommendations set out in the NRC Phase 1 review (NRC, 2008). Four project teams have been awarded funding, with ARRA funding providing support to two of the teams (Cummins–Peterbilt and Daimler). With 50/50 cost sharing between government and industry, the total engine and vehicle funding for the project teams is $77.7 million for Cummins–Peterbilt; $79.1 million for Daimler Trucks North America; $76.2 million for Navistar; and $38 million for Volvo, for a total of about $284 million. Two teams, Cummins–Peterbilt and Daimler, aimed at completing their projects late 2014/early 2015 and their demonstration vehicles. The Navistar and Volvo teams will wind up in 2016. It should be emphasized that the teams have numerous companies, national laboratories, and universities working with them. With this funding for comprehensive demonstration vehicles incorporating many technologies, the teams are addressing areas such as engine efficiency, hybridization, aerodynamics, rolling resistance, idle reduction, and lightweight materials that prior to SuperTruck were only addressed through the core VTO projects.

The four project teams were awarded projects under the SuperTruck program and given the same basic targets, along with a requirement to maintain "comparable vehicle performance":

- Achieve 50 percent BTE from the engine at a cruise operation speed and load point,
- Demonstrate a path to 55 percent BTE from the engine, and
- Demonstrate a 50 percent increase in freight efficiency, measured in freight ton-miles per gallon, on a long-haul drive cycle.

In addition to these targets, the Cummins and Daimler teams added a target to measure the effectiveness of their auxiliary power unit (APU) systems, which handle hotel loads when the vehicle is parked: Demonstrate a 68 percent increase in freight efficiency on a 24-hour duty cycle (drive cycle plus overnight hotel load).

Finding 8-1. Overall, the committee finds the SuperTruck program to be a great success and finds that the system integration aspect of SuperTruck was a key to the program's success. The SuperTruck program drove technology development at a faster pace than industry would have achieved on its own. SuperTruck teams used the program to do the following:

- Increase both test and analysis capabilities, and improve the correlation between test and analysis;
- Use simulation results to drive improved experimental techniques, and use experimental results to help improve simulation techniques;
- Integrate combinations of technologies that had never been tested on a complete vehicle;
- Learn about opportunities, issues, and trade-offs with fuel saving technologies in real-world vehicle testing; and

[5] Second-generation systems will be able to detect stationary threat objects in the roadway through the fusion of radar and vision systems, while third-generation systems will have more aggressive automated braking deceleration, achieving 0.6 g.

- Understand the challenges that must be overcome in order to make certain technologies cost effective.

Finding 8-2. The Cummins and Daimler SuperTruck teams have met the goal of an engine with 50 percent brake thermal efficiency (BTE) at the cruise power point, and the other two teams are working to meet this goal. The Cummins and Daimler teams have also exceeded by a wide margin the goal of a 50 percent increase in freight efficiency (33 percent reduction in load-specific fuel consumption [LSFC]) over a long-haul drive cycle. The other two teams are working to meet or exceed the program goal in 2015 (Volvo) and early 2016 (Navistar).

Finding 8-3. The Cummins–Peterbilt SuperTruck team has comfortably exceeded a self-imposed goal of a 68 percent increase in freight efficiency (40.5 percent reduction in LSFC) over a 24-hour long-haul duty cycle. It achieved an 86 percent increase in freight efficiency (46 percent reduction in LSFC). The Daimler team demonstrated a 115 percent increase in freight efficiency (53.5 percent reduction in LSFC) on a different 24-hour duty cycle. This 24-hour goal does not apply to the Volvo program, and Navistar's status is to be determined.

Finding 8-6. Using the results available to date, about 26 to 44 percent of the total vehicle fuel savings are due to engine efficiency improvements, while about 56 to 74 percent are due to vehicle power demand reduction. In the Cummins–Peterbilt project, 42 percent of fuel savings are due to the engine and waste heat recovery, 14 percent to tractor aerodynamics, 28 percent to trailer aerodynamics, and 15 percent to tire and driveline improvements. In the Daimler SuperTruck, engine improvements account for 26 percent of the total fuel savings while 74 percent is a result of vehicle power demand reductions, including the effect of the hybrid system.

Finding 8-7. SuperTruck project results show a limited potential benefit on long-haul duty cycles for hybrid systems using currently available technology. Much of the benefit of a hybrid system can be captured with much less expensive and heavy alternatives, such as a GPS-based cruise control that uses the vehicle as a kinetic energy storage device. Microhybrid systems (smart control of auxiliary power demand, possibly combined with limited energy storage to handle auxiliary and/or hotel loads) may prove to be a more promising hybrid approach for long-haul trucks.

Finding 8-9. The SuperTruck vehicles incorporate technologies with a wide range of production readiness: Some will go into production soon; some will never become cost-effective with technology that is now known. The outstanding fuel savings achieved in this program thus need to be treated carefully. Actual production vehicles achieving SuperTruck fuel savings may not be cost-effective for several decades unless fuel costs increase substantially.

Recommendation 8-1. The SuperTruck demonstration vehicles represent a huge investment. DOE should consider ways of extracting additional research results from this investment by using the trucks that have been built to evaluate additional technologies. Some possibilities include these:

- Evaluation of additional technologies, such as microhybrid;
- Comparison of SuperTrucks on identical test cycles, with additional work to help understand any differences in performance;
- Vehicle evaluation of hardware resulting from future system or subsystem research projects;
- Exploration of a range of routes and payloads to determine the sensitivity of technologies to various applications.

Recommendation 8-2. Because of the great value demonstrated by the SuperTruck program, DOE should be working on at least one vehicle integration project at any given time. Owing to likely funding limitations, it will not be possible to have three or four similar projects running. A range of integration projects are possible, including these:

- A regional haul SuperTruck,
- A heavy-duty vocational SuperTruck (refuse, dump, etc.),
- A SuperTrailer program to help trailer manufacturers build engineering capability, and
- A delivery truck of Class 3, 4, 5, or 6.

Finding 8-12. Although it did not conduct a detailed safety analysis, the committee believes that it is unlikely that most of the efficiency technologies under consideration in the SuperTruck program will have a negative impact on safety.

Recommendation 8-5. It is important for the 21CTP, probably through DOT, to monitor and analyze in detail the technologies implemented in the SuperTruck projects to verify that they do not have a negative effect on safety, since one or more of the technologies may be considered for future production vehicles.

Finding 8-13. DOE is still using fuel economy (FE) in miles per gallon and freight efficiency in ton-miles per gallon for their fuel use metric, while the NHTSA regulations that were published 5 years ago use fuel consumption (FC) in gallons per 100 miles and load-specific fuel consumption (LSFC) in gallons per 1,000 ton-miles.

Recommendation 8-6. DOE should use FC and LSFC in its studies in order to be consistent with EPA/NHTSA regulations and to provide in the literature the percent improvements in magnitudes that relate to the metrics used in the regulations. Also, DOE needs to be a leader in changing the culture so that FC and LSFC become accepted metrics by industry.

REFERENCES

21CTP (21st Century Truck Partnership). 2013. Roadmap and Technical White Papers. U.S. Department of Energy Office of Energy Efficiency and Renewable Energy. https://www1.eere.energy.gov/vehiclesandfuels/pdfs/program/21ctp_roadmap_white_papers_2013.pdf.

NRC (National Research Council). 2008. *Review of the 21st Century Truck Partnership*. Washington, DC: The National Academies Press.

NRC. 2012. *Review of the 21st Century Truck Partnership, Second Report.* Washington, DC: The National Academies Press.

1

Introduction and Background

INTRODUCTION

In March 2014, the National Research Council (NRC) appointed the Committee on Review of the 21st Century Truck Partnership, Phase 3 (called the committee in this report) to conduct an independent review of the 21st Century Truck Partnership (21CTP) (see Appendix A for biographical information on committee members). The results of the committee's review are presented in this report. This NRC Phase 3 review follows on two previous reviews, the first conducted in 2007 resulting in the NRC Phase 1 report, issued in 2008 (NRC, 2008), and the second review, conducted in 2010-2011, resulting in the NRC Phase 2 report issued in 2012 (NRC, 2012). The Partnership's responses to the recommendations in the NRC Phase 2 report are contained in Appendix C of the present report. Given the extensive background presented in the NRC Phase 1 and Phase 2 reports and in other related reports, the committee refers to these other reports as appropriate.

The 21CTP is a cooperative research and development (R&D) partnership including four federal agencies—the U.S. Department of Energy (DOE); the U.S. Department of Transportation (DOT); the U.S. Department of Defense (DOD), more specifically the U.S. Department of the Army; and the U.S. Environmental Protection Agency (EPA)—and 15 industrial partners: Allison Transmission, ArvinMeritor, BAE Systems, Caterpillar, Cummins Inc., Daimler Trucks North America (which includes Freightliner and Detroit Diesel Corporation), Eaton Corporation, Honeywell International, Navistar, Mack Trucks, NovaBUS, Oshkosh Truck, PACCAR, and Volvo Trucks North America (see Figure 2-3 in Chapter 2 for the Partnership organization).[1] The Partnership was formed in 2000 and announced on April 21, 2001, at a press event in Romulus, Michigan.[2]

The Partnership is not a program in the formal sense of an R&D program managed by a director with lines of authority to its partners and a specific budget item appropriated by Congress. Rather, the Partnership is a means of exchanging information and coordinating ongoing activities that are occurring at the various agencies and private-sector companies to contribute to national goals of reducing fuel usage and emissions while improving heavy vehicle safety. (Chapter 2 addresses the organization of the Partnership and how it operates.) The 21CTP vision is "that our nation's trucks and buses will safely and cost-effectively move larger volumes of freight and greater numbers of passengers while emitting little or no pollution and dramatically reducing the dependency on foreign oil" (21CTP, 2013). The focus of the R&D is on medium- and heavy-duty vehicles (MHDVs), which range from Class 3 trucks greater than 10,000 lb to larger commercial trucks such as delivery vans, garbage trucks, and on up to tractor-trailer combinations ranging up to 80,000 lb or greater in some special cases (see Table 1-1 for different size categories; also NRC, 2010a, 2012).

The Partnership addresses the following "national imperatives":

> (a) Transportation in America supports the growth of our nation's economy both nationally and globally. (b) Our nation's transportation system supports the country's goal of energy security. (c) Transportation in our country is clean, safe, secure, and sustainable. (d) America's military has an agile, well-equipped, efficient force capable of rapid deployment and sustainment anywhere in the world. (e) Our nation's transportation system is compatible with a dedicated concern for the environment (21CTP, 2013).

[1] In this report, Daimler or Daimler Trucks will be used interchangeably with Daimler Trucks North America; Detroit Diesel will be used interchangeably with Detroit Diesel Corporation, and Volvo will be used interchangeably with Volvo Trucks North America.

[2] For further details of the history, see 21CTP (2006) and NRC (2000, 2008, 2012).

TABLE 1-1 Comparing Classes of Medium- and Heavy-Duty Vehicles

Class	Applications	Gross Weight Range (lb)	2006 Fleet Registrations (millions)	Typical Miles per Gallon Range 2007	Typical Fuel Consumed (gal/1,000 ton-mi)	Annual Fleet Fuel Consumption (billion gal)	Annual Fleet Miles Traveled 2006 (billion mi.)	Share of Annual Miles (%)	Share of Fuel Use (%)
2b	Large Pick-Up, Utility Van, Multi-Purpose, Mini-Bus, Step Van	8,501-10,000	6.2	10-15	38.5	5.5	93	29.5	11.6
3	Utility Van, Multi-Purpose, Mini-Bus, Step Van	10,001-14,000	0.69	8-13	33.3	1.46	12	3.8	3.1
4	City Delivery, Parcel Delivery, Large Walk-in, Bucket, Landscaping	14,001-16,000	0.29	7-12	23.8	0.53	4	1.3	1.1
5	City Delivery, Parcel Delivery, Large Walk-in, Bucket	16,001-19,500	0.17	6-12	25.6	0.26	2	0.6	0.5
6	City Delivery, School Bus, Large Walk-in, Bucket	19,501-26,000	1.71	5-12	20.4	6.02	41	13.0	12.7
7	City Bus, Furniture, Refrigerated, Refuse, Fuel Tanker, Dump, Tow, Concrete, Fire Engine, Tractor-Trailer	26,001-33,000	0.18	4-8	18.2	1.93	9	2.9	4.1
8a	Dump, Refuse, Concrete, Furniture, City Bus, Tow, Fire Engine (straight trucks)	33,001-80,000	0.43	2.5-6	8.7	3.51	12	3.8	7.4
8b	Tractor-Trailer: Van, Refrigerated, Bulk Tanker, Flat Bed (combination trucks)	33,001-80,000	1.72	4-7.5	6.5	28.1	142	45.1	59.4

NOTE: The *Transportation Energy Data Book* (Davis et al., 2013) estimates that in 2011 light trucks (gross vehicle weight <10,000 lb, including Classes 1, 2a, and 2b) used 7.24 quadrillion British thermal units (quads) of gasoline and 0.344 quads of diesel fuel; trucks of Classes 3-6 consumed 0.536 quads of gasoline and 0.727 quads of diesel; and Classes 7 and 8 combination trucks consumed 0.047 quads of gasoline and 4.468 quads of diesel.

SOURCE: Adapted from NRC (2010a, Table 2-1), with estimates based on data from 2006 and 2007. Note that total annual fleet miles amounts to about 315 billion miles and total fuel use 47.31 billion gallons.

This report builds on the NRC Phase 1 and 2 reviews and reports and also, as part of its charge, comments on changes and progress since the Phase 2 report was issued in 2012. The strategic approach of the Partnership includes the following elements as laid out in the 2013 21CTP roadmap and white papers, which evolved from the 2006 21CTP roadmap (21CTP, 2006, 2013):

- Develop and implement an integrated vehicle systems R&D approach that validates and deploys advanced technology necessary for both commercial and military trucks and buses to meet the aforementioned national imperatives.
- Conduct research for engines, powertrains, combustion, exhaust aftertreatment, fuels, and advanced materials to achieve both significantly higher efficiency and lower emissions.
- Conduct research focused on advanced heavy-duty hybrid propulsion and auxiliary power systems that will reduce energy consumption and pollutant emissions.
- Conduct research to reduce vehicle power demands (also referred to as parasitic losses) to achieve significantly reduced energy consumption.
- Support research on the development of technologies to improve truck safety, resulting in the reduction of fatalities and injuries in crashes involving trucks.

- Support research on the development and deployment of technologies that substantially reduce energy consumption and exhaust emissions during idling.
- Conduct the validation, demonstration, and deployment of advanced truck and bus technologies, and improve their reliability to the point where they can be adopted in the commercial marketplace.
- Research, validate, and deploy technologies and methods that save fuel through the more efficient operation of trucks and transportation systems, targeting an overall improved freight efficiency (DOE, 2013).

As is discussed in more detail in this report, the Partnership has been evolving and making some changes since the Phase 1 and 2 reviews. For example, since 2006 the roadmap and a series of white papers have been revised and updated.

NATIONAL CONCERNS

The federal government, including DOE, has addressed in varying degrees the economic, energy security, and environmental aspects of energy supply, distribution, and use for many decades, and the focus of efforts has changed from time to time. Supporting R&D for vehicle technologies that would reduce fuel consumption and emissions has been a cornerstone of federal R&D for decades and has complemented a number of National Highway Traffic Safety Administration and Environmental Protection Agency (NHTSA/EPA) regulations. Developing vehicle technologies to reduce fuel consumption helps to reduce demand for petroleum-derived gasoline and diesel fuel, which addresses concerns about energy security and U.S. dependence on petroleum imports, and addresses fuel affordability and price concerns by applying downward pressure on fuel demand. The United States has also implemented policies to replace petroleum-based fuels with fuels derived from domestic feedstocks, such as biofuels (NAS-NAE-NRC, 2009a,b; NRC, 2011). Concerns about air quality and the effects of pollutants on human health have led to a number of stringent regulations, significantly reducing exhaust emissions, such as oxides of nitrogen (NO_x) and particulates, for both light-duty vehicles (LDVs—e.g., cars, vans, and light trucks) and MHDVs. These regulations have stimulated the development of technologies to meet these regulations. In addition, concerns about climate and emissions of greenhouse gases (GHGs) from human activity have increased interest in developing and deploying vehicle technologies to reduce fuel consumption and GHG emissions such as carbon dioxide (CO_2).

Fuel Consumption

An extensive discussion of policy initiatives to reduce fuel consumption and emissions from vehicles can be found in previous NRC reports and will not be repeated here (NRC, 2008, 2012, 2014). LDVs have been regulated for decades with regard to both (1) emissions that contribute to air pollution and threaten health and (2) fuel economy from the standpoint of energy security. The Obama administration has been moving forward on regulatory measures to reduce GHG emissions as well as petroleum consumption and is focused on improving energy security concerns, with a goal of reducing oil imports by one third by 2025. Both increasing domestic production of petroleum and other fuels and improving the fuel efficiency of vehicles contribute to meeting this goal.[3] The most recent ruling on fuel economy for LDVs was promulgated in a combined fuel economy and GHG emissions rule by NHTSA and EPA; the rule calls for a GHG CO_2 level of 163 g/mi, which is equivalent to a 54.5 mile per gallon (mpg) corporate average fuel economy (CAFE) standard by 2025 (EPA/NHTSA, 2012).

Partially as a result of the promulgation of these LDV fuel economy regulations, the Energy Information Administration (EIA, 2014) forecasts in its reference case that energy consumption by LDVs in the United States will decline by an average of about 0.8 percent/year between 2012 and 2040, from about 8.41 million bbl/day (oil equivalent) to 6.38 million bbl/day, respectively. On the other hand, MHDVs, which in 2012 consumed about 25 percent of the petroleum used by on-road vehicles in the U.S. transportation sector, are expected to increase their fuel consumption by about 40 percent between 2012 and 2040, from about 2.8 million bbl/day to 3.91 million bbl/day. About 70 percent of the fuel used by MHDVs is used by Class 6 and Class 8 trucks, where diesel engines are the dominant technology (DOC, 2002; NRC, 2012). Table 1-1 provides estimates of the fuel consumed by various classes of trucks.[4] U.S. refineries have traditionally been set up to maximize gasoline output. However, with diesel fuel demand projected to increase while gasoline demand decreases, U.S. refineries will need to change to be able to supply the diesel fuel needed as the demand for gasoline decreases. It is only recently that MHDVs have been regulated in the United States with regard to fuel consumption, with a Phase I rule promulgated in 2011 covering vehicles beginning in model year 2014 and extending through model year 2018. The CO_2 standard in g/ton-mi for 2017 vocational vehicles is 6 to 9 percent below that for a model year 2010 vehicle and up to 23 percent below that for combination tractors and the engines installed in them.[5] A second rule for MHDVs is currently under development,

[3] P. Davis, DOE, "Vehicle Technologies Program Overview," Presentation to the Phase 3 committee on May 14, 2014.

[4] Note that for the global economy, commercial transport (trucks, ships, planes, and trains) energy demand is projected to grow by 70 percent between 2010 and 2040; an important component of this will be attributable to trucks (ExxonMobil, 2014). Worldwide diesel fuel use is projected to increase from 18 million bbl/day in 2014 to 30 million bbl/day in 2040 (IEA, 2014). Also, see http://www.dieselforum.org/news/the-global-fuel-forecast-is-sunny-for-diesel, November 18, 2014. Accessed December 8, 2014.

[5] Vocational vehicles cover a wide range of vehicles, including delivery trucks, dump trucks, cement trucks, buses, cranes, bucket trucks, and others. They are typically sold as an incomplete chassis, with multiple

and a Notice of Proposed Rulemaking was announced in June 2015 that will be enacted in 2016 and presumably lead to further reductions in fuel consumption for such vehicles beyond model year 2018.[6]

U.S. petroleum production today is greater than it has been in more than 25 years. During September 2014, daily U.S. oil production exceeded 8.8 million barrels per day, the most since early 1986 (Shenk, 2014). The EIA stated on September 14, 2014, that it expected daily oil production to reach over 9.5 million barrels per day in 2015, the most since 1970 (EIA, 2014). The greatest reason for this increased production has been the significant amounts of oil generated from shale deposits using advanced extraction techniques developed during the last several decades. As a result of this increased production, it is expected that petroleum-based fuels will continue to be the major source of transportation fuels well into the 21st century. Alternative fuel sources will contribute some portion of the transportation fuel pool, but petroleum-based diesel fuel, perhaps with modified properties or blended with other components, will be the primary energy source for heavy-duty trucks (particularly Classes 7 and 8). This increased U.S. energy production, together with improvements in vehicle fuel consumption, has improved the U.S. energy security position with regard to dependence on imports of petroleum. Increased domestic natural gas production and the associated sharp decline in natural gas prices have also stimulated interest in the use of natural gas in certain applications in the transportation sector, another trend that can contribute to improving energy security. Nonetheless, EIA, as noted above, forecasts a significant increase in fuel use by MHDVs, mostly diesel, in the next few decades, with natural gas playing a relatively small role as a percentage of the total transportation fuel consumption (see Table 1-2). Increasing use of biomass-based fuels has also helped with domestic production, but it remains to be seen to what extent cellulosic-based biofuels will contribute in the future.[7] EIA also forecasts that the U.S. net import share of petroleum and other liquids will decline from 2012 to 2040, but net expenditures (in constant 2012 dollars) for these will increase (Table 1-2). These trends, if they continue, will ameliorate U.S. dependence on imports but still represent significant expenditures and, given U.S. and global projections for increased fuel use by MHDVs, reducing the fuel consumption of MHDVs can help to further improve energy security.

Environmental Concerns

Added to the concern over imported petroleum and energy security is the concern about climate change. Nations around the world are beginning to exert more stringent control over human-made emissions, especially GHGs such as carbon dioxide (CO_2). The European Union aims to reduce GHG emissions by 2020 to levels 20 percent lower than in 1990, and the European Commission announced in May 2014 that it will develop a strategy to reduce CO_2 emissions from trucks, buses, and coaches. Numerous discussions have taken place in the U.S. Congress about climate change, and many pieces of climate change legislation have been proposed, although at present (2014 as this is being written) and on into 2015, it is unlikely that any major legislation would be forthcoming, albeit the NHTSA/EPA Phase 2 Notice of Proposed Rulemaking for MDHVs was announced in June 2015. In November 2014, according to the U.S.-China Joint Announcement on Climate Change, the United States "intends to achieve an economy-wide target of reducing its emissions by 26-28% below its 2005 level in 2025 and to make its best efforts to reduce its emissions by 28%."[8] The administration's regulations to decrease the fuel consumption of both LDVs and MHDVs are also aimed at reducing GHGs from the transportation sector. It is estimated that the transportation sector accounted for about 28 percent of the total anthropogenic CO_2-eq emissions in the U.S. economy in 2012 (EPA, 2014).[9] The total on-road emissions in 2012 from the use of gasoline, diesel, and alternative fuels are estimated to have been about 1.48 billion metric tons of CO_2-eq emissions. In 2012, MHDVs are estimated to account for about 0.41 billion metric tons of CO_2-eq emissions. Of these emissions, gasoline accounts for about 10 percent, diesel approximately 89.5 percent, and natural gas and liquid petroleum gas approximately 0.5 percent (EPA, 2014).

As for future trends, EIA (2014) forecasts that between 2012 and 2040, CO_2 emissions from the transportation

"outfitters"—such as an engine manufacturer, a body manufacturer, and an equipment manufacturer.

[6] See President Obama's National Fuel Efficiency Policy at http://www.whitehouse.gov/the-press-office/president-obama-directs-administration-create-first-ever-national-efficiency-and-em. A Notice of Proposed Rule Making for MDHVs was issued on October 26, 2010. Final standards issued by EPA and DOT's NHTSA on September 15, 2011, applied to model year 2014 (EPA/NHTSA, 2010, 2011). EPA/NHTSA issued a Notice of Proposed Rulemaking, Phase 2 vehicle fuel efficiency and greenhouse gas emissions, on June 19, 2015 (see http://www.nhtsa.gov/fueleconomy).

[7] Since the 1970s, Congress has supported legislation that requires increasing the production of fuels from renewable, bio-based sources and other alternative fuels as part of efforts to reduce petroleum-based fuel consumption. The Energy Independence and Security Act (EISA) of 2007 (Public Law 110-140) includes a subtitle that amended the Renewable Fuel Standard (RFS) contained in the Energy Policy Act of 2005 (EPAct 2005, Public Law 109-58) and substantially increased the volumes of renewable fuels to be phased in to the fuel supply. The mandated volumes of renewable fuels to be used begin with 9 billion gallons in 2008 and reach 36 billion gallons in 2022. These fuels are anticipated to include corn-based ethanol, cellulosic-based ethanol, and biodiesel made from vegetable oils (e.g., from soybeans), animal fats, and cellulose. Much R&D is occurring to develop, demonstrate, and commercialize the advanced biofuels that would be made from cellulose, but costs and technology performance are still uncertain (NAS-NAE-NRC, 2009b; NRC, 2011).

[8] See http://www.whitehouse.gov/the-press-office/2014/11/11/us-china-joint-announcement-climate-change. Accessed December 8, 2014.

[9] EPA considers emissions of the GHGs CO_2, methane (CH_4), and nitrous oxide (NO_2) and converts them to CO_2-equivalents (CO_2-eq), accounting for their different warming potentials in the atmosphere.

TABLE 1-2 Historical and Projected U.S. Energy Use and CO_2 Emissions by the Transportation Sector

Item	Units	2012	2040
Natural gas, use in transportation	quad	0.04	0.86
	million bbl oil equivalent	6.9	148.2
Total transportation energy use	quad	26.72	25.5
	million bbl oil equivalent	4,606	4,396
Net import share of petroleum and other liquids	percent[a]	40.3	32
	billions, constant 2012$	314	385
CO_2 emissions			
Transportation sector	million metric ton	1,815	1,700
Commercial light trucks[b]	million metric ton	35.6	35.6
Buses	million metric ton	16.1	15.8
Freight trucks	million metric ton	358	503

[a] See EIA (2014), table on petroleum and other liquid supply disposition.
[b] Gross vehicle weight of 8,500 to 10,000 lb.
SOURCE: EIA, 2014.

sector as a whole will decline, the result of decreased emissions from LDVs. Emissions from commercial light trucks (8,500 to 10,000 lb) remain about the same, buses decline somewhat, and freight trucks increase substantially because an increased quantity of freight will need to be moved as a consequence of the growing economy (see Table 1-2).

Emissions of oxides of nitrogen (NO_x) and particulate matter (PM) from heavy-duty vehicles have been significantly reduced by PM standards that went into effect in 2007 and NO_x standards that were phased in between 2007 and 2010.[10] In order to meet lower NO_x requirements, the trade-offs in engine and emissions control designs led to some decline in the brake thermal efficiency of diesel engines as NO_x standards phased in during the first decade of the 21st century. With NO_x requirements having been stabilized by 2010, engine manufacturers are now more focused on thermal efficiency improvements (NRC, 2012). It is uncertain when more stringent NO_x emission standards will be promulgated (e.g., by California) in the coming years and, if they are, how they may affect fuel consumption improvements.

Thus, for economic and environmental reasons and for energy security, the transportation sector is a key sector for consideration and a focus for policy, and MHDVs are a significant and increasingly important component. The 21CTP can play an important role is this regard. The public sector—through advanced R&D, and especially in partnering with the private sector, where the ultimate decisions will be made to deploy and commercialize new technology—is an important complement to regulatory and market-pull requirements. In this vein, the Partnership's fostering of technology that can reduce fuel consumption and emissions by MHDVs has gained in importance in recent years.

[10] A summary review of these emissions standards and changes can be found in the NRC Phase 1 and Phase 2 reports (NRC, 2008, 2012), as well as in references in this chapter (Ehlmann and Wolff, 2005; Johnson, 1988).

AREAS OF INTEREST AND LEVELS OF OVERALL FUNDING

As a means of providing focus and a set of goals and objectives for itself as a whole, the Partnership developed a roadmap and supporting technical white papers, which have evolved since 2006 (21CTP, 2006, 2010, 2011, 2013). The technical areas covered by the white papers include these:

(1) Engine systems;
(2) Advanced heavy-duty hybrid propulsion systems;
(3) Vehicle power demands (sometimes called parasitic losses, including, for example, losses due to aerodynamics, tire rolling resistance, and the like);
(4) Idle reduction;
(5) Vehicle safety;
(6) Operational efficiency, or efficient operations; and
(7) Additional infrastructure considerations, which were not included in previous versions of the roadmap.

These areas and the associated goals are discussed in further detail in the remaining chapters of this report. In addition, four major cost-shared contracts were awarded to four industry teams to carry out R&D and demonstrate for a complete long-haul tractor-trailer a freight efficiency improvement of 50 percent in ton-miles per gallon of fuel. These contracts were awarded under the SuperTruck program, which is a part of the 21CTP and which is addressed in Chapter 8.

DOE provides the central leadership for the 21CTP through its Vehicle Technologies Office (VTO), which is within DOE's Office of Energy Efficiency and Renewable

Energy (EERE). The VTO has the primary role in DOE for pursuing the development of advanced vehicle technologies both for LDVs and MHDVs. The LDV activities are carried out in the U.S.DRIVE (Driving, Research, and Innovation for Vehicle Efficiency and Energy Sustainability) partnership; the MHDV activities are carried out in the 21CTP. The U.S.DRIVE programs include work on combustion and emissions control, fuel cells, hydrogen storage, batteries, lightweight materials, power electronics, and vehicle systems. In terms of the baseline DOE budget, the LDV program activities during the past 10-15 years have been much larger than efforts directed toward MHDVs. However, there is some overlap between work that is done for LDVs and MHDVs—for example, in areas such as the understanding and modeling of advances in combustion, advances in lightweight materials, or advances in electrochemistry and battery technologies—and such overlapping areas are all managed under the VTO to support both LDV and MHDV technologies, as appropriate. Consequently, advances made in technical areas that are characterized and budgeted as part of the U.S.DRIVE could benefit MHDVs. DOE also contracts work out to the private sector and involves the 21CTP industry partners in cost-shared contracts, and it supports R&D in the national laboratories and universities. It also plays an important role by hosting its Annual Merit Review (AMR) at which all VTO projects are peer reviewed and evaluated. It has also traditionally hosted annually the Directions in Energy Efficiency and Emissions Research (DEER) conference, which brings together professionals in the engine community to share the latest advances in combustion engine R&D; however, the last DEER conference was held in 2012 (see Chapter 2 for further discussion).

The EPA has an interest in reducing emissions and works with the private sector and promotes and provides information on various technologies for the reduction of fuel consumption and of GHG emissions through its SmartWay program. DOD also is very interested in improving the fuel efficiency and reducing the fuel consumption of its noncombat vehicles; for combat vehicles it is interested in increased power density and low heat rejection. DOT is focused on safety issues, including the use of advanced technology and regulations that can improve highway safety, as well as on the overall system and infrastructure for moving freight efficiently and economically and on not compromising safety in order to reduce fuel consumption.

Since 21CTP is a partnership and not a formal program with a specific budget line item, it has been difficult to quantify the level of effort for the different areas of interest. While levels of funding are available for DOE's VTO, which is the lead organization for 21CTP, a distinction has to be made, as noted above, between efforts directed toward LDVs as opposed to MHDVs. In the Phase 1 and 2 reviews, DOE estimated what portion of VTO's budget was applicable to the 21CTP. Congress in its appropriations breaks its budget down not by LDVs and heavy-duty vehicles but by technical areas. Also of note is that the Congress now requires DOE to fund projects up front and not yearly, as was the practice in previous years; thus, for multiyear projects, DOE would have to fund the project completely at the beginning. The other agencies (DOD, DOT, EPA) associate their own existing programs or projects that are relevant to the goals of the 21CTP under the 21CTP umbrella, but budgets are not clearly associated with 21CTP, and during previous reviews there has been no specific budget information from these other agencies. The efforts of the private sector associated with 21CTP are also not available except for specific projects that individual companies may be jointly funding with a government agency—for example, DOE. The other factor that makes budgets and projects involved in the 21CTP unclear is that the different agencies receive their budget appropriations from different committees in Congress and are managed by the individual agency program managers.

The DOE estimated the part of the DOE VTO budget that could be attributed to 21CTP activities to be about $87 million in 2002. That amount then declined to about $46 million in 2010 as efforts on LDVs became a higher priority (NRC, 2012); note that the total VTO budgets in FY 2010 and FY 2013 were about $304 and $303 million, respectively. These past estimates for heavy-vehicle work did not appear to include work on energy storage (e.g., batteries), which could be associated with hybrid and electric drive technologies for MHDVs. At the time of this Phase 3 review, the budget request, $359 million, for the FY 2015 appropriations indicated a significant increase from 2013-2014 for VTO, but Congress appropriated only $280 million. What proportion of VTO's FY 2015 appropriations will be directed toward MHDVs remains uncertain. The other three agencies have their own, separate projects and budgets that can be associated with helping the Partnership to meet its goals.

Nevertheless, as recommended during the Phase 2 review, the 21CTP leadership, at the urging of the Phase 3 committee, put together a list of projects and associated funding that 21CTP estimates contribute to the 21CTP effort (see Appendix D). As far as the committee can tell, the projects are not all R&D projects, with some addressing demonstration, deployment, field testing, facilities, and the like. These are not strictly what the committee considered to be the focus of the technical areas in the statement of task, namely, R&D activities in areas directly relating to heavy-duty trucks. These estimates by the 21CTP leadership resulted in Figure 1-1, which depicts the estimated funding contributed by the four agencies. DOE estimated that about $116 million of FY 2014 funds for the four agencies are contributing to 21CTP goals (see Figure 1-1). These funding levels do not include funding from the American Recovery and Reinvestment Act of 2009 (ARRA, better known as "the stimulus") or efforts by the private sector.

The ARRA injected a significant amount of funding into activities, including R&D, on vehicles. Although this funding was a one-shot infusion and is not included as part of

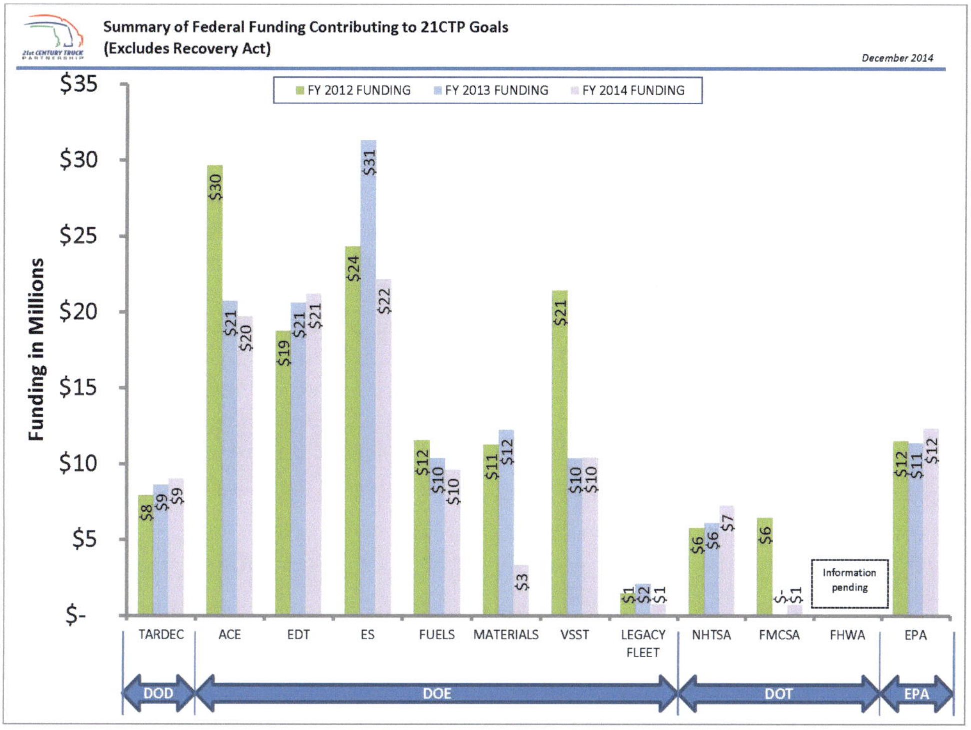

FIGURE 1-1 Summary of estimated federal funding contributing to 21CTP goals. Provided to the committee by the 21st Century Truck Partnership in December of 2014. ACE, advanced combustion engines; EDT, electric drive technologies; ES, energy storage; FHWA, Federal Highway Administration; FMCSA, Federal Motor Carrier Safety Administration; VSST, vehicle systems simulation and testing.

the congressional appropriations for the agencies, it allowed the initiation of a number of LDV and MHDV activities that helped to promote technologies for reducing fuel consumption. For example, approximately $1.5 billion was provided to accelerate the manufacturing and deployment of the next generation of U.S. batteries, $500 million to manufacture electric-drive components, and $400 million for transportation electrification. Such efforts, for example, can help to promote the more rapid development of battery technologies and help to stimulate the demonstration and deployment of hybrid vehicles. The 21CTP estimates that ARRA funding[11] contributed about $206 million to efforts supporting the 21CTP, including the SuperTruck projects.

ARRA funding also allowed a solicitation to be announced and funded called Systems Level Technology Development, Integration, and Demonstration for Efficient Class 8 Trucks (SuperTruck) and Advanced Technology Powertrains for Light-Duty Vehicles (ATP-LD). The heavy-vehicle part of this solicitation has a goal "to develop and demonstrate a 50-percent improvement in overall freight efficiency on a heavy-duty Class 8 tractor-trailer measured in ton-miles per gallon."[12] Four SuperTruck industry teams have been funded, generally about 5-year contracts, with ARRA funding of about $86 million contributing to two of the teams. The total funding for the four SuperTruck vehicle and engine projects is estimated at about $284 million, which includes ARRA funding, DOE funding, and private sector funding (see Chapter 8).

In summary, it is difficult to have a complete picture of the funding for the Partnership since there is not a budget item appropriated by Congress. The 21CTP leadership has estimated the level of effort by making judgments about which projects are associated with meeting 21CTP goals, but this exercise has not been complete for all four agencies (see Figure 1-1). The DOE estimated that the funding for its projects associated with 21CTP goals was about $87 million in FY 2002 but steadily declined to about $45 million in FY 2010 (NRC, 2008, 2012). In recent years, and with inclusion of some of the other agency projects, the most recent estimate for DOE, DOT, and EPA is about $116 million per year (Figure 1-1). The ARRA funding injected an additional $86 million of federal funding spread over 5 years or so for the SuperTruck projects, whereas the private sector contributed about $138 million. Since some of the SuperTruck projects were addressing vehicle power demands and engine idle reduction, funding for individual projects in those areas was reduced.

[11] P. Davis, DOE, "Vehicle Technologies Program Overview," Presentation to the Phase 2 committee on September 8, 2010.

[12] See http://www07.grants.gov/search/search.do?&mode=VIEW&flag2006=false&oppId=47867.

ORIGIN AND SCOPE OF THIS STUDY

In response to a request from the director of the DOE's Office of Vehicle Technologies, the NRC appointed the committee to fulfill the following statement of task:

(1) Review the high-level technical goals, targets, and timetables for R&D efforts, which address such areas as heavy vehicle systems; hybrid electric propulsion; advanced internal combustion engines (ICEs); and materials technologies.
(2) Review and evaluate progress and program directions since the inception of the Partnership towards meeting the Partnership's technical goals, and examine on-going research activities and their relevance to meeting the goals of the Partnership.
(3) Examine and comment on the overall balance and adequacy of the 21st Century Partnership's research effort, and the rate of progress, in light of the technical objectives and schedules for each of the major technology areas.
(4) Examine and comment, as necessary, on the appropriate role for federal involvement in the various technical areas under development.
(5) Examine and comment on the Partnership's strategy for accomplishing its goals, which might include such issues as (a) program management and organization; (b) the process for setting milestones, research directions, and making Go/No Go decisions; (c) collaborative activities within DOE, other government agencies, the private sector, universities, and others; and (d) other topics that the committee finds important to comment on related to the success of the program to meet its technical goals.
(6) Examine and comment on the response of the Partnership to the recommendations made in previous NRC reviews and reports of the 21st Century Truck Partnership.
(7) Write a report documenting its Phase 3 review of the 21st Century Truck Partnership with conclusions and recommendations.

The statement of task contains a number of standard elements that the NRC has used to review a number of DOE R&D programs since it is general enough to allow a committee to make an assessment either narrowly, broadly, or both, as appropriate. As noted in the Phase 2 report, in an ideal world, every technical area would have well-defined projects, budgets, milestones, and targets against which to assess progress. But in reality, given the multiagency and multi-industry nature of the 21CTP, the identification of such well-defined projects that can fall under the 21CTP umbrella is not uniform across the various areas and agencies (see Chapter 2). The Partnership has coalesced around six technical areas in its roadmap and has white papers and goals for each of those areas. In some instances there are precise targets against which to measure progress; in others there are not, and committee judgment has been used. The assessments of the committee are contained in the respective technical chapters, which correspond to the areas addressed by the white papers. In some cases, such as in hybrid propulsion, the review has been made complicated because the goals and targets have been undergoing revision. The SuperTruck projects are in various stages of completion, and the committee's comments on this important component of the 21CTP are in Chapter 8 for the results that are available. The situation is not dissimilar to that during the Phase 1 and 2 reviews, whose recommendations helped to focus some of the 21CTP efforts; the committee anticipates that the current report's recommendations also will help the Partnership with its focus over the next few years.

ROLE OF THE FEDERAL GOVERNMENT

The role of the federal government in R&D varies depending on the administration and the Congress and the issues that they deem important for the nation to address.[13]

An extensive economics literature on the subject points to the importance of R&D in promoting technical innovation, especially for the kinds of research where the private sector finds it difficult to capture the return on its investment; this is especially true for basic research, the results of which can be broadly used. Such innovation, if successful, can foster economic growth and productivity and lead to improvements in the standard of living (Bernanke, 2011). Furthermore, in the energy area, the government generally has to confront issues of national security, environmental quality, or energy affordability. Many of these issues are addressed through policy initiatives or regulations, which place a burden on private firms to achieve. Thus there is a role for the federal government in supporting R&D not only to help the private sector achieve these policy goals but also to help U.S. firms remain competitive in the face of international competition.

The committee believes that the federal government plays an important role in the development of technologies that can help to address government policies and regulations aimed at reducing emissions and fuel consumption from MHDVs. There are similar reasons for the government's playing a role in R&D for light-duty vehicles as well. The Partnership for a New Generation of Vehicles (PNGV), the FreedomCAR and Fuel Partnership, the U.S.DRIVE partnership, and the 21CTP are examples of public-private efforts to support R&D and to develop advanced technologies for vehicles (NRC, 2001, 2010a,b, 2013). These partnerships generally entail a variety of efforts (fundamental research, development, demonstration, and—in some cases—deployment).

[13] This section is repeated from the Phase 2 report since an important component of the Phase 3 committee's statement of task is the appropriate role of the federal government in the various technical areas and this view underlies a number of the committee's recommendations (NRC, 2012).

The federal government can support fundamental research through the national laboratories and universities, and industry can focus on development. The importance of having government–industry collaboration is that the private sector can help to transform improvements from research into cost-effective and marketable products. Generally, the contracting that is engaged in with the private sector is cost-shared, and those research contracts more closely associated with fundamental or basic research will have a majority of federal funding, whereas contracts with a strong development or product component will have significant support from the private sector. According to Section 988 of the EPAct of 2005, DOE-wide cost sharing requirements are 20 percent cost share for R&D, with an exemption for basic or fundamental R&D, and a 50 percent cost share for demonstration and commercial application activities (Public Law 109-58; also see Chapter 2). In its recommendations in each of the technical areas, the committee has considered which activities are most appropriate for the 21CTP to support. Implicit in all the recommendations that relate to the support of additional research is the committee's belief that the federal government has a role to play in the R&D.

STUDY PROCESS AND ORGANIZATION OF THE REPORT

The committee held meetings to collect information through presentations on 21CTP activities by representatives of the four federal agencies involved in the Partnership as well as individuals outside the program (see Appendix B for a list of the presenters and their topics). The committee reviewed the 21CTP roadmap and white papers, including a list of related projects and funding; submitted questions to the 21CTP leadership and received informative answers; and considered DOE's annual reports issued in the various technical areas. Subgroups of the committee also made site visits to the Cummins Technical Center in Columbus, Indiana; Daimler Corporate Facilities in Portland, Oregon; Daimler subsidiary Detroit Diesel in Detroit, Michigan; and Volvo in Greensboro, North Carolina for the SuperTruck projects. Committee subgroups also visited the Oak Ridge National Laboratory and the U.S. Army Tank Automotive Research, Development and Engineering Center (TARDEC) to understand the relationship between the Army and 21CTP R&D activities following on the DOE and DOD Advanced Vehicle Power Technology Alliance (AVPTA) partnership, entered into on July 18, 2011.[14] The committee also reviewed papers on the various DOE projects under 21CTP at the 2013 and 2014 DOE Annual Merit Review; in fact, some committee members attended and served as reviewers of the Annual Merit Review.[15] The committee was not in a position to review every project that the 21CTP said was associated with the Partnership, but based on the 21CTP presentations on various projects and on the committee's own review of projects presented at the Annual Merit Review, it believes it received sufficient information to make judgements on the activities associated with the various technical areas. The committee's findings and recommendations are based on the information gathered during the study and on the expertise and knowledge of committee members.

Chapter 2 addresses the overall management strategy and priority setting of the Partnership. Chapter 3 addresses work on engines and related activities on aftertreatment, fuels, and propulsion materials. Chapter 4 focuses on hybrid vehicles. Chapter 5 addresses vehicle power demands, including such areas as aerodynamics, tire rolling resistance, friction losses in the drivetrain, auxiliary loads, and weight reduction. Chapter 6 addresses idle reduction technologies for reducing fuel consumption and emissions during truck idle time. Chapter 7 addresses safety, which comes mostly under DOT. Chapter 8 addresses the four SuperTruck projects and Chapter 9 the area of efficient operations.

Appendix A presents biographical sketches of the committee members. Appendix B lists all of the public presentations at the committee's four meetings. Appendix C contains the list of findings and recommendations from the NRC Phase 2 report as well as the 21CTP responses to them. Appendix D is an inventory of 21CTP projects. Appendix E lists abbreviations and acronyms used in the report.

[14] See, for example, www.dtic.mil/cgi-bin/GetTRDoc?AD=ADA554222. Accessed March 6, 2015.

[15] The DOE Annual Merit Review papers can be accessed at http://energy.gov/eere/vehicles/vehicle-technologies-office-annual-merit-review-presentations.

REFERENCES

21CTP. 2006. 21st Century Truck Partnership Roadmap and Technical White Papers. 21CTP-003. December. Washington, D.C.: Office of FreedomCAR and Vehicle Technologies.

21CTP. 2010. Energy.gov. Office of Vehicle Technologies, Office of Energy Efficiency and Renewable Energy. http://www1.eere.energy.gov/vehiclesandfuels/about/partnerships/21centurytruck/21ct_goals.html. Accessed October 4, 2010.

21CTP. 2011. Updated 21st Century Truck Partnership Technical White Papers. Working draft, February 25. Washington, D.C.: Office of Vehicle Technologies.

21CTP. 2013. 21CTP Roadmap and Technical White Papers (February). Washington, D.C.: Office of Vehicle Technologies. http://energy.gov/eere/vehicles/downloads/roadmap-and-technical-white-papers-21st-century-truck-partnership. Accessed October 10, 2014.

Bernanke, B.S. 2011. Promoting research and development: The government's role. *Issues in Science and Technology*, a publication of the National Academies and the University of Texas-Dallas. Summer. http://www.issues.org/27.4/.

Davis, S.C., S.W. Diegel, and R. G. Boundy. 2013. *Transportation Energy Data Book*, Edition 32. Center for Transportation Analysis, Oak Ridge National Laboratory, Tenn. http://cta.ornl.gov/data.

DOC (U.S. Department of Commerce). 2002. *2002 Vehicle Inventory and Use Survey; Class 2B*. Oak Ridge National Laboratory, Tenn. March.

Ehlmann, J., and G. Wolff. 2005. *Automobile Emissions–The Road Toward Zero*. January. Pittsburgh, Pa.: Air and Waste Management Association.

EIA (Energy Information Administration). 2014. AEO 2014 [Annual Energy Outlook]. http://www.eia.doe.gov/forecasts/aeo/. Accessed September 30, 2014.

EPA (U.S. Environmental Protection Agency). 2014. Inventory of U.S. Greenhouse Gas Emissions and Sinks: 1990-2012. EPA 430-R-14-003. April 15. http://www.epa.gov/climatechange/ghgemissions/usinventoryreport.html#fullreport. Accessed September 30, 2014.

EPA/NHTSA (U.S. Environmental Protection Agency/National Highway Traffic Safety Administration). 2010. Greenhouse Gas Emissions Standards and Fuel Efficiency Standards for Medium- and Heavy-Duty Engines and Vehicles. EPA-HQ-OAR-2010-0162 and NHTSA-2010-0079, October 25. http://www.regulations.gov.

EPA/NHTSA. 2011. Greenhouse Gas Emissions Standards and Fuel Efficiency Standards for Medium- and Heavy-Duty Engines and Vehicles, Final Rules, September 25. http://www.nhtsa.gov/fuel-economy.

EPA/NHTSA. 2012. Final Rulemaking for 2017-2025 Light-Duty Vehicle Greenhouse Gas Emission Standards and Corporate Average Fuel Economy Standards. EPA-420-R-12-901. http://www.epa.gov/otaq/climate/regs-light-duty.htm.

ExxonMobil. 2014. The Outlook for Energy: A View to 2040. http://corporate.exxonmobil.com/en/energy/energy-outlook/charts. Accessed October 9, 2014.

IEA (International Energy Outlook). 2014. World Energy Outlook 2014. http://www.worldenergyoutlook.org/.

Johnson, J. 1988. Automotive Emissions. In: *Air Pollution, the Automobile, and Public Health.* Washington, D.C.: The National Academies Press.

NAS-NAE-NRC (National Academy of Sciences, National Academy of Engineering, National Research Council). 2009a. *America's Energy Future: Real Prospects for Energy Efficiency in the United States.* Washington, D.C.: The National Academies Press, pp. 39-75.

NAS-NAE-NRC. 2009b. *Liquid Transportation Fuels from Coal and Biomass: Technological Status, Costs, and Environmental Impacts.* Washington, D.C.: The National Academies Press.

NRC. 2000. *Review of the U.S. Department of Energy's Heavy Vehicle Technologies Program.* Washington, DC: The National Academies Press.

NRC. 2001. *Review of the Research Program of the Partnership for a New Generation of Vehicles.* Washington, D.C.: The National Academies Press.

NRC. 2008. *Review of the 21st Century Truck Partnership.* Washington, D.C.: The National Academies Press.

NRC. 2010a. *Technologies and Approaches to Reducing the Fuel Consumption of Medium- and Heavy-Duty Vehicles.* Washington, D.C.: The National Academies Press.

NRC. 2010b. *Review of the Research Program of the FreedomCAR and Fuel Partnership, Third Report.* Washington, D.C.: The National Academies Press.

NRC. 2011. *Renewable Fuel Standard: Potential Economic and Environmental Effects of U.S. Biofuel Policy.* Washington, D.C.: The National Academies Press.

NRC. 2012. *Review of the 21st Century Truck Partnership, Second Report.* Washington, D.C.: The National Academies Press.

NRC. 2013. *Review of the Research Program of the U.S. DRIVE Partnership, Fourth Report.* Washington, D.C.: The National Academies Press.

NRC. 2014. *Reducing the Fuel Consumption and Greenhouse Gas Emissions of Medium- and Heavy-Duty Vehicles, Phase Two: First Report.* Washington, D.C.: The National Academies Press.

Shenk, M. 2014. U.S. Oil Output Surges to Highest Since 1986 on Shale. Bloomberg, September 17. http://www.bloomberg.com/news/print/2014-09-17/u-s-oil-output-surges-to-highest-since-1986-on-shale.html.

2

Management Strategy and Priority Setting

INTRODUCTION

As part of its Phase 3 review of the 21st Century Truck Partnership (21CTP), the committee received presentations from the four participating agencies—Department of Energy (DOE), Department of Transportation (DOT), Department of Defense (DOD), and Environmental Protection Agency (EPA)—and the 21CTP industrial partners. These presentations included detailed responses to the concerns about the program's overall effectiveness, funding mechanisms, priority setting, Partnership coordination and performance, and other 21CTP issues raised in the National Research Council's Phase 1 (NRC, 2008) and Phase 2 (NRC, 2012) reports. The committee also collected information by reviewing documents and formulating questions to which the 21CTP provided detailed responses and by making site visits to several of the key partners. In addition, the 21CTP provided responses to the recommendations in the Phase 2 report (please see Appendix C).

In this chapter the committee reviews each of these areas of concern and reports its findings and recommendations. For background on the structure of the Partnership, the chapter also includes and summarizes relevant information from the NRC Phase 2 report (NRC, 2012).

PROGRAM MANAGEMENT

As noted in Chapter 1, overall management of the Partnership currently resides with the DOE's Vehicle Technologies Office—VTO, formerly the Office of Freedom CAR and Vehicle Technologies (FCVT)—in the Office of Energy Efficiency and Renewable Energy (EERE).

DOE personnel publish Partnership goals through white papers and roadmaps (21CTP, 2006, 2013), maintain the information-flow infrastructure, and organize meetings and conference calls. The management of individual projects under the 21CTP umbrella rests with the individual agencies that have funded the work. These agencies communicate with one another through the 21CTP information-sharing infrastructure in an attempt to coordinate their efforts and to ensure that valuable research results are shared and that any overlap of activities among their respective efforts is minimized.

Figure 2-1 illustrates the relations among the key participants in developing and conducting 21CTP research programs. Government agencies request funding from Congress through the administration and work with the industrial partners and research organizations, including universities and government laboratories, to establish research programs that meet national priorities and the interests of industry. However, final funding levels are determined by congressional appropriations, with each agency overseen by different

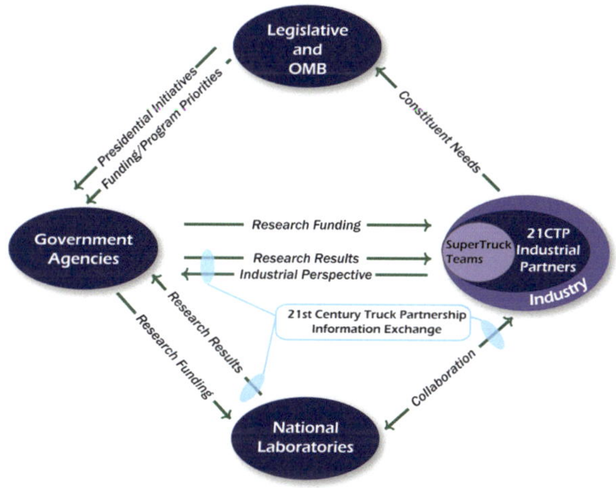

FIGURE 2-1 Relations between 21CTP participants. SOURCE: Submitted to the Committee on Review of the 21st Century Truck Partnership, Phase 2, by the DOE Office of Vehicle Technologies (January 29, 2011).

congressional committees. This makes prioritization of all of the 21CTP projects across the four agencies extremely difficult, if not impossible.

This limitation was discussed in the NRC Phase 1 and Phase 2 reports, and there are signs of some improvement in the coordination of projects in the subsequent 3 years, but the overall structure and funding of the government agency side of the Partnership remains a limitation to efficient program management.

In the NRC Phase 1 and 2 reviews, the previous committees asked for a specific list of projects within each agency deemed to fall under the 21CTP umbrella, the associated line-item funding, and overall agency budgets for 21CTP. While DOE was able to provide this information for its own projects, neither the Phase 1 nor the Phase 2 review was able to secure this information from DOT, DOD, or EPA.

Even in the case of DOE, light-duty and heavy-duty vehicle work often overlaps, in such areas as combustion or lightweight materials, for example. Accordingly, there is difficulty in parsing exactly which projects are part of 21CTP, although leveraging appropriate research activity across the entire vehicle spectrum, regardless of where the research program resides, is desirable wherever possible.

In addition, with the exception of the SuperTruck projects, it was difficult for the committee to ascertain the level of resources contributed by the private sector participants. Note, however, that DOE 21CTP activity (which accounts for 78 percent of total 21CTP funding) falls under the Energy Policy Act of 2005, which requires a 20 percent cost share for R&D projects and a 50 percent cost share for demonstration and commercial application projects (such as SuperTruck), so that the private sector's contribution is at least 20 percent in all such activities (DOE, n.d.).

The situation improved somewhat by the time of this Phase 3 review: Led by DOE, the Partnership provided an inventory of projects categorized as falling under 21CTP (although detailed EPA information was still missing[1]). It also provided the estimated associated funding levels (see Appendix D) and a partial summary of federal funding (excluding American Recovery and Reinvestment Act [ARRA] funds) contributing to 21CTP goals (see Figure 1-1 in Chapter 1).

In DOE vehicle research, which specifically addresses the national issue of energy security and the increasing pressures from the rising global consumption of oil, the VTO has involved the affected industries in planning the research agenda and identifying technical goals that, if met, will provide the basis for commercialization decisions. The government's approach is intended to allow industry-wide collaboration in precompetitive research, which is then followed by competition in the marketplace.

[1] EPA data were presented as simply two line items, of which the larger, approximately $10 million, was that portion of the EPA budget estimated to relate to heavy-duty vehicle testing and certification, and not R&D.

The Partnership provides a forum for the exchange of technical information among the industry and government partners involved in heavy-duty transportation. While this exchange of information provides an opportunity for coordination of relevant initiatives, what it does not, and cannot, do is provide single-point management or direction of such activities, or enable a "combined portfolio approach" to managing collaborative research efforts.

The four agencies involved in 21CTP have many areas of common interest, illustrated in Figure 2-2.

The Partnership points to examples of successful collaboration between agencies, such as the Advanced Vehicle Power Technology Alliance (AVPTA) between DOD and DOE, work between DOT, EPA, and DOE on the relationship between heavy-truck fuel efficiency regulations and 21CTP research, and collaboration between EPA and DOE on hydraulic hybrid research (since dropped) and the EPA SmartWay Transport Partnership.

The overall management structure of the Partnership remains largely unchanged since its inception in 2001 as a virtual network (see Figure 2-3). The industry side is led by an executive committee comprising one representative from each of the three major industry sectors involved: engines, truck original equipment manufacturers (OEMs), and hybrids. At the time of this review, those representatives were from Cummins, Volvo, and Eaton.

In Finding S-1 of the NRC Phase 2 report, it was noted that the Partnership needs to review whether additional partners—such as major truck and component manufacturers that are not currently members but that could contribute to the R&D program—should be recruited. No changes have been

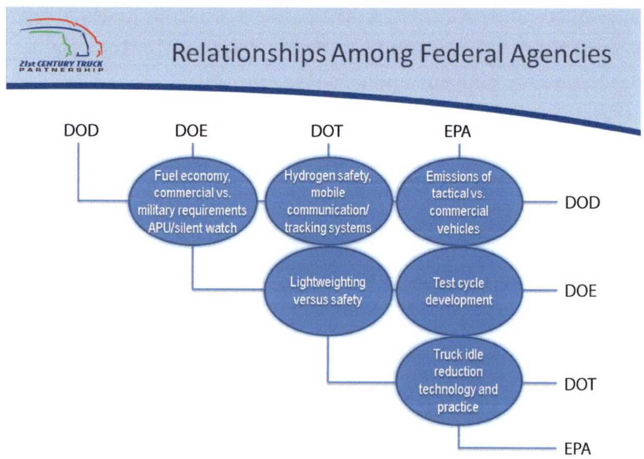

FIGURE 2-2 Some areas of common interest among government agencies participating in 21CTP. SOURCE: K. Howden, DOE, "21CTP Overview," Presentation to the committee on May 14, 2014.

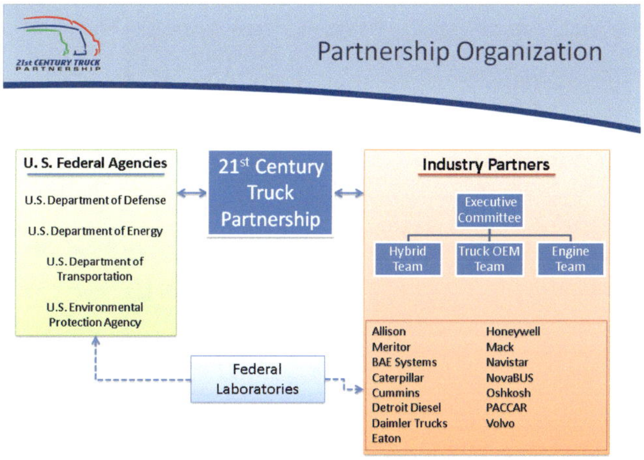

FIGURE 2-3 Partnership organization. SOURCE: K. Howden, DOE, "21CTP Overview," Presentation to the committee on May 14, 2014.

made to the Partnership structure since 2001, and meanwhile at least two industrial members have exited the on-road truck business, and consolidation in the industry has reduced six of the members to two separate corporate entities.

The executive committee meets by teleconference once every month and is chaired by the 21CTP director. The purpose of these meetings is to discuss 21CPT activities for the month and make any appropriate planning decisions that are best facilitated by a small representative group.

The Partnership also holds regular teleconference meetings of its full membership once a month. The purpose of these meetings is to facilitate open communication among the federal and industry partners about research activities, industry business, and technical accomplishments related to the Partnership's goals and objectives. The meetings are chaired by the 21CTP director and separated into three distinct activities: a government-only discussion, a government-industry discussion, and an industry-only discussion. Participation in these calls involves all of the federal agency and industry partner organizations, and attendance ranges from 20 to 40 people per call. In addition to the regular teleconferences, the Partnership holds at least one, and usually two, in-person meetings a year at a partner location.

Agendas and minutes of these meetings are archived on the Partnership's internal website. The committee has reviewed examples of these meeting minutes and found the meetings to be comprehensive, well attended, and productive.

The descriptions of the overall program management process, originally published in the NRC Phase 1 and Phase 2 reports, have been updated in the present report to reflect current Partnership practices. They reflect the Partnership's responses to questions from the committee during this Phase 3 review, dated August 28, 2014.

The original partnership structure was judged in the Phase 1 and Phase 2 reviews to be far from ideal. Accepting in the Phase 2 review that the form of centralized management and control preferred by the committee was simply not feasible within the prevailing government and congressional structure, the committee recommended (Appendix C, Recommendation 2-1) as follows:

> DOE is urged to continue to improve the functioning of the 21CTP "virtual" management structure in every way possible. Such improved functioning would include strengthening interagency collaboration…and documenting and publishing specific 21CTP activity within all four agencies.

In its response, the Partnership argued that its informal virtual organization offers some advantages but also noted it is exploring new communication methods and continuing its efforts to strengthen interagency partnerships, among other initiatives.

Overall, the committee accepts that the Partnership is striving to operate as effectively as possible despite a less than ideal organizational structure. In this regard, illustrated in Figure 2-4, the DOE has a dedicated leader assigned to 21CTP matters, even though his responsibilities are listed as "Aftertreatment" in the Advanced Combustion Engines group. It would be most helpful if this person reported directly to the VTO director for 21CTP matters and, furthermore, had designated 21CTP counterparts at the other three agencies.

The other program management recommendation in the NRC Phase 2 report dealt with the lack of a clearly defined inventory of projects and budgets included under the 21CTP umbrella, and recommended a brief annual report documenting the projects funded and the progress made (Appendix C, Recommendation 2-2). The initial response from the Partnership concurred that no such report exists, and promised to consider the development of such a document as a nonduplicative complement to other 21CTP-related reports.

Subsequently, in responses to follow-up questions from the committee, the Partnership indicated its intention to strengthen the coverage of 21CTP activity on the DOE website (which currently contains only the 2006 and 2013 roadmaps and white papers) and provided the committee with a first draft of a proposed periodic report of recent Partnership accomplishments with digital links to further detailed information. The committee considers these actions to be fully responsive to the prior recommendations (21CTP, 2013).

In addition to the committee's hearing a DOD presentation on May 14, 2014, a committee subgroup visited the U.S. Army Tank Automotive Research, Development, and Engineering Center (TARDEC) in Warren, Michigan, on December 5, 2014, to review 21CTP-related projects. Although the DOD places a high priority on reduced energy consumption, it is of necessity focused on a totally different operating environment, one that is exempted from emissions

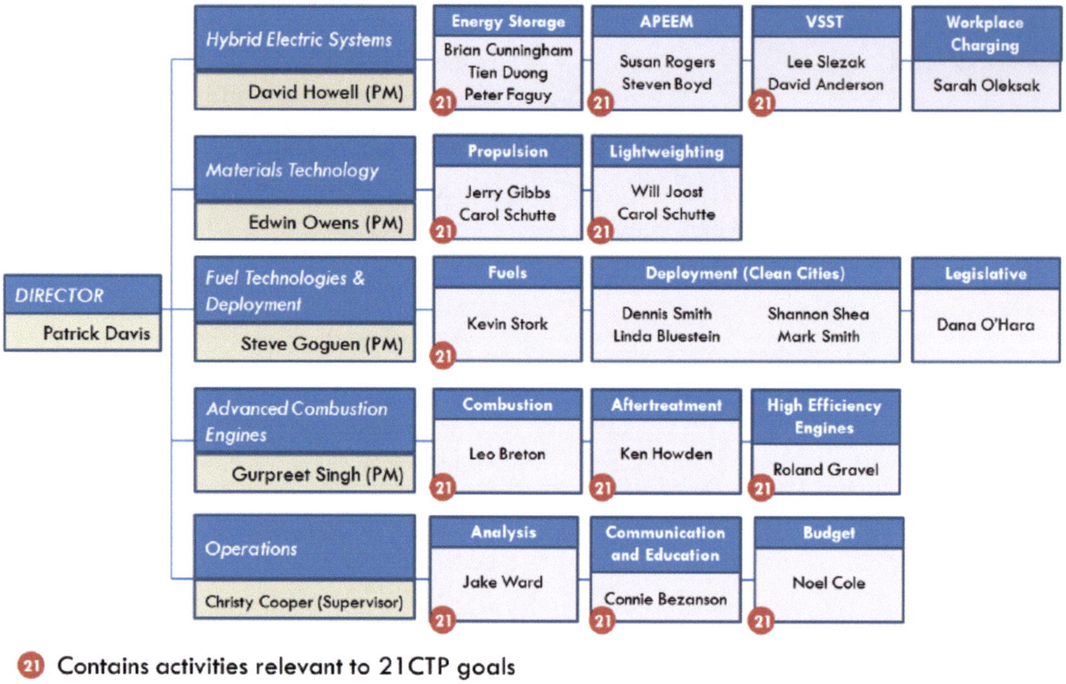

FIGURE 2-4 DOE VTO organization. SOURCE: Submitted to the committee by the DOE VTO, October 7, 2014. NOTE: The "21" notations indicate some 21CTP activity within the indicated groups. APEEM, advanced power electronics and electric motors; VSST, vehicle systems simulation and testing.

regulations and emphasizes high power density and JP-8 fuel. Consequently, there is little synergy with the needs of the commercial heavy-truck industry or with the stated 21CTP vision of "making trucks and buses safer, cleaner, and more efficient,"[2] given the unique nature of DOD's goals and their dissimilarity to those of 21CTP. Consequently, as shown in Figure 1-1 in Chapter 1, the Partnership lists DOD funding related to 21CTP goals as approximately $9 million, or only 6 percent of total 21CTP federal spending.

The national laboratories conduct many DOE programs synergistically with the 21CTP. Examples of such projects include advanced combustion research, fuels and lubricants, exhaust aftertreatment, lightweight materials, simulation software for combustion and engine operation, hybrid systems, and vehicle parasitic losses, among many others. A committee subgroup visited Oak Ridge National Laboratory on October 22, 2014, to review the activities related to 21CTP. Participation in 21CTP fosters ongoing technical interchange with industry at the working level, facilitating collaboration between the national laboratories, the government agencies, universities and industry, thereby ensuring that the national laboratories know industry's needs and priorities. It also expands the awareness across industry of activity at the national laboratories beyond that reported at the DOE Annual Merit Review. Data provided by DOE in response to committee questions shows that DOE funding related to 21CTP was split fairly equally between industry and the national laboratories: In FY 2014 it was 49 percent each, and over the last three fiscal years, it was 43 percent for industry and 55 percent for the national laboratories. Universities received the remaining 2 percent. While the university share appears to be small, it does not include the fundamental research sponsored under joint National Science Foundation/DOE programs or the many instances of universities partnering with national laboratories or industry on their projects. The importance of maintaining connections to universities suggests that DOE needs to determine whether the funding levels are sufficient. These connections are important for encouraging students to work on research projects related to the automotive industry so that there will be a next generation of engineers educated for government and industry. Unfortunately, no similar breakdown was available for the other three agencies.

Using the 21CTP to leverage the enormous capabilities in the national laboratories to address the specific research needs of industry adds considerable value and will facilitate the timely transition of promising technologies from the laboratories into the marketplace.

[2] K. Howden, DOE, "21CTP Overview," presentation to the committee, May 14, 2014.

As noted in Chapter 1, another important information-sharing activity supported by DOE was the Directions in Energy and Emissions Research (DEER) conference. This conference brought together professionals in the engine community and addressed research and projects related to the 21CTP. The conference has traditionally been an important meeting for bringing these professionals together for in-depth discussions of engines and issues related to their emissions, including a wide variety of experts not necessarily involved directly with the 21CTP. In this way, the researchers involved in the 21CTP projects can learn from a wide variety of engine and emission control experts. Unfortunately, the last DEER conference was held in 2012.

PRIORITIZATION OF PROJECTS

The organizational structure of 21CTP precludes any systematic prioritization of research projects for the total program across all the participants. Each of the four agencies included in 21CTP has its own separate budgets and priorities, and the industrial partners also have their own needs, priorities, and resources.

The Partnership provided an inventory of 162 21CTP projects in October 2014 and then provided an updated inventory on December 29, 2014 (see Appendix D); the committee asked for clarification of which projects were considered "key" and how key projects are prioritized. On October 24, the Partnership provided the response shown in Box 2-1.

The NRC Phase 1 report recommended the creation of "a portfolio management process that sets priorities and aligns budgets among the agencies and industrial partners" (NRC, 2008; Recommendation 2-2). The Partnership responded that the recommendation "will be considered . . .[but] the ability to directly align budgetary decisions across the agencies, however desirable, may be outside the scope of this voluntary collaborative organization."

Since that time, the DOE has alluded periodically to the adoption of an Advanced Research Projects Agency-Energy (ARPA-E)-like portfolio management approach to its projects, and the committee asked if that approach had been adopted and how it operates. The response shown in Box 2-2 was received on October 24, 2014.

In addition, upon request, DOE provided examples of two projects that had encountered difficulties in achieving their respective goals and were substantially renegotiated and redesigned.

While these responses are clearly focused on DOE activities, they represent a welcome move in the direction of more aggressive program management and a portfolio approach, and the committee would like to see a similar approach taken at the other participating agencies.

A major component of 21CTP in recent years has been the SuperTruck initiative, enabled by the American Recovery and Reinvestment Act (ARRA) of 2009. SuperTruck was discussed in the NRC Phase 2 report and is covered here in detail in Chapter 8. Four teams have been awarded cost-sharing contracts to develop prototype Class 8 trucks employing many of the technologies being pursued by 21CTP, with very specific performance goals and time tables. Two teams are supported by ARRA funds and two teams benefit from DOE internal funds redirected to this purpose. The committee fully endorses this reprioritization of DOE funds to enable the SuperTruck projects to proceed and applauds SuperTruck's emphasis on applying a total systems approach to evaluating and demonstrating the candidate technologies in real-world vehicle applications against stringent test criteria. The SuperTruck teams have made significant vehicle and engine progress, as described in chapters 3 and 8, and in so doing have shown the value of carefully designed demonstration programs to complement component and system technology research. As the four SuperTruck projects approach their conclusion, it is not too soon to develop proposals to build on their momentum and success and to prioritize the next set of objectives in a future resource-constrained environment.

BOX 2-1
Partnership's Response to Committee Questions on Project Prioritization

The federal agencies that are members of 21CTP do not explicitly "prioritize" their ongoing projects after the portfolio has been established to identify a subset of them as being "key." The federal agencies do broadly prioritize their research efforts when they are building their research portfolio, based on agency mission, funding levels, and other factors, however. The federal agencies prioritize their technical focus areas through their strategic planning and goal setting work, and then use a variety of mechanisms to identify performers (industry, academia, laboratories) to complete the work. The mechanisms are frequently competitive in nature, effectively prioritizing the proposed projects for each technical focus area as they are selected for funding.

For DOE, the industry-led projects chosen in response to a funding opportunity announcement are competitively selected (or "prioritized") by technical review committees and DOE technical staff, who are selecting projects that are likely to be best able to contribute successfully to the relevant DOE goals. The laboratory-led projects are selected through a different process that still prioritizes the projects most likely to be successful and achieve critical DOE goals. DOE technical staff works with the laboratories to select the appropriate mix of projects, based on available funding and technical priorities. In both cases (industry and lab), the portfolio selections are reviewed each year at the Annual Merit Review, and changes to portfolio components can be made as a result of that feedback.

Box 2-2
Partnership's Response to Committee Questions on Management Approach

The DOE Office of Energy Efficiency and Renewable Energy's (EERE's) program management approach is broadly similar to the ARPA-E approach, in that it involves establishing and tracking of critical project milestones, and close coordination with the project performers to ensure that milestones are met and any issues are identified and resolved in a timely manner (see the ARPA-E Strategic Vision explanation of their management approach at http://arpa-e.energy.gov/sites/default/files/ARPA-E_Strategic_Vision_Report_101713.pdf).

EERE maintains a Project Management Center (PMC, viewable at https://www.eere-pmc.energy.gov/) to assist DOE HQ technology managers and project performers in managing their projects, and to provide a common framework and business practices across the diverse group of EERE offices. The PMC supports EERE through two field offices in Golden (Colorado) and NETL (Pittsburgh and Morgantown). DOE EERE technology managers and representatives from the PMC work together with the project performers to negotiate project awards, establish and track milestones, actively review project progress on a quarterly basis (either via webinar or in-person meetings), and gather project management data using centralized EERE software tools to track progress. Complete details on how EERE program management is conducted may be found at http://energy.gov/eere/about-us/eere-program-management-guide. As with any project management system, the EERE system is continually being reviewed and refined to keep pace with current best practices and lessons learned.

FINDINGS AND RECOMMENDATIONS

In summary, the 21CTP continues to operate as a virtual network of government agencies, industry, and national laboratories, led by DOE using a relatively flat and informal management structure to discuss research priorities, communicate research successes, and provide feedback on future trends.

The Partnership has responded positively to prior NRC recommendations to improve the functioning of this virtual management structure, particularly across the other three agencies involved; publish an inventory of Partnership projects and associated budgets; and consider dedicated communication, such as a brief annual report, of Partnership activities and accomplishments.

Finding 2-1. The 21CTP remains a virtual organization facilitating communication among four government agencies, the national laboratories, and industry, led by DOE but with no single-point authority over its activities, priorities, or budgets. While far from optimal, this structure is necessitated by the separate reporting and budgeting mechanisms for each agency. The Partnership has made good progress in adapting to this reality by improving communications, coordination and collaboration among the partners, and documenting most of the projects and budgets under the 21CTP umbrella.

Recommendation 2-1. The DOE is urged to continue to improve by maintaining and publishing the inventory of projects and budgets across all four agencies, tying those projects into the specific 21CTP goals and promoting a portfolio management approach or the DOE's Office of Energy Efficiency and Renewable Energy's Project Management Center (EERE PMC) equivalent within the other agencies. Furthermore, EPA, DOT, and DOD should appoint a dedicated counterpart to DOE's designated 21CTP leader, who in turn should report directly to the director of the Vehicle Technologies Office on 21CTP matters.

Recommendation 2-2. The Partnership should develop and adopt criteria for including projects under the 21CTP umbrella, such as "Does the project clearly address one of the specific goals of 21CTP?" and "Does the project fall within the R&D interests of the member partners of the 21CTP?" The committee recognizes that there will be at least two levels of projects—those tightly connected to specific 21CTP goals and a supporting set of projects that have longer term impact. Better definition of the criteria for including a project and at what level would assist in evaluating and increasing the effectiveness of the Partnership.

Finding 2-2. While many projects deemed to fall under the 21CTP umbrella are reviewed in their own right at the annual DOE Merit Review and the Directions in Engine Efficiency and Emissions Research (DEER) conferences (until the latter ended in 2012), and the SuperTruck projects have an annual reporting requirement, there remains no dedicated report on 21CTP activities, in any medium. In response to prior NRC recommendations, DOE has proposed the development of such a report and given the committee a first draft proposal.

Finding 2-3. The annual DEER conferences sponsored by DOE for over 25 years were an excellent way to share research results among industry, national laboratory, government, and university personnel, but they were not held in 2013 and 2014.

Recommendation 2-3. The DOE should publish a brief annual report on 21CTP activities and accomplishments, such as the first draft provided to the committee on October 24, 2014, with references to published technical reports from all four agencies and the national laboratories. This would be a great help to Congress and to future review committees and the public in understanding the work and scope of the Partnership.

Recommendation 2-4. Because the DEER conference was an excellent approach to communicating research results of the 21CTP, the Partnership should consider holding it in 2015 and each year thereafter. If funding constraints prevent an annual meeting, it should be held at the very least every other year.

REFERENCES

21CTP (21st Century Truck Partnership). 2006. 21st Century Truck Partnership Roadmap and Technical White Papers. 21CTP-003. December. Washington, D.C.: Office of FreedomCAR and Vehicle Technologies.

21CTP. 2013. Roadmap and Technical White Papers. Department of Energy, Office of Energy Efficiency and Renewable Energy. https://www1.eere.energy.gov/vehiclesandfuels/pdfs/program/21ctp_roadmap_white_papers_2013.pdf.

DOE (Department of Energy). n.d. Chapter 35.2, Cost Sharing in Research and Development Contracting. http://energy.gov/sites/prod/files/35.2_Cost_Sharing_in_Research_and_Development_Contracts_0.pdf. Accessed March 18, 2015.

National Research Council (NRC). 2008. *Review of the 21st Century Truck Partnership.* Washington, D.C.: The National Academies Press.

NRC. 2012. *Review of the 21st Century Truck Partnership, Second Report.* Washington, D.C.: The National Academies Press.

3

Engine Systems, Aftertreatment, Fuels, Lubricants, and Materials

INTRODUCTION

Heavy-duty truck engines, emissions control technology, and fuels are central to all aspects of the 21st Century Truck Partnership's (21CTP's) vision of improved thermal efficiency, reduced oil dependency, low-exhaust emissions, lower cost, and improved safety. Although diesel engines used in new trucks are among the most efficient and clean on-road transportation power plants available today, in the opinion of the committee there are still opportunities for making them better. In some heavy-duty applications, gasoline spark-ignition engines are also used. Despite the fact that they are not as efficient as diesel engines, the lower cost of the engine system (which includes emissions control) and the lower cost of the fuel results in lower ownership and operating expenses relative to diesel power plants for their specific applications. However, spark-ignition engines being developed today are incorporating technologies that make them look more diesel-like (e.g., turbocharged direct-injection engines). Consequently many of the fundamental issues that need to be addressed to facilitate reduced fuel consumption in diesels are also of value for spark-ignition engines, and vice versa (e.g., spray characterization, vaporization and mixing phenomena, autoignition, combustion and emission kinetics, and cost-effective lean-emissions control systems).

This chapter covers the 21CTP programs in diesel engines, fuels and lubricants, aftertreatment systems, high-temperature materials, and health concerns raised by diesel engine emissions. In the mobility domain being addressed within 21CTP, the internal combustion engine and its associated emissions control systems and their fuels represent continuously improving, state-of-the-art transportation technologies, offering the lowest life-cycle costs for near-term propulsion technologies (21CTP, 2013, p. 33).

Commensurate with the evolution of heavy-duty (HD) engine and emissions control technologies during the first 15 years of the 21st century, vehicle fuel and lubricant technology is also changing. Petroleum-based diesel fuel regulations have been updated (e.g., lower sulfur limits) to allow for advanced emissions control components, a variety of biofuels have been developed for the purpose of extending transportation fuel supplies from renewable sources, and synthetic hydrocarbons have been produced from natural gas, recycled plastics, and organic refuse. In addition, new lubricant formulations have provided increased fuel efficiency for light-duty vehicles. Research on development of new fuel-efficient lubricant formulations for heavy-duty vehicles is in progress. It is also important to note that the supply of petroleum-based fuels from within the United States has increased significantly owing to the development of new hydraulic fracturing ("fracking") and directional drilling techniques.

Nonpetroleum diesel fuels can be produced from renewable resources such as seed oils and animal fat, as well as synthesized from natural gas, biomass, oil sands, coal, and other resources. Cellulosic ethanol production facilities are being brought online using technology developed in part by Department of Energy (DOE) laboratories, although the volumes produced will be small. Facilities for the production of renewable diesel fuel from biomass resources continue to be developed, and the production and sale of biodiesel is growing in the United States at a modest rate.[1] The use of syncrudes from tar sands in Canada has also grown. Fischer-Tropsch (F-T) diesel fuel, synthesized from natural gas, has been studied in conventional diesel engine tests in many laboratories to quantify its beneficial impact on emissions. Natural gas has also been described as a potential replacement for liquid petroleum fuels. This application is discussed in detail in a recent National Research Council (NRC, 2014) report. Future expanded use in medium- and heavy-duty vehicles will depend on lowering the cost of on-board fuel storage, as well as the cost of dispensing facilities. Lubricant properties and composition can have a beneficial effect on

[1] See Biodiesel Production Statistics at http://www.biodiesel.org/production/production-statistics.

vehicle fuel efficiency by reducing engine and driveline friction. Conversely, some engine oil components can adversely affect vehicle emissions by reducing the durability of exhaust emissions control devices. The sulfur, phosphorus, and ash content of lubricants needs to be minimized to prevent degradation of all types of catalytic devices.

Integral to the industries' efforts to increase efficiency is the push to operate the engine at higher peak cylinder pressures. As oxides of nitrogen (NO_x) aftertreatment systems continue to improve, there is a tendency for the industry to also push the in-cylinder temperatures higher because the improved aftertreatment can reduce the increased NO_x. Consequently, propulsion materials are required that can withstand higher pressures and temperatures. These advanced materials are an enabler for cost-effective fuel savings. Given the long timeline for the identification, development, and implementation of new materials, it is essential that R&D continues without interruption.

ENGINE SYSTEMS PROGRAM: STATE OF TECHNOLOGY AND GOALS

As with any power generation device using chemical reactions to provide energy, both diesel and gasoline engines have thermodynamic constraints associated with the combustion process. These engines have additional practical constraints as well:

- Impracticality of extremely large expansion ratios,
- Inability to capture all of the useable energy in the heat rejection and exhaust flow,
- Inability to totally eliminate pumping work,
- The presence of friction due to rubbing contacts, and
- The work consumed in driving auxiliaries and accessories.

All of these constraints add up to limit the efficiency that can be obtained from practical, economical engines. Figure 3-1-1 in Box 3-1 delineates the partitioning of energy within the engine between that for the engine and that for the vehicle. The fundamental causes of these limitations are known, and the 21CTP works to coordinate and advise the federally-funded programs focused on minimizing these limitations with technologies that would be viable in the market. The specific goals of the Partnership within the Engine Systems and Fuels area are the following (21CTP, 2013, p. 33):

(1) Develop and demonstrate an emissions compliant engine system for Class 7-8 highway trucks that achieves 50 percent brake thermal efficiency (BTE) in an over-the-road cruise condition, improving the engine system fuel efficiency by about 20 percent (from approximately 42 percent thermal efficiency today) (by 2015).

(2) Research and develop technologies which achieve a stretch thermal efficiency goal of 55 percent in prototype engine systems in the laboratory. (This efficiency gain would be equivalent to an additional 10 percent gain in over-the-road fuel economy when prototype concepts are fully developed for the market.) (by 2015).

(3) Through experiments and models with FACE (Fuels for Advanced Combustion Engines) fuels and other projects, determine the most essential fuel properties, including renewables, to help achieve 55 percent engine brake efficiency (by 2014).

Implicit in all of the above goals is meeting emission regulations. In the NRC Phase 2 review this was a significant focus because new emissions regulations had just come into effect (NRC, 2012). These standards were met through the development and integration of new emissions control technologies into the engine system. This exemplified a transition to a new combustion system development paradigm—synergistic integration of engine and fuel combustion development to most effectively utilize aftertreatment system capabilities.

Exhaust Emissions

The considerable effort and research funding focused on improving emissions control systems is complementary to the development of engine combustion processes. To meet US2007 HD regulations, exhaust gas recirculation (EGR) combustion strategies were the primary NO_x reduction technology, while diesel oxidation catalysts (DOCs) and diesel particulate filters (DPFs) were the primary hydrocarbon (HC), carbon monoxide (CO), and particulate matter (PM) reduction technology (EPA, 2000). A transition occurred to meet the US2010 HD regulations since EGR was not sufficient to efficiently meet the required 0.2 grams per brake horsepower-hour (g/bhp-hr) NO_x tailpipe emissions levels. Selective catalytic reduction (SCR) was added to medium- and heavy-duty vehicles (MHDVs) to meet this new NO_x requirement, while some engines were sold without SCR because some manufacturers used credits to meet an average NO_x standard. Because the SCR can effectively remove NO_x, the US2010 engines could then operate at higher engine-out NO_x levels and run more efficiently. For Class 8 truck engines, this resulted in about a 5 percent reduction in fuel consumption for US2010 engines compared to US2007 engines. However, the trade-off required about 2 to 3 percent urea (contained in the diesel exhaust fluid [DEF]) relative to fuel (Charlton, 2010).

The requirements were adjusted again in 2013-2014, when onboard diagnostics (OBD) (2013) and the first phase of HD greenhouse gas (GHG) regulations (2014) were introduced. In conjunction with this, engine and emissions control systems were further optimized for better performance and/or reduced cost.

Now, emissions control technology needs further understanding to optimize performance, to reduce cost and fuel consumption, and to meet any future regulatory tightening.

BOX 3-1
Typical Energy Flows in an Engine

Figure 3-1-1 (Figure 2 from Delgado and Lutsey, 2014) shows a typical partitioning of the energy flows within the engine as a result of the different phenomena associated with burning the fuel to producing work. Such displays of the energy flow are very useful when interpreted from the thermodynamic perspective that different forms of energy have different potentials to produce work. Practically speaking, all of the energy in a typical hydrocarbon fuel is useful, so the fuel energy input on the left-hand side of the figure could, theoretically, be converted into useful work. Thus it is a good reference against which to evaluate engine performance.

The second column from the left shows the energy flows associated with phenomena occurring in the cylinder. Within the cylinder the energy flow is partitioned between leaving the cylinder at the piston face as indicated work, the desired outcome, and leaving the cylinder via heat transfer or within the exhaust gas. It is also known that any energy transformation process, like combustion, that is used to release the energy bound within the fuel will have losses, or irreversibilities, associated with it. For the case of chemical reactions this irreversibility is a degradation of useable fuel energy into energy that can no longer be converted into work. This nonusable energy becomes part of the heat transfer and exhaust flow leaving the cylinder. (Approximate proportions of this nonuseful energy within the heat transfer and exhaust have been marked on the figure.[a]) Thus, even though there is significant energy flow leaving the cylinder as heat transfer and exhaust flow, it is not possible to convert all of that energy into work using additional energy conversion devices such as waste heat recovery, which use the heat transfer and/or exhaust flow as the energy input.

This is well understood by the researchers and engineers in the technical community and is instrumental in determining the cost-effectiveness and technical viability of incorporating work- producing devices onto the engine that use the heat transfer or exhaust flow as energy inputs.

The final three columns in the figure show what happens to the Indicated work that leaves the piston. Some of the work transferred from the cylinder into the piston must be used to affect the gas exchange within the engine (pumping), some must be used to drive auxiliaries and accessories, and some is dissipated as friction. Even though these are relatively small, any reduction in the amount of work expended on these processes represents work that stays on the shaft and makes it to the drivetrain. The work that is ultimately delivered to the flywheel is called brake work. An analogous statement can be made for the energy transfer through the drivetrain to the wheels, the fourth column. Consequently the work that makes it through the drivetrain to the wheels is called drive work.

Finally, the last column shows how the drive work is used to move the vehicle once it makes it to the wheels. The extent to which the categories in this column can be reduced directly impacts the amount of work necessary to move the vehicle.

This figure helps to put in perspective the research activities associated with the 21CTP. In regard to the Engine, Aftertreatment, Fuels and Lubricants, and Materials subprograms in 21CTP, the Indicated and Brake columns are the relevant energy partitions. In general, one would like to maximize the work that is obtained by the expansion process, and minimize the uncontrolled transfer of energy from the cylinder via heat transfer and exhaust flow. Within the engine itself, minimizing the necessary expenditure of work for pumping and driving auxiliaries and accessories is important and yields immediate benefit in terms of brake work, as does any reduction in the friction.

[a] Approximations determined from exergy balances on the energy transformation processes within an internal combustion engine (see Caton, 2000).

Engine Systems Program

The industry has demonstrated great technical competence in reducing fuel consumption and meeting the 2010 emission standards while making a product that meets customers' reliability and cost of operation goals and also the Environmental Protection Agency's (EPA's) 2014 GHG standards. What facilitates the industry's achievements is the continued progress in understanding the technical subtleties of the thermodynamic, chemical, and physical processes involved in the conversion of the fuel energy into power. As the demands for higher efficiency and low emissions grow, so too does the need for an increasingly deeper understanding of the fundamentals. This is the principal deficit that the Engine Systems program of the 21CTP is facing as it works to achieve its efficiency goals (21CTP, 2013, pp. 40-41). The challenges specifically relating to engine system fundamentals are these:

- Inadequate understanding of the thermodynamic, chemical, and physical fundamentals of combustion and the consequent inability to incorporate them into robust simulation capabilities, especially across the full range of combustion approaches, from conventional diesel combustion to new low-temperature combustion (LTC) regimes.
- Inability to optimize in-cylinder combustion processes for efficiency via synergistic coupling of enhanced aftertreatment system performance.
- Lack of exploration and development of innovative engine processes and architectures.

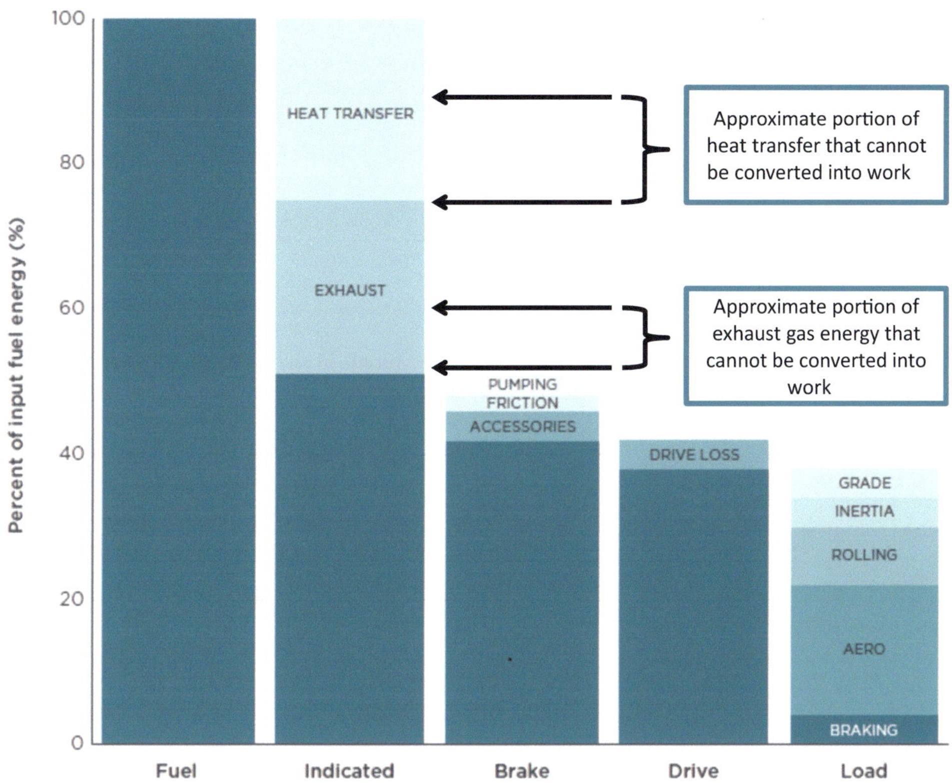

FIGURE 3-1-1 Hypothetical energy audit for Class 8 truck. SOURCE: Delgado and Lutsey (2014). Attribution-ShareAlike 3.0 Unported.

The above challenges motivate the fundamental research and development projects in the 21CTP.

Although SuperTruck is discussed at greater length in Chapter 8, its engine development activities in SuperTruck are discussed here separately, before the individual DOE and Department of Defense (DOD) engine programs. SuperTruck's engine programs interface closely with the 21CTP fundamental engine projects and are synergistic with them in achieving two of the Engine Systems' program goals: the 50 percent BTE demonstration and the technical roadmap to 55 percent BTE. The discussion as a whole is separated into activities directed toward demonstrating 50 percent BTE on the road in a truck (Goal 1), which is a SuperTruck engine accomplishment, and then covers the activities showing a technical pathway to 55 percent BTE (Goal 2). The discussion of the SuperTruck engine teams' work toward achieving Goal 2 segues into the discussion of the individual DOE and DOD programs nicely because of the extent to which the SuperTruck programs will be relying on the advancements made within the individual DOE and DOD projects to achieve the 55 percent goal. This is the sequence in which the research activities are discussed.

Research Budgets for the SuperTruck Program

Funding for the SuperTruck engine and vehicle program comes from two sources: the American Reinvestment and Recovery Act (ARRA) and the DOE Vehicle Technologies Office (VTO). The Cummins-Peterbilt and Daimler programs are funded through the ARRA, while the Volvo and Navistar programs are funded through the DOE VTO. There is no specific budget line item for engine development

TABLE 3-1 Estimated Federal Budgets for SuperTruck Engine Research (millions of dollars)

SuperTruck Team	Engine Research Budget	Total Federal Budget for SuperTruck
Cummins-Peterbilt	15.5 (estimate by 21CTP)	38.8
Daimler	15.8 (estimate by 21CTP)	39.5
Navistar	12.7	37.3
Volvo	7.6	17.7

within the ARRA-funded programs, Cummins-Peterbilt and Daimler. However the Partnership did provide estimates to the committee. These estimates and the engine research budgets for Volvo and Navistar, along with the total budgets of federal dollars, are given in Table 3-1.

Progress Toward Meeting Engine Goals 1 and 2

The demonstration of 50 percent BTE in a truck on the road involved integrating laboratory-proven technologies into a vehicle powertrain system, with an eye toward assessing the viability of those technologies for commercial introduction. The SuperTruck program was developed around this goal. Accomplishments in this effort are highlighted here via a table of technologies used, which has been extracted from the more comprehensive Table 8-1 given in Chapter 8. The extension of this effort to reach 55 percent BTE entails a fundamental research program to explore and quantify the potential of using advanced combustion/fuel and engine technologies that are currently being explored within research laboratories, with an eye on showing technical potential. The activities of the SuperTruck teams are closely aligned with, and will depend on the results of, the individual research programs within 21CTP. The approaches each team is taking toward this goal are summarized in the paragraphs below.

Goal 1: Develop and demonstrate an emissions-compliant engine system for Class 7-8 highway trucks that achieves 50 percent brake thermal efficiency in an over-the-road cruise condition.

Status: The Partnership has successfully achieved this goal. As shown in Table 3-2, two of the four SuperTruck teams have successfully demonstrated brake thermal efficiencies greater than 50 percent in on-road tests using commercial, ultra-low-sulfur diesel fuel.

TABLE 3-2 Achievement of Goal 1, 50 Percent Engine BTE at Cruise, by the Four SuperTruck Teams

SuperTruck Team	Status	Complete?
Cummins-Peterbilt	51% engine + WHR BTE demonstrated	Yes
Daimler	50.2% engine + WHR BTE demonstrated	Yes
Volvo	48% engine + WHR BTE demonstrated	TBD
Navistar	47.4% BTE engine only, WHR being considered	TBD

NOTE: WHR, waste heat recovery.

Both Cummins-Peterbilt and Daimler are in the final stage of their SuperTruck program. Their programs end in 2015, whereas Volvo and Navistar are in earlier phases of their programs. Both the Volvo and Navistar projects have completion dates in 2016. Their not having achieved Goal 1 is attributed to not being as far along in their programs as the other two teams. It is expected that they will meet their goal of 50 percent by the time they will have completed their work.

Table 3-3, which is extracted from Table 8-1, shows the engine and combustion technology that the respective SuperTruck teams are using in the 50 percent BTE engine. Each of the technologies being used can be categorized in terms of the second and third columns of the energy partitioning Figure 3-1-1. Each technology is used either to enhance the work extraction or to reduce the work expenditure for pumping, friction, accessories, and auxiliaries.

Achieving 50 percent BTE in a truck on the road, Goal 1 represents the successful integration of laboratory-proven technologies into a complex vehicle powertrain system, and the committee congratulates the Partnership and SuperTruck teams for this accomplishment.

Goal 2: Research and develop technologies that achieve a stretch thermal efficiency goal of 55 percent in prototype engine systems in the lab (by 2015).

Status: To date this goal has not been achieved, but progress has been good. It is anticipated that by the end of each of the respective SuperTruck programs, the teams will have developed a technology pathway for achieving 55 percent BTE. It is expected that the pathway will be combinations of individual technologies that are either demonstrated in the laboratory or simulated via advanced computational fluid dynamic (CFD) models.

Approaches to Meeting Goal 2

During the review, the committee heard presentations from the SuperTruck teams at its meetings, and made site visits to Cummins (August 28, 2014), Daimler (November 24, 2014), and Volvo (December 5, 2014). Achieving 55 percent BTE with a Class 7 or a Class 8 HD truck engine will be extremely challenging. The aggressiveness of the SuperTruck research programs is consistent with the high risk approach that needs to be pursued if this goal is to be

TABLE 3-3 Summary of Engine Technologies Used by the Four SuperTruck Engine Teams in Their Efforts to Achieve 50 Percent BTE on the Road in a Heavy-Duty Truck

Technology	Cummins- Peterbilt	Daimler	Volvo	Navistar
Base engine	15 L inline 6, no downsizing	10.7 L inline 6, downsized from 15 L baseline	11 L inline 6, downsized from 13 L baseline	12.6 L, I-6 baseline and SuperTruck
rpm @ 65 mph	~1,180	~1,300	Data not provided	~1,050 or 1,125
Engine efficiency features	High-efficiency turbo, low friction seals, lower power oil pump, low viscosity. Oil, cylinder kit friction reduction, higher PCP, cal. optimization, overall 30% FMEP reduction	Turbo match, optimized liner cooling, variable speed water pump, low viscosity. Oil, piston friction reduction, 15% higher PCP, cal. optimization	High-efficiency turbo, variable coolant and oil pumps, reduced friction pistons, rings, and liners, low viscosity oil, improved thermal management	High-efficiency turbo, elevated coolant temperature, low friction power cylinder, thermal insulation, reduced air flow restrictions
Fuel system	HPCR with reduced parasitic fuel pump	Amplified HPCR	HPCR (converted from unit injector baseline)	Amplified HPCR
Combustion refinement	Very high CR, piston bowl, injector match, 4.3 g/hp-hr engine-out NO_x, conventional diffusion burn	High CR, piston bowl, low EGR, injector match, conventional diffusion burn, higher engine-out NO_x, model- based controls	Increased CR, advanced piston bowl design, conventional diffusion burn, same engine-out NO_x as US2010	Looking at 6 g engine-out NO_x, higher injection press, revised piston bowl and high CR, evaluating diesel and dual fuel options, low swirl
Electric drive components		Electric HVAC	Electric dual-zone HVAC	Electric HVAC, 48 V
Waste heat recovery	Rankine cycle, R245 working fluid, mechanical drive, uses EGR and exhaust heat, turbine expander	Rankine cycle, ethanol working fluid, electric drive, uses EGR and exhaust heat, scroll expander	Turbocompound plus Rankine cycle with ethanol working fluid, mechanical drive, uses EGR and exhaust heat	Turbocompound, Rankine cycle, and e-turbo are being evaluated
Aftertreatment	High conversion efficiency, low back pressure	High conversion efficiency, low back pressure	High conversion efficiency, low back pressure	High conversion efficiency, low back pressure
Turbo technology	High efficiency VG	Asymmetric	High efficiency	Possible e-turbo
EGR loop	Reduced flow and restriction HPL	HPL	Reduced flow HPL	Reduced flow and restriction HPL
Variable valve actuation	No	No	No	Being evaluated
Cooling system	Conventional cooling package, engine-driven fan	Angled cooling package, hydraulic motor fan drive, active grill shutters	Variable speed engine-driven fan, variable-speed cooling pump	3-speed engine-driven fan, electronic stat, high coolant temp., variable-speed cooling pump, variable coolant pressure
Accessory power demand		Clutched air compressor with active controls, clutched power steering pump with reservoir, cab insulation, solar reflective paint	Clutched air compressor with active controls, low-energy power steering, look-ahead smart alternator, LED lighting, cab insulation	Variable displacement oil pump, clutched air compressor with intelligent dryer control, accessories run on deceleration/coasting

NOTE: FMEP, friction mean effective pressure; PCP, peak cylinder pressure; HPCR, high-pressure common rail; CR, compression ratio; HVAC, heating, ventilation, and air conditioning; LED, light-emitting diode; EGR, exhaust gas recirculation; VG, variable geometry; HPL, high-pressure loop.

achieved. From a generic perspective, there is similarity in the overall approach being followed by the four teams. All of the programs are pursuing continued reduction in friction and pumping, more effective air boosting systems, smaller auxiliary and accessory loads, and improvements in their waste heat recovery (WHR) systems and are investigating advanced low-temperature combustion (LTC) approaches. All of the teams are also engaged with the aftertreatment technical community looking to capitalize on further improvements in exhaust gas treatment of criteria pollutants that will allow further optimization of the engine-aftertreatment combination. However, the details of how these technologies will be applied and the gains from each differ from program to program.

The basic premise of LTC processes stems from an understanding of the energy flow partitioning presented in Box 3-1 at the beginning of this chapter. If the in-cylinder temperatures can be kept low during the closed portion of the cycle, the thermal efficiency will increase. This is explained by the dependence of the closed-cycle efficiency of the engine on the ratio of the specific heats of the gases ($\gamma = c_p/c_v$) in the cylinder.[2] Lower temperatures and leaner mixtures within the cylinder result in values of gamma (γ) that are larger than when the temperatures are higher or the mixtures are stoichiometric. A larger average gamma results in more work being extracted during the closed cylinder portion of the engine's mechanical cycle, which subsequently decreases the amount of useful energy leaving the energy in the exhaust. Furthermore, lower in-cylinder temperatures also result in less heat transfer from the cylinder.

The challenge with trying to drive the in-cylinder temperatures down is that the burning velocity of the fuel and air mixture decreases as the temperature decreases, and in trying to push this concept to the limit, the time necessary to complete combustion gets too long and engine efficiency and emissions suffer. The overview of the individual teams' programs shows that shortening the combustion interval is an important aspect of achieving the 55 percent target. The general approach in LTC strategies is to keep combustion durations short by minimizing the need for flame propagation through volumetric combustion via autoignition. Achieving this type of combustion is highly dependent on the chemical and physical characteristics of the fuel and requires very precise control of the thermokinetic state of the air–fuel mixture within the cylinder. Understanding the fundamentals of these phenomena is prerequisite to success and is a principal focus of the individual DOE engine combustion research projects in 21CTP.

Researchers have proposed many different approaches for achieving LTC. They will typically name their specific approach with an acronym, such as HCCI (homogeneous-charge compression ignition), PPCI (partially-premixed-charge compression ignition), RCCI (reactivity-controlled compression ignition)—often more generically referred to as dual-fuel combustion—and many more. One advantage of the dual-fuel approach is that using varying ratios of two fuels with different degrees of reactivity gives the operating system an additional and powerful combustion phasing control lever. Indeed the dual-fuel approach is being investigated by most of the SuperTruck teams, although to date they have not divulged the specific fuel combinations they are currently exploring.

If successfully achieved, LTC strategies yield higher closed-cycle efficiency that minimizes heat loss from the cylinder and exhibits low NO_x and particulate emissions, but at the same time introduces concern about unburned hydrocarbon and carbon monoxide (CO) emissions. Consequently it is likely that aftertreatment will still be required, if not for PM and NO_x, then for HC and CO, and the aftertreatment systems will most likely need to operate at lower temperatures than current systems today.

Cummins

The Cummins approach to the 55 percent BTE requirement is described in its 2014 DOE Annual Merit Review (AMR) presentation (Project ACE057) (Koeberlein, 2014). Two basic combustion strategies are under evaluation. Both approaches will pursue downspeeding the engine and operating at higher loads to get the requisite power. The first approach, which uses relatively conventional diesel combustion, is summarized in Figure 3-1.

The approach embodied in the technologies listed in Figure 3-1 represents a continued effort at improving conventional diesel combustion. The optimized bowl, injector, and heat-transfer efforts represent combustion improvement. The team is performing simulation and experiments with the objective of shortening the combustion interval as much as possible to maximize the work from expansion and minimize heat loss. As shown in Figure 3-1, the objective is to gain approximately three percentage points improvement in engine BTE through this combustion improvement.

In addition to trying to further improve conventional diesel combustion, the Cummins-Peterbilt team is also pursuing a dual-fuel LTC strategy it calls alternate fuel compression ignition (AFCI). Simulation and laboratory results indicate that there is sufficient potential for improvement from this combustion strategy to merit further investigation. The

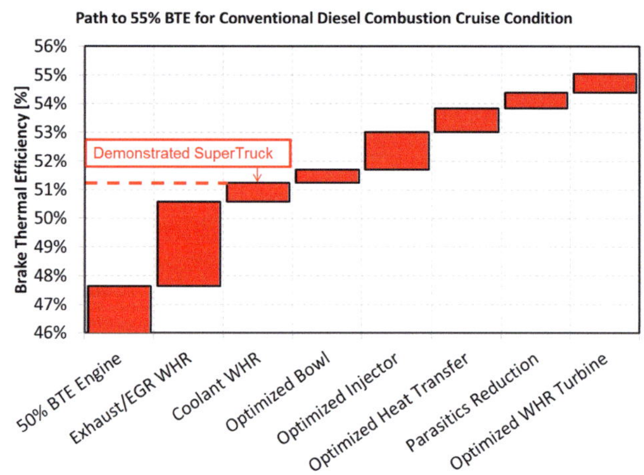

FIGURE 3-1 Cummins-Peterbilt SuperTruck team's projected incremental gains to get from its current 50 percent BTE engine to 55 percent BTE. SOURCE: L. Kocher, "SuperTruck 55% BTE Update Technology and System Level Demonstration of Highly Efficient and Clean, Diesel Powered Class 8 Trucks," presentation to the committee, Columbus, Indiana, August 28, 2014, slide 20. SuperTruck Annual Merit Review Presentations, Cummins, Inc.

---

[2] C_p, specific heat at constant pressure; C_v, specific heat at constant volume.

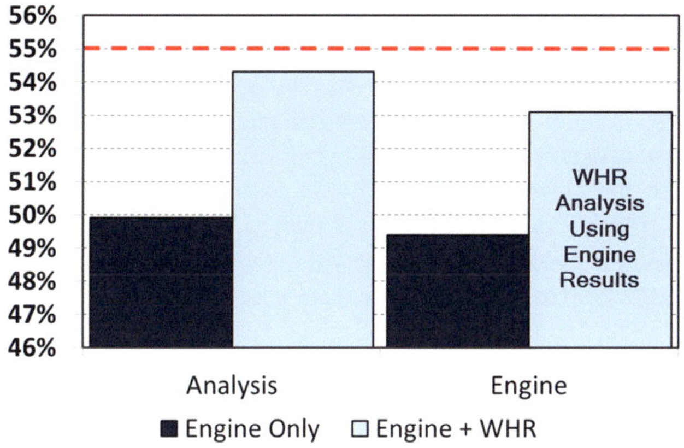

FIGURE 3-2 Cummins-Peterbilt SuperTruck team's analysis and results using AFCI at 1,000 rpm and 10 bar. SOURCE: L. Kocher, "SuperTruck 55% BTE Update Technology and System Level Demonstration of Highly Efficient and Clean, Diesel Powered Class 8 Trucks," presentation to the committee, Columbus, Indiana, August 28, 2014, slide 24. SuperTruck Annual Merit Review Presentations, Cummins, Inc.

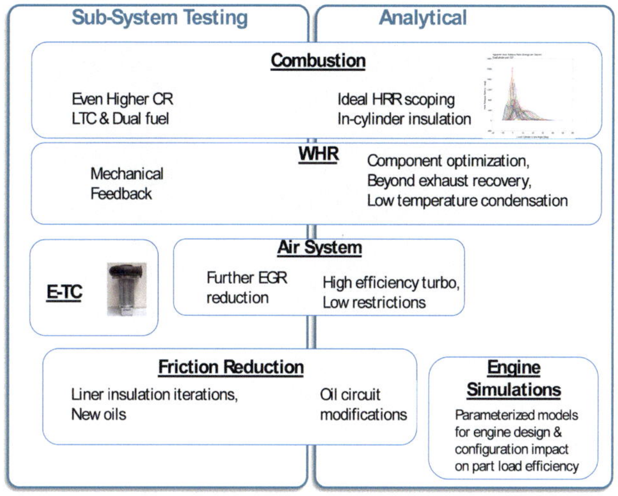

FIGURE 3-3 Overview of Daimler SuperTruck team's approach to achieving 55 percent BTE. SOURCE: Singh (2014). NOTE: E-TC, electronic turbocharger; CR, combustion ratio; LTC, low-temperature combustion; HRR, heat release rate; WHR, waste heat recovery; EGR, exhaust gas recirculation.

team's current results comparing both the predictions and actual engine results are shown in Figure 3-2.

The simulation does a reasonably good job of predicting the engine-only performance. However, because the exhaust energy was lower with AFCI, the work output from the WHR system in the exhaust was lower than that from conventional diesel combustion. The Cummins AFCI approach shares challenges with other LTC strategies, including the difficulty of running the engine above 10 bar brake mean effective pressure (BMEP). Cummins plans additional work to find ways of increasing the BMEP limit. It expects the efficiency of the engine to improve as it achieves higher loads with AFCI.

Daimler–Detroit Diesel[3]

The Daimler approach to the 55 percent BTE requirement is described in the team's most recent AMR presentation (Project ACE058) (Singh, 2014). The approach plans to use both downspeeding and downsizing of the engine. An overview of its 55 percent BTE scoping activities is presented in Figure 3-3.

Additional work on liner cooling optimization is ongoing with the Massachusetts Institute of Technology (MIT), as well as the development of new lubricants and optimization of the oil pump and lube circuits to reduce lube system power demand and further upgrades to the WHR system. Daimler-Detroit Diesel believes it may be possible to improve the WHR system contribution from 2.4 points of engine BTE (achieved in the final SuperTruck demonstration engine) to 3.6 points of BTE but acknowledges that this level of performance may prove impractical with currently available technology. As with other teams it is exploring the potential of LTC for shorter combustion intervals and lower heat loss. The team is working with the Oak Ridge National Laboratory (ORNL) on a dual-fuel engine approach, using natural gas as one of the fuels.

Volvo

To achieve the target of 55 percent BTE, Volvo is pursuing different engine architectures, alternative combustion cycles, and fueling optimization (Project VSS081) (Amar, 2014). These approaches will be pursued in a downsized and downspeeded engine.

New combustion concepts like PPCI and RCCI have demonstrated very high indicated efficiencies as well as low engine-out emissions. However, these kinds of combustion are significantly more difficult to simulate than normal diesel diffusion combustion. So, enhancements to the simulation capabilities are under way. A transported probability distribution function (PDF) combustion model has been developed to address this challenge, which is backed up by extensive testing. A cetane ignition device equipped with optical access is used for testing of fuels and validation of spray and chemical kinetics submodels. Figure 3-4 shows a stack chart of where Volvo believes the improvements in engine processes can be made to achieve 55 percent BTE.

The Volvo SuperTruck program is leveraging Sandia National Laboratory's Engine Combustion Network to validate the CFD subprograms it is developing. The team stated

[3] Detroit Diesel Corporation is a subsidiary of Daimler Trucks North America.

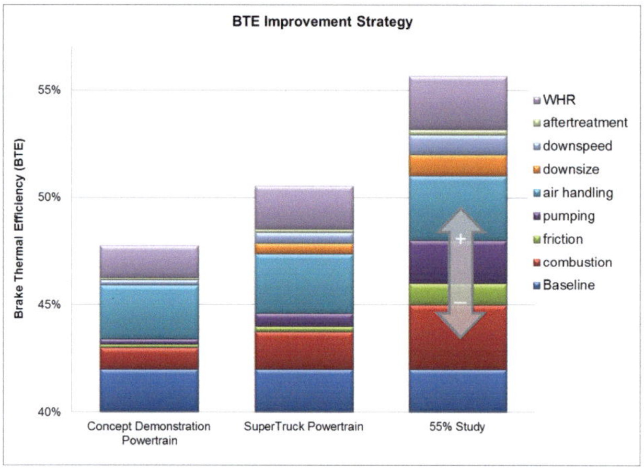

FIGURE 3-4 Volvo SuperTruck team's stack chart of the incremental improvements in engine technology to reach 55 percent BTE from a concept demonstration powertrain. SOURCE: A. Greszler, "SuperTruck Development and Demonstration of a Fuel-Efficient Class 8 Highway Vehicle," presentation to the committee, May 14, 2014, slide 14. Volvo Group Truck Technology.

that it is exploring a different engine architecture and running its heavy-duty engine on gasoline-like fuels, simulated as an 87 octane primary reference fuel blend.

Navistar

Navistar is also pursuing an aggressive research path in the technologies it has identified to achieve 55 percent BTE. Its approach includes downspeeding. Navistar's simulation predicts that through continued improvements of the aftertreatment system, which will allow more efficient combustion phasing, advanced turbomachinery, thermal barrier coating, dual-fuel combustion with variable valve actuation, continued reduction in friction and parasitic losses, and incorporating an advanced organic Rankine cycle (ORC) as a WHR system, it will be able to achieve the 55 percent BTE target. An overview of the incremental gains it expects from these technologies is shown in Figure 3-5.

Progress and Fundamental Programs Toward Overcoming Technical Barriers to Achieving 55 Percent BTE (Goal 2)

As seen in the descriptions of the SuperTruck team's activities, advanced combustion strategies and sophisticated CFD modeling are essential parts of their technical roadmaps to achieving 55 percent BTE. The requisite understanding of the fundaments of advanced combustion strategies, like LTC, and incorporation of that understanding into usable CFD codes is the focus of the individual research programs within 21CTP. Success in achieving Goal 2 will depend on the advancements being made within the individual 21CTP research programs.

The Partnership has increased its emphasis on incorporating their research results into simulations or conceptual models that can be used by stakeholders for either predictive simulation or for comparative analysis with laboratory results to gain an understanding of the data that is not achievable through routine analysis. Additionally, the development of phenomenological models, which conceptually model the different processes occurring within the engine, helps others in the field to understand the differences between processes that appear to be similar globally but are fundamentally different, e.g., different LTC approaches relative to conventional diesel or direct ignition combustion.

The 21CTP has been successful in its engine research efforts to increase BTE. Advanced CFD is being used extensively by all of the SuperTruck engine teams. Optimization of the combinations and interactions of the myriad of parameters that affect the efficiency and emissions of the complex engine systems could not have been done without advanced CFD. Details of the fuel spray breakup, how it depends on what is occurring inside the nozzle, and how the fuel then mixes with the combustion chamber gases and is impacted by the fluid motion within the chamber, along with the influence of the composition of the chamber gases on the nature of the energy conversion process, all impact the efficiency and emissions of the engine. As the industry tries to push the limits of efficient engines by lowering the engine-out emissions, an understanding of these details, as well as other phenomena like the localized boundary layer heat transfer, the state of thermal gradients within the chamber, and the evolution of the fuels' reactivity, becomes critical. CFD simulations with accurate submodels of the thermodynamic, chemical, and physical processes occurring in the engine and the aftertreatment systems enable these activities, and will need to be advanced further to facilitate achieving Goal 2, 55 percent BTE.

DOE programs have made significant contributions to the capabilities of the CFD programs. The continued development of more accurate, high-fidelity, kinetic routines for different fuel mixtures has been an important contribution. An industrial collaborator, Convergent Science, has licensed a kinetic solver developed through DOE research programs and is now using its code to do simulations with these advanced kinetic routines and fluid mechanic models for many in the engine industry. Argonne National Laboratory (ANL) is preparing to launch a program called the Virtual Engine Research Institute and Fuels Initiative (VERIFI),[4] an organization that will be available to industry and that integrates high-performance computing, fuel chemistry, and combustion science and engine performance with some of the world's fastest supercomputers. This organization will facilitate simulations that industry does not have the capital resources to do but that are important to achieving the goals

[4] For more on the VERIFI program at ANL, see http://verifi.anl.gov/.

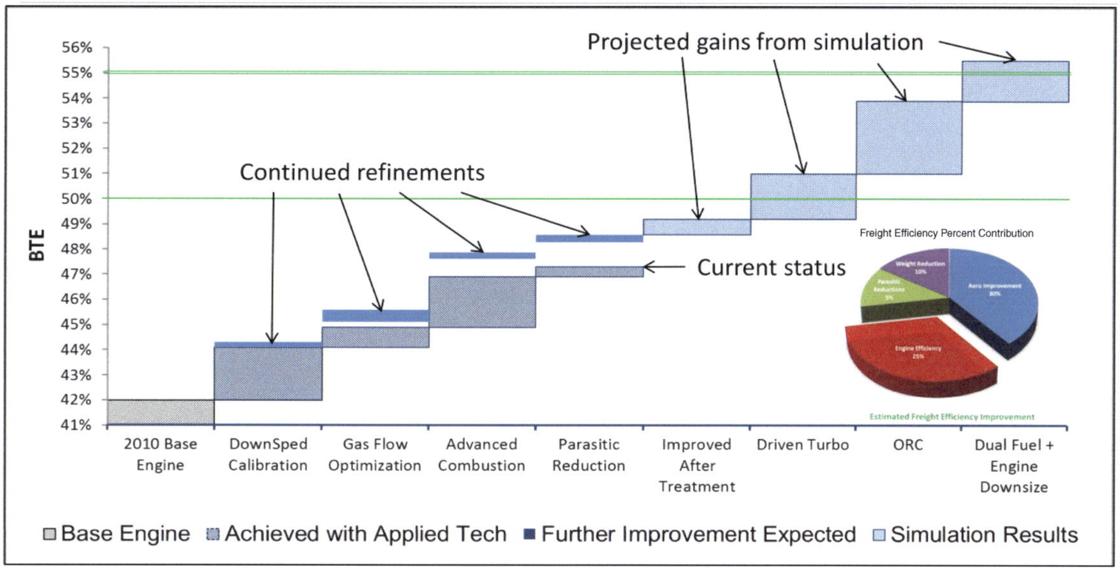

FIGURE 3-5 Navistar SuperTruck team's projected improvements in the engine technologies that will enable them to achieve 55 percent BTE. SOURCE: R. Nine and R. Zukouski, "SuperTruck–Development and Demonstration of a Fuel-Efficient Class 8 Tractor & Trailer Vehicle System," presentation to the committee, November 18, 2014, slide 12.

set for engine systems relating to reduced fuel consumption and lower emissions.

Because of the importance of CFD as a development tool, along with the rapid pace at which computing technology is changing, DOE held a high-performance computing workshop on August 19-21, 2014. The purpose of the workshop was to get feedback from users and stakeholders on the best way to integrate new research results into simulations. Discussion was on topics such as these: Is there a preferred platform by which newly developed subprograms can be publically demonstrated and critiqued? What is the appropriate role of the government in this arena? The outcomes of that workshop were not available at the writing of this report, so they could not be included in this report.

Finally, an anecdote about the success of the 21CTP: During a site visit to the Tank Automotive Research, Development and Engineering Center (TARDEC) on November 24, 2014, a subgroup of the committee was told that results from the 21CTP have been influential in production decisions on light- and medium-armored vehicles and on decisions about power plants and powertrains.

Individual DOE and DOD Engine Systems Projects and Funding Levels

As part of the review process, the Partnership supplied to the committee a listing of the projects in its research portfolio. The projects from that list that the committee has interpreted as falling within the advanced engines arena are listed in Table 3-4. They include the TARDEC Automotive Research Center, run by the University of Michigan, and the DOE programs that are focused on advancing the fundamental understanding of engine processes. As mentioned above, a committee subgroup visited TARDEC as part of the fact-finding effort. During that visit, the subgroup was told about an advanced engine program in which innovative combustion processes and an alternative engine architecture are being assessed as a high-power-density engine with improved efficiency. The presentation on this work indicated that the opposed-piston two-stroke diesel engine met or exceeded program requirements for improved BTE, heat rejection to coolant, power density, and a 50-hr durability test. In the request for proposals for the next phase of this development program, DOD specifically required the opposed-piston architecture. The results of the proposal evaluations have now been made public. TARDEC awarded two contracts, one to the Achates-Cummins team and the other to the AVL group, to design, build, test, and evaluate advanced single-cylinder (SC) opposed-piston engine technology for potential future combat vehicle applications. Both teams were given the same combat engine performance parameter targets that are representative of multicylinder engine performance expectations. The performance parameters include these:

(1) Specific heat rejection of 0.45 Kw/kw, which includes charge air cooling, water jacket cooling, and engine oil cooling,
(2) A best brake specific fuel consumption point of 0.32 lb/bhp-hr,
(3) A rated speed air:fuel ratio not to exceed 30:1,
(4) A targeted rated speed of 2,600 rpm,

TABLE 3-4 Research Projects Identified by 21CTP as Part of the Engine Systems Program (dollars)

Public Review Project No./Title	Recipient	Funding 2012	Funding 2013	Funding 2014
DOD017 Automotive Research Center	University of Michigan	←	Funding not provided	→
ACE001 Heavy-Duty Low-Temperature and Diesel Combustion and Heavy-Duty Combustion Modeling	Sandia National Laboratories (SNL)	815,000	805,000	825,000
ACE004 Low-Temperature Gasoline Combustion (LTGC) Research	SNL	760,000	740,000	720,000
ACE005 Spray Combustion Cross-Cut Engine Research	SNL	730,000	740,000	950,000
ACE007 Large Eddy Simulation (LES) Applied to Low-Temperature and Diesel Engine Combustion Research	SNL	500,000	450,000	200,000
ACE010 Fuel Injection and Spray Research Using X-Ray Diagnostics	Argonne National Laboratory (ANL)	1,100,000	1,000,000	850,000
ACE012 Model Development and Analysis of Clean and Efficient Engine Combustion	Lawrence Livermore National Laboratory (LLNL)	520,000	740,000	475,000
ACE013 Chemical Kinetic Models for Advanced Engine Combustion	LLNL	620,000	600,000	550,000
ACE014 2014 KIVA Development	Los Alamos National Laboratory (LANL)	720,000	763,000	695,000
ACE015 Stretch Efficiency for Combustion Engines: Exploiting New Combustion Regimes	Oak Ridge National Laboratory (ORNL)	350,000	350,000	300,000
ACE052 Neutron Imaging of Advanced Transportation Technologies	ORNL	200,000	200,000	200,000
ACE054 Collaborative Combustion Research with Basic Energy Sciences	ANL	400,000	320,000	325,000
ACE075 Advancement in Fuel Spray and Combustion Modeling for Compression Ignition Engine Applications	ANL	350,000	500,000	350,000
ACE076 Improved Solvers for Advanced Engine Combustion Simulation	LLNL	340,000	340,000	475,000
ACE077 Cummins ORNL\FEERC Combustion CRADA: Characterization and Reduction of Combustion Variations	ORNL	300,000	300,000	300,000
Total federal dollars		7,705,000	7,848,000	6,520,000

NOTE: Information provided includes project number, title, lead organization, and federal dollars supporting the program.

(5) A rated speed power of 250 bhp and a minimum peak torque of 500 ft-lb,
(6) Military fuel use compatibility encompassing jet and diesel fuels, and
(7) Steady-state and transient smoke targets not to exceed visible limits. Both efforts also include a conceptual multicylinder engine study that targets representative combat vehicle claim space.

Although not officially categorized as part of the 21CTP program, this opposed-piston engine technology is worthy of mention because exploration of nonconventional engine architectures is an area of interest for the 21CTP.

The total 2014 federal budget for all of the DOE Engine Systems projects listed in Table 3-4 is $6.52 million. The committee was not given the federal dollar budget for the DOD-funded Automotive Research Center at the University of Michigan, so a sum total of all the engine research activities aside from the SuperTruck engine program is not known. Investigators for each of the engine systems projects with DOE funding are required to submit quarterly progress reports, participate in semiannual research progress meetings, and give a presentation at the DOE AMR. Each AMR presentation states the project's relevance, the budget, milestones for the project, the technical approach, accomplishments, lists of collaborators, and future work. The presentations and the reviewers' comments are available to the public.[5]

Brief Summary of DOE Individual Programs

The Engine Systems research programs listed in Table 3-4 span a range from fundamental experimental work, to kinetic mechanism development, to mechanism evaluation and simplification, to development of advanced numerical methods, and further development of the computational codes. The program is well managed and there is good collaboration and synergy between the individual DOE 21CTP engine projects. Brief summaries highlighting the accomplishments for each of the DOE projects shown in Table 3-4, along with links to the 2014 Annual Merit Review presentations, are given below.

Project ACE001. Using in-cylinder optical imaging, Musculus has developed a conceptual model of direct injected LTC, and the bridging between conventional combustion and LTC (2014). The model along with the in-cylinder imaging shows the spatial and temporal evolution of soot precursors. These soot evolution histories compared favorably with simulations that were performed as part of a collaboration with the University of Wisconsin. Musculus's research has also shown how injection rate shapes affect postinjections and how piston bowl geometry affects multiple injections.

Project ACE004. Dec and collaborators General Motors, Cummins, LLNL, University of California-Berkeley, University of Melbourne, and Chevron have demonstrated a peak indicated thermal efficiency of 49.8 percent and were able to explore the maximum load that could be achieved using homogeneous charge and direct injection partially stratified charge compression ignition of gasoline-like fuels (Dec, 2014). Maximum loads in excess of 16 bar BMEP were achieved, and it was concluded that significant noise reduction could be achieved with a minimal loss of thermal efficiency.

Project ACE005. Through the research efforts of Pickett and his research team, the Spray Combustion Cross-Cut Engine Research Network continues to grow (Pickett and Skeen, 2014). This network represents a collaboration among approximately 20 international laboratories, industries, and universities dedicated to a coordinated experimental and computational evaluation of engine-relevant spray conditions for the purpose of developing predictive computational tools that can be used by industry. The dissemination and collaboration is done through Sandia's Engine Combustion Network.[6]

Project ACE007. In conjunction with the modeling efforts taking place as part of the Engine Combustion Network, Oefelein and his research team are continuing the development of large eddy simulation (LES) to facilitate more accurate spray and fluid mixing simulations (Oefelein et al., 2014). Trying to understand the underlying causes of cycle-to-cycle variation, correctly simulating the differences between gasoline and diesel sprays, and predicting the effects of internal nozzle geometry on the spray processes in the cylinder is currently outside the precision and fidelity of the models. Such work provides a link between the DOE Office of Science and the VTO.

Project ACE010. Dr. Powell and his research team at ANL are using the laboratory's unique Advanced Photon Source (APS) to perform x-ray measurements of near-nozzle and intra-nozzle phenomena on production-type fuel injectors (Powell, 2014). They have been able to make detailed measurements of the internal nozzle needle wobble that occurs during injection and have measured the cavitation of the fuel inside the nozzle and the effects these phenomena have on injection and on injection variation. These results have been incorporated into the work of the Engine Combustion Network. Collaborators in this work include Delphi, Caterpillar, the University of Massachusetts-Amherst, and computational colleagues at ANL.

Project ACE012. The simulation of advanced compression ignition combustion processes, often generically referred to as LTC, requires detailed high-fidelity kinetic

[5] See Vehicle Technologies Office: Annual Merit Review and Peer Evaluation at http://energy.gov/eere/vehicles/vehicle-technologies-office-annual-merit-review-and-peer-evaluation.

[6] See the Engine Combustion Network at http://www.sandia.gov/ecn/.

representation of the fuel and the in-cylinder fluid mechanics. Whitesides and his research team at LLNL are working on developing faster and more accurate combustion solvers to facilitate these calculations and evaluate the results using these solvers (Whitesides et al., 2014). The emphasis of this project is to use the advanced solver as a means of validating detailed engine and combustion modeling tools through simulation of LTC results from a variety of collaborators.

Project ACE013. In conjunction with the more efficient solvers, it is important to also have comprehensive, high-fidelity kinetic routines to simulate the in-cylinder combustion process. This is the focus of Dr. Pitz and his research team's efforts (Pitz et al., 2014). They are developing predictive chemical kinetic models for gasoline, diesel, and next-generation fuels by creating surrogate fuels, which are fuel blends in which the number of components in the fuel is computationally manageable. They are developing models for FACE, which include blends that have been specified for researchers to represent a matrix of fuels in which the properties vary over a range that might be expected in the future as feedstocks change. The diesel surrogate fuel under investigation currently has nine components, a mixture of selected n-alkanes, isoalkane, cycloalkane, one- and two-ring aromatics, and a napthoaromatic. Their current gasoline surrogate has 10 components. Their work involves collaboration with many who are performing experiments and simulations to assess the representativeness of the surrogate fuels to actual fuels, and to assess the accuracy of their kinetic routines.

Project ACE014. At LANL, DOE is supporting Dr. Carrington to write the next version of KIVA, the open source software program that has been used extensively as the framework for past engine CFD simulations (Carrington, 2014). Dr. Carrington is collaborating with the University of New Mexico, Purdue University, Calumet Specialty Products Partners, L.P., the University of Nevada-Las Vegas, and many KIVA users. The new version, KIVA 4, uses a high-performance finite element method in a modular object-oriented parallel processing code. This is being coupled to faster grid generation capabilities. The general topic of advanced CFD program development, which includes the KIVA 4 program, was the focus of DOE's High Performance Computing Workshop.

Project ACE015. At ORNL, DOE is supporting analysis of fundamental thermodynamic strategies and implementation methods that could provide an increase in efficiency that would be revolutionary rather than evolutionary. Daw and his colleagues are analyzing the potential of reformate assisted dilute combustion through thermochemical recuperation (Daw et al., 2014). They are looking at steam reforming the fuel—octane, ethanol, or methanol, for example—to maximize the fuel's exergy while facilitating highly dilute combustion, which would reduce heat transfer and improve the working properties of the gas. Catalyst performance experiments have been performed and an engine test is under development. The research team is collaborating with SNL, the Gas Technology Institute, Cummins, the University of Michigan, and Pennsylvania State University.

Project ACE052. Toops and his research team are exploring the use of nondestructive neutron imaging to visualize the internal flow dynamics in fuel injectors and the buildup of soot and ash in diesel and gasoline particulate filters (Toops et al., 2014a). Images can be obtained at a single cross section or a complete reconstruction can be constructed to provide a cross section of the entire sample at a resolution on the order of 10-20 microns. Voids in the nozzle's fuel reservoir can be detected. This work is in collaboration with the DOE Office of Basic Energy Sciences, the University of Tennessee, MIT, the University of California, GM, and NGK Spark Plugs.

Project ACE054. Goldsborough and colleagues at ANL are using their rapid compression machine to acquire fundamental data that will be used to develop and evaluate kinetic routines for transportation-relevant fuels at conditions representative of advanced combustion regimes (Goldsborough et al., 2014). This work is being done in collaboration with DOE Basic Energy Sciences, LLNL, King Abdullah University of Science and Technology (KAUST) and Chevron, the University of Wisconsin, and the DOE working groups on HCCI and diesel engines. The behavior of FACE and the comparison to the predictions of that behavior using kinetic mechanisms from the surrogate models is one of their projects. They are also evaluating the impact of fuel additives such as ethylhexyl nitrate (EHN).

Project ACE075. Som and his colleagues, also of ANL, are pursuing advances in fuel spray and combustion modeling for compression ignition engine applications (Som et al., 2014). This is a comprehensive program with collaborations among other groups at ANL, Convergent Science, Caterpillar, Cummins, LLNL, the Sandia Engine Combustion Network, the Advanced Engine Combustion Working Group, the University of Connecticut, and the Politecnico di Milano and University of Perugia. Simulations are being done for flow inside the nozzle tip, with 50 million computational cells, using the data from the images of the internal nozzle and needle tip motion obtained in the advanced photon source research. The simulations show the complexity of the flow inside the nozzle, how it is impacted by needle wobble, and how it impacts the spray behavior when the flow enters the cylinder. These are the capabilities that will be made available to participants in the VERIFI program, mentioned above.

Project ACE076. McNenly is the principal investigator of the program at LLNL, which is focused on the development of improved solvers for advanced engine combustion simulation (McNenly et al., 2014). Colleagues of this research team are also evaluating the effectiveness of the developed solvers to validate detailed engine and combustion models for a variety of LTC engine results, see discussion above. In the development of the solver, better algorithms and applied mathematics are being coupled with new Graphical Processor Unit (GPU) computing architecture to facilitate inclusion of improved physical submodels for better accuracy and smaller error. This is especially important as the number of species in the simulation grows. Currently, the team has demonstrated a 4.8-fold speedup over a conventional modern code for a simulation containing 2,000 species. Collaborators in this work include Cummins, Ford, Volvo, Bosch, GE Research, Convergent Science, Nvidia, ANL, National Renewable Energy Laboratory (NREL), SNL, FACE, the Advanced Engine Combustion working group, and multiple universities.

Project ACE077. Finally, in a cooperative research and development agreement (CRADA) between ORNL and Cummins, Partridge and colleagues are developing and applying an advanced EGR probe to help characterize and reduce combustion variations (Partridge et al., 2014). The probe is a laser-based multiplex EGR probe that measures CO_2. Measurements are being made in the intake manifold to assess charge components and fluctuations by measurement of the residual gas backflow and external EGR. When combined with models, the nature of the residual gas in the cylinder can be predicted. This in turn will facilitate control of advanced combustion strategies. The team is also developing a multicolor, multispecies EGR probe that measures CO_2, water, and the temperature of the cylinder charge components. Other collaborators include the Cummins SuperTruck engine team and the University of Central Florida. This project was given a 2013 R&D 100 Award.

As seen in the brief summaries given above, the Engine Systems research programs range from fundamental experimental work, to kinetic mechanism development, evaluation, and reduction, to advanced numerical methods, and to further development of the computational codes. The committee believes that the program is well managed and there is much collaboration and synergy between the individual projects of the DOE 21CTP engine projects. And, the program is addressing some of the important technical barriers standing in the way of achieving 55 percent BTE, Goal 2.

Approaches to Goal 3

Goal 3: Through experiments and models with FACE and other projects, determine the most essential fuel properties, including renewables, to help achieve 55 percent engine brake efficiency (by 2014).

(When asked for clarification of the intent of Goal 3, Kevin Stork, DOE, responded that the fuels research in the 21CTP was to "support experimental and modeling work to determine the impacts of fuel properties on enhancing (or hindering) attainment of advanced combustion modes, such as LTC, over a greater portion of an engine map.")

Status: A more detailed discussion of the fuels and lubricant research within the 21CTP is given in a separate section later in the chapter, so only general comments will be made here. The committee feels that fuel research is much more important than what is stated or implied in Goal 3. As mentioned in the preceding summary of DOE engine research, FACE provides researchers with the ability to perform experiments with fuels of known characteristics, having property ranges that are within the range of variations that might be seen in future fuels. This is superior to running specific blends of research grade fuels that are not representative of what an engine will experience in the field. Using FACE also helps with the kinetic model development being pursued in the surrogate fuel simulation program. Researchers can now test their advanced kinetic models against realistic, but known, fuels in real engines, an important step in developing simulation capabilities for predictive behavior. The committee believes a more detailed understanding of the impact of fuel characteristics on engine operation and potential facilitation of advanced combustion will also enable the high-level objective of maximizing the utility of our fossil fuels, thus reducing their use.

Partnership Responses to NRC Phase 2 Recommendations: Engine Systems

NRC Phase 2 Recommendation 3-1. The 21CTP fundamental research program should continue to provide important enablers for the 55 percent BTE goal, and DOE should continue to look for leverage opportunities with other government- and industry-funded projects.

21CTP Response: The Partnership agrees with the need to continue research toward the 55 percent thermal efficiency goal, and has included this as a research goal for the SuperTruck partners (with technology scoping toward this goal being the major activity).
The Partnership will continue to look for new opportunities to work together: one possible new collaborative arena is the recently announced partnership between DOE and the U.S. Army (the Advanced Vehicle Power and Technology Alliance). DOE is working with the U.S. Army to identify areas of common interest that could result in collaborative research efforts.

NRC Phase 2 Recommendation 3-2. The DOE should ensure that the engine R&D for the goal of 50 percent BTE at over-the-road cruise conditions and the stretch goal of 55 percent BTE in an engine in a laboratory that will now be carried out under the SuperTruck program receive the appropriate share of the SuperTruck funding and benefit extensively from the DOE-funded research programs in advanced engine combustion.

21CTP Response: Participating SuperTruck companies are also involved in the rest of the VTP R&D program (the advanced combustion MOU, the advanced engine crosscut team, and the Annual Merit Review), and are thus made aware of the DOE-funded advanced engine combustion programs. DOE's Annual Merit Review included the SuperTruck team members as active participants, and presented the entire research portfolio to them. This ensures that SuperTruck teams are aware of the portfolio and can harvest breakthrough results for their use

NRC Phase 2 Recommendation 3-3. The DOD and the DOE should increase their awareness of one another's programs and look for opportunities to share technologies on areas of joint interest, such as thermal efficiency. One way to encourage interaction is for the DOE to invite DOD program participants to present their findings at the DEER (Diesel Engine-Efficiency and Emissions Research) Conference.

21CTP Response: In 2011, DOE and the U.S. Army announced the formation of a research collaboration, the Advanced Vehicle Power and Technology Alliance. DOE is working with the U.S. Army to identify areas of common interest that could result in collaborative research efforts. This partnership should enhance the interaction between these federal departments: some areas of collaboration have already been identified. The U.S. Army also participates in meetings of the Diesel Crosscut Team and the light-duty USCAR partnership with DOE and industry partners. Incorporation of DOD presentations at the yearly DEER meeting will also be considered: DOD has presented papers at DEER in the past, and DOE's role as the chair for the meeting will ensure that DOD can have access to presenter slots as needed.

Committee Comment on 21CTP Responses

The committee is pleased with the responses by 21CTP to the NRC Phase 2 recommendations. Unfortunately there has not been a DEER Conference since the Phase 2 review, so DOD participation has not been possible. The committee commends DOE and DOD for the formation of the Advanced Vehicle Power and Technology Alliance.

Findings and Recommendations: Engine Systems

Finding 3-1. The 21CTP has successfully met Goal 1, to develop and demonstrate an emissions-compliant diesel engine system for Classes 7 and 8 highway trucks that achieves 50 percent brake thermal efficiency in an over-the-road cruise condition. The engine uses a waste heat recovery system.

Finding 3-2. The projects in the engine systems portion of 21CTP represent a closely coordinated set of research activities that are pursuing a better fundamental understanding of processes critical to efficient engine operation. Fundamentals associated with fuel injection, sprays, gas exchange, in-cylinder flows, advanced combustion processes, plus comprehensive yet robust kinetic routines for realistic fuels are being investigated. The learning from these activities is being incorporated into models, both detailed and phenomenological, which serve as tools for advanced engine development. Integral to this effort is the continued advancement of the base computer program itself and the solvers that facilitate rapid computational turnaround time. The program is well managed and interfaces well with industry stakeholders.

Finding 3-3. The 21CTP has realized the importance of transferring the new knowledge generated in its research programs into the stakeholder community and is active in disseminating this learning via appropriate forms and forums, such as the development of computer submodels that can be used by other researchers in the field, and through user groups such as the Engine Combustion Network, to maximize leverage and learning obtained from the research by encouraging broad base participation within the scientific community.

Finding 3-4. Increased emphasis has been placed on issues such as numerical algorithm development, advanced computer architectures, and CFD code development. The Partnership's awareness of the importance of these activities was evinced by the high-performance computing workshop DOE sponsored in August 2014.

Recommendation 3-1. With the increased importance of advanced CFD for developing the engines and operating scenarios necessary for minimum fuel consumption and in light of DOE's role in the generation of new knowledge that gets incorporated into these CFD codes as submodels, a critical review of the Partnership's program to develop the next-generation code (KIVA 4) should be performed. Feedback from the participants of the high-performance computing workshop should be matched against the current code development activities, and the adequacy of the current program should be assessed. If necessary, the next-generation code development should be adjusted.

Finding 3-5. Achieving Goal 2, 55 percent BTE in a laboratory engine, will be very challenging. This is a high-risk, high-reward fundamental research program. It is an important stretch goal because it will facilitate identifying the potential of different advanced engine, fuel, and combustion concepts for increased engine efficiency, even though these concepts may not be commercially viable in the near future.

Recommendation 3-2. The fundamental diesel engine research program pursuing advanced technologies and combustion processes and engine architectures to achieve 55 percent BTE should continue to be a focus of the 21CTP engine activities. However, the experiments and modeling should maintain a focus on dynamometer R&D, as opposed to attempting to build a demonstration vehicle. The achievement of this goal should be extended from 2015 to 2020, in order to have sufficient time to carry out R&D on this stretch goal. Also, this activity should not be at the expense of efforts to reduce load-specific fuel consumption via system integration and road load reductions.

AFTERTREATMENT SYSTEMS

Introduction

HD diesel aftertreatment systems have evolved worldwide as separate systems. Europe was developing and optimizing the SCR systems to meet Euro IV and V regulations (2005, 2008 respectively), while Japan and the United States were developing diesel particulate filter (DPF) technology. Both technologies came together to meet the US2010 and Euro VI (2013) HD regulations. New U.S. regulatory requirements went into effect in 2013 and 2014, when OBD (2013) and the first phase of the medium- and heavy-duty vehicle fuel efficiency and GHG regulations (2014) were introduced.

More-efficient SCR systems allow higher engine-out NO_x, resulting in further reduction in fuel consumption and low engine-out PM levels. The NO_2 levels coming out of the diesel oxidation catalyst (DOC) are sufficient in many 2013 engines to oxidize the PM retained on the filter without the need for high-temperature active regeneration. This resulted in filters with less PM mass and lower back pressure. An example of a modern emission control system architecture is shown in Figure 3-6.

The required OBD system adds significant complexity, with upwards of 18 control points, as illustrated in Figure 3-7 (Stanton, 2013). The OBD system is needed to diagnose deficiencies in the emissions control system and allow the defective parts to be identified to facilitate remediation. Major industry efforts are being expended on OBD, and emissions control system choices are always made in the context of OBD requirements.

The results of these efforts to date are quite impressive. In many cases the tailpipe concentration of fine particles is less than that of ambient air. NO_x reductions are approaching 98 percent from engine-out levels. In Europe, trucks have lower NO_x emissions per kilometer than modern diesel cars (Bergmann, 2013).

California is now independently considering another 90 percent reduction of the HD NO_x tail pipe standard for around 2020 (CARB, 2015). EPA may consider following with similar tightening depending on the level of the new National Ambient Air Quality Standard (NAAQS) ozone standard, proposed in December 2014 to be in the range of 65 to 70 ppb. To have minimal impact on fuel consumption, these new tail pipe NO_x levels (~0.02 g/bhp-hr) will require nominally 99.5 percent NO_x reductions on the hot federal test procedure (FTP) cycle and 96 percent reductions on the cold FTP cycle, both of which depend on additional innovations in emissions control technology. The California initiative will stimulate new approaches to HD NO_x aftertreatment, particularly related to cold start emissions. Some technologies being considered are SCR filters (SCR catalyst coated on DPF) to

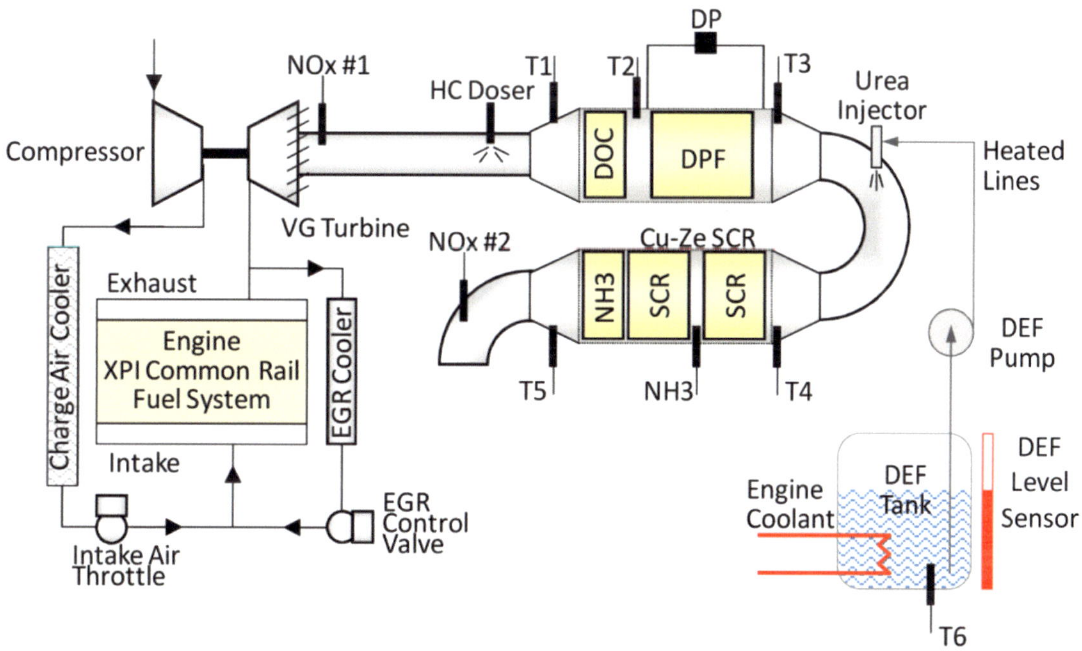

FIGURE 3-6 Layout of a modern HD diesel emission control system. SOURCE: D. Stanton, "Systematic Development of Highly Efficient and Clean Engines to Meet Future Commercial Vehicle GHG Regulations," *SAE Int. J. Engines* 6(3): 1395-1480. Reprinted with permission from SAE paper 2013-01-2421Copyright © 2013 SAE International.

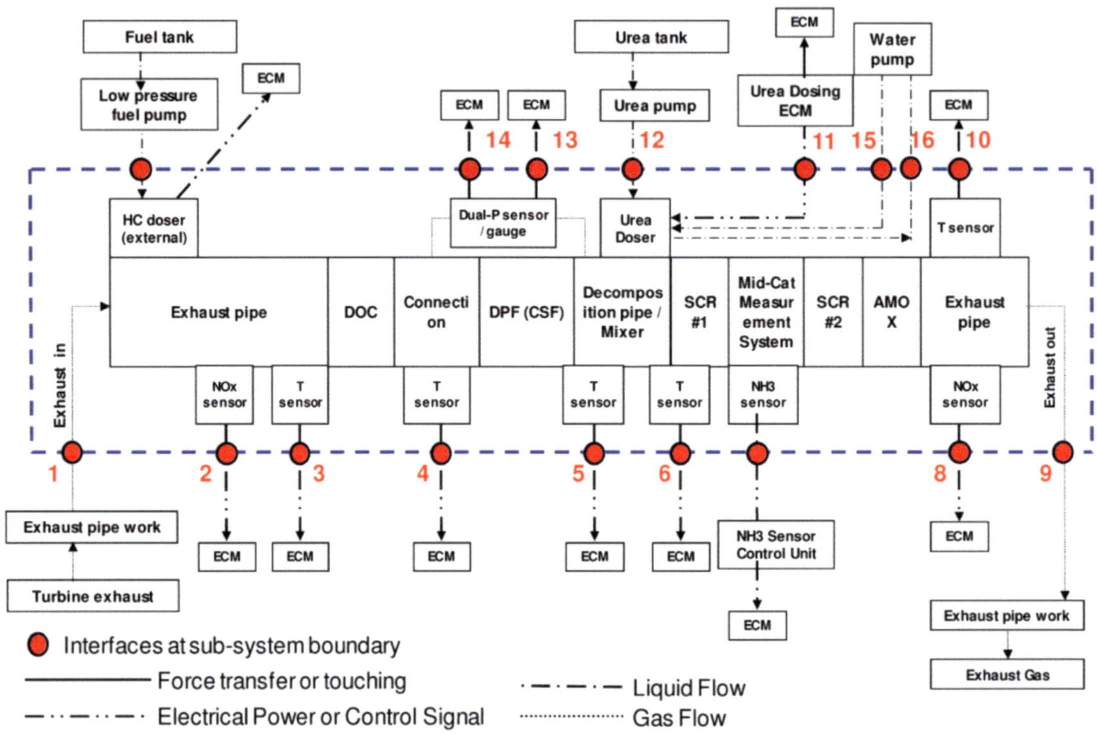

FIGURE 3-7 Example of OBD layout for a 2013 HD aftertreatment system. SOURCE: D. Stanton, "Systematic Development of Highly Efficient and Clean Engines to Meet Future Commercial Vehicle GHG Regulations," *SAE Int. J. Engines* 6(3): 1395-1480. Reprinted with permission from SAE paper 2013-01-2421 Copyright © 2013 SAE International.

add SCR catalyst closer to the engine, and low-temperature NO_x adsorbers that release the NO_x at higher temperatures when the SCR is functional. Further, at such high deNO_x efficiencies, proper management of the diesel emission fluid (ammonia) will be critical to prevent the formation of N_2O, a powerful greenhouse gas. Expected OBD requirements at these very low tail pipe NO_x levels are not achievable with today's sensor and modeling technology.

Although 21CTP has no specific aftertreatment goals, in the February 2013 Roadmap (21CTP, 2013, p. 45), 21CTP listed several aftertreatment elements to the overall technical strategy:

- High-efficiency SCR.
- Resolve remaining issues on DPF regeneration, ash loading and removal, and aging.
- Mitigate sulfur effects.
- Improve the catalyst materials and systems for lean NO_x catalysis with urea and other reductants for performance over a wider temperature range while minimizing reductant slip.
- Develop monitors and thresholds for sensors in controls and diagnostics in conjunction with OBD. Develop and use fundamental knowledge of catalysts and sensors for OBD methods.
- Materials for catalysts and filters that have high efficiency, low back pressure, and minimal space requirements for at least 1 million miles of durability.
- Robust sensors with direct sensing of emissions constituents (e.g., PM, N_2O).

Aftertreatment Projects

The NRC Phase 2 report (2012) put the total spending during the previous 7 years of 21CTP heavy-duty truck aftertreatment work (through FY 2010) at about $37 million. Spending FY 2011 through FY 2014 was about $13 million, for a total of about $50 million of aftertreatment-related funding over 11 years ($4.5 million per year average). Table 3-5 describes the expenditures on active aftertreatment projects reported to the committee by 21CTP since the NRC Phase 2 review.

The aftertreatment research and development community is quite active, with upwards of 400 technical papers and presentations annually presented worldwide on industry- and government-funded work. In the opinion of the committee, the body of work sponsored by 21CTP ought to complement, not duplicate, the industrial programs. Following is a summary of the 21CTP project progress with comments on corollary work from outside the DOE projects.

TABLE 3-5 Expenditures on 21CTP Aftertreatment Projects (dollars)

Public Review Project No./Title	Recipient	2012 Funding	2013 Funding	2014 Funding	Note
ACE022 Joint development and coordination of emissions control data and models (CLEERS analysis and coordination)	ORNL	350,000	712,000 (700,000)*	558,000 (650,000)*	*According to 2014 AMR
ACE023 CLEERS aftertreatment modeling and analysis	Pacific Northwest National Laboratory (PNNL)	750,000	750,000	750,000	
ACE026 Enhanced high- and low-temperature performance of NO_x reduction materials	PNNL	300,000	300,000	300,000	Funding matched by Cummins in CRADA
ACE028 Experimental studies for CPF and SCR model, control system, and OBD development for engines using diesel and biodiesel fuels	Michigan Technological University	607,000			Project completed in FY 2012; 323,000 matched funding
ACE032 Cummins/ORNL-FEERC CRADA: NO_x control and measurement technology for heavy-duty diesel engines, self-diagnosing smart catalyst systems	ORNL	450,000	595,000 (400,000)*	232,000 (350,000)*	*According to 2014 AMR; funding matched by Cummins in CRADA
ACE089 Development of radio frequency diesel particulate filter sensor and controls for advanced low-pressure drop systems to reduce engine fuel consumption	Filter sensing technologies	487,000	386,000*	836,000*	*From 2014 AMR; total private share adds 565,000
Totals		2,944,000	2,743,000	2,676,000	

Crosscut Lean Exhaust Emission Reduction Simulation (CLEERS) Program

The Joint Development and Coordination of Emissions Control Data and Models (ACE022 and ACE023) is a project managed by ORNL with subprojects managed by the Pacific Northwest National Laboratory (PNNL). The core activities are to support and coordinate emissions control research, which evolves with DOE priorities and industry needs. Efforts are communicated to the 22 industrial partners, 11 universities (including three in Europe), and two national laboratories through monthly teleconferences and an annual workshop that is open to the public. The 2014 workshop had more than 100 attendees, 39 technical papers, and 12 posters. Topics most pertinent to the 21CTP included diesel particulate characterization and filtration; SCR catalysts, reaction mechanisms, and modeling of urea spray; oxidation and reforming catalysts; passive adsorbers and traps; multifunctional catalysts and aftertreatment devices; emissions controls and engine integration; low-temperature catalysis; interpretation of experimental aftertreatment measurements; development of microkinetic and global reaction mechanisms; drive-cycle simulations of conventional and hybrid vehicles; and engine exhaust speciation. Examples of recent accomplishments of CLEERS are the provision of basic data in support of vehicle systems aftertreatment modeling; the establishment of a new online database for references relevant to modeling of emissions control devices; the analysis and reporting of results from a 2013 industry priority survey; the measurement of NH_3 storage isotherms on a commercial small pore Cu zeolite; the development and application of analytical techniques for extracting adsorption enthalpies from isotherm data; and the development of reaction mechanisms for NO SCR reactions that are consistent with reaction rate measurements and diffuse reflectance infrared spectroscopy (DRIFTS) observations. Future work will continue mechanistic investigations into small pore Cu zeolite and candidate NO_x adsorber materials, with emphasis on low-temperature operating conditions and will initiate the characterization of passive adsorber materials and protocols for their development.

Emissions Projects

Project ACE026. The CRADA project "Enhanced High and Low Temperature Performance of NO_x Reduction Materials" focuses on determining factors that limit low- and high-temperature NO_x performance, including mechanisms for deactivation for candidate materials due to hydrothermal aging and poisoning mechanisms. NO_x adsorber work that ended in 2014 shows enhanced potassia-titania high-temperature NO_x storage catalysts deactivated through irreversible reaction of the two oxides. Work is now focused on preparing and modeling three emerging SCR catalysts with improved low-temperature and high-temperature performance. Model Cu/SAPO-34, Fe/SSZ-13, and SSZ-13 with various Si/Al ratios have been prepared for a number of studies of low- and high-temperature performance of commercial Cu-chabazite (CHA)-based SCR catalysts. These studies led, in part, to the identification of SCR catalyst materials with significantly lower (up to 20°C lower) "light-off" temperatures than the contemporary Cu-SSZ-13 catalyst. Future work will focus on limitations of low- and high-temperature performance of CHA-based SCR catalysts.

Project ACE028. The project "Experimental Studies for CPF and SCR Model, Control System, and OBD Development for Engines Using Diesel and Biodiesel Fuels" was completed in September 2012. A core aspect of the project was communication and collaboration between stakeholders to facilitate the achievement of emissions regulations with minimal fuel penalty for a wide range of engines, including those operating on diesel or biodiesel fuel. The project developed DOC, catalyzed particulate filter (CPF), and SCR reduced-order models and estimator strategies that were validated on engines for use in calibration efforts. Importantly, an industrial consortium was formed in 2014 to continue this work.

Project ACE032. The Cummins/ORNL-FEERC CRADA, "NO_x Control & Measurement Technology for Heavy-Duty Diesel Engines, Self-Diagnosing Smart Catalyst Systems," is aimed at developing diagnostics to promote the understanding of both the SCR catalyst and the impact of aging on catalyst performance, focusing on distributed NH_3 storage and NO_x conversion performance. Accomplishments include assessment of impacts of hydrothermal laboratory aging on commercial SCR catalyst functions of NH_3 capacity, the SCR reaction, parasitic NH_3 oxidation, NH_3 oxidation characterization, and determining that the common approach using capillary sampling was noninvasive. Future work will be to extend the work to field aging and assess aging impacts via experimental correlations and comparison to catalyst models.

Project ACE089. "Development of Radio Frequency Diesel Particulate Filter Sensor and Controls for Advanced Low-Pressure Drop Systems to Reduce Engine Fuel Consumption," has the objectives of developing radio frequency (RF) sensors and controls for direct, in situ measurements of DPF soot and ash levels; quantifying associated fuel savings; exploring additional efficiency gains with advanced combustion modes, alternative fuels, and advanced aftertreatment via RF sensing and control; and developing production designs and commercialization plans. The investigators completed preproduction RF sensors and antennas; demonstrated combined DPF soot and ash measurements; benchmarked RF transient response with an established microsoot sensor; evaluated RF performance over 240,000-mile equivalent DPF aging; and quantified fuel savings potential via extended regeneration intervals and reduced regeneration duration of about 50 percent relative to stock original equipment manufacturer (OEM) controls in a fleet test. Future work will focus on developing optimized calibrations and controls to quantify performance relative to baseline conditions in a wide range of engine and vehicle applications.

Materials Work at ORNL Related to Catalysts/Emissions

ORNL makes use of capabilities that are hard to maintain at universities and difficult to justify in industry, given the need for experienced researchers to operate and to maintain state-of-the-art equipment. One example is the aberration-corrected electron microscope (Project ID 18865), which provides atomic-level imaging to better than 1 Å resolution. Samples can be heated in situ up to 1,000°C and follow a catalytic reaction in a controlled atmosphere. Basic research studies have been followed using catalysts such as Au/CeO_2 and Pt/Rh on a perovskite. In other work (Narula et al., 2010) scientists at ORNL are using theoretical models to explore catalyst materials via first principles for low-temperature operation. Materials being explored are bimetallic zeolites such as CuFe ZSM-5.

Project PM055. In this project, "Biofuels Impact on Aftertreatment Devices," ORNL is investigating the impact of biodiesel fuel on aftertreatment devices. Impurities (Na, Ca, and Mg) in biofuels have been found to accumulate on the aftertreatment devices. The sources of these impurities are NaOH and KOH from the transformation of the feedstock and Ca/Mg from the washing.

Project PM009. This project, "Materials Issues Associated with EGR Systems," concerns soot in the exhaust that can deposit and interfere with the EGR system, causing the engine to lose BTE. Advanced engines EGR will be required to operate over a wider range of engine speed and loads. Low-temperature combustion will increase this problem and also hinder waste heat recovery. One approach is to identify the optimum operating temperature for the system. Imaging technologies are being applied to characterize the deposits. Strategies being explored include deposit removal techniques. U.S. diesel engine manufacturers are collaborating on this project with ORNL.

Project PM010. In a CRADA with Cummins, "Durability of Diesel Engine Particulate Filters," ORNL is characterizing properties of ceramic diesel particulate filters and developing tools to assess durability and reliability. One goal is to be able to reduce the fuel economy penalty associated with the DPF by 25 percent relative to the baseline 2009 vehicle. The regeneration of the filter that is the focus of this work must be reliable and the filter durable. A new zeolite-based support with a finer pore structure is being investigated. The Cummins test rig is being used to do simulated measurements of filter lifetimes. Data generated will be used as input to models to predict the behavior of the DPF.

Project PM049. ORNL worked on this project, "Catalyst Characterization and Deactivation Mechanisms," in two separate stages from 2009 through 2013. In one it partnered with Cummins and on the other with Ford, University of Michigan, and Protochips. The overall objective was cost-effective emission control using new engine operating conditions that minimize emissions. The approach was to increase understanding of the deactivation mechanisms and address durability requirements for light-duty diesel aftertreatment: ammonia oxidation (AMOX) and SCR materials. Hydrothermal aging was done at elevated temperatures for lifetime prediction and to evaluate degradation mechanisms. Transmission electron microscopy (TEM) provided atomic resolution of rhodium nanoparticles on a $CaTiO_3$ support over a wide temperature range and in an oxidizing-reducing atmosphere.

Other Emissions-Related Work Outside 21CTP

Fundamental or characterization work outside the 21CTP has been reported on many pertinent aspects of emissions control. Improvements in SCR catalyst formulations have been reported on Mn zeolites (Kim et al., 2012), improved Cu zeolites (Walker, 2012; Reith et al., 2013), SCR filter catalysts (Rohe et al., 2012), and advanced substrates and LT urea injection methods (Strots et al., 2014). SCR system durability and aging studies are reported by SwRI (Bartley et al., 2012), and Cummins (Yezerets et al., 2014; Chen et al., 2013; Kumar et al., 2013). Daimler and Milano Politecnico investigators showed NO_x can adsorb on SCR catalysts upon engine start-up, until water reaches the catalyst (Schmeisser et al., 2013). Dioxin emissions are minimal and not a concern, as reported by EPA (Laroo et al., 2013). Tenneco reported engineering and fundamental work on silver-alumina catalysts that use E85 as a reductant instead of urea (Patel, 2012). The University of Wisconsin (Viswanathan et al., 2012) and MIT (Kamp et al., 2013; Sappok et al., 2013) reported fundamental characterization of ash-soot interactions in DPFs. Soot regeneration by NO_2 on DPFs with SCR catalyst coatings is characterized by Liebherr (Hohl, 2014) and modeled by BASF (Tang et al., 2013). PGM-free DOCs were reported by Honda (Ishizaki et al., 2012) and Nanostellar (Wang et al., 2012). Heesung (Kim et al., 2013) and University of Pennsylvania (Cargnello et al., 2012) described new LT methane catalysts. Regarding technologies pertinent to California's low-NO_x regulatory initiative, Theis and Lambert (2014) reported some work on the performance of low-temperature NO_x adsorbers, showing that the NO_x is stored on the base-metal storage material as a nitrite (from NO) using palladium as a catalyst, and the device is relatively resistant to sulfur poisoning and durable to 740°C. Not much is known about fundamental reaction mechanisms or the materials used in these components, and calibration studies are lacking on how they should be implemented. Michigan Technological University recently completed a thorough literature review on SCR+DPF system integration (Song et al., 2014). They identified several needs, including developing a testing protocol and a better understanding of ash-soot-catalyst interactions as they pertain to soot regeneration and SCR performance.

There are gaps in the literature pertaining to understanding or even reporting secondary or unregulated emissions, such as CH_4 or N_2O. Fundamental understanding on the formation mechanisms of the greenhouse gas N_2O from advanced combustion and emission control systems is lacking. This will become increasingly important as the GHG regulations tighten. Also, developments ought to be put into the context of the tightening HD regulations emerging in California and perhaps the EPA/NHTSA rulemaking in 2015.

Health-Related Studies

"Advanced Collaborative Emissions Study (ACES)" was a $15.5 million 7-year consortium program (2007-2013) in which the DOE was one of many partners. Completed in FY 2013, it characterized the emissions of US2007- and US2010-compliant HD engines and health effects of 2007 engines using a rat model. The emissions results are impressive. Particle number (PN) emissions on the FTP certification cycle for US2010 engines were reduced 99.9 percent from 2004 levels and 40 percent from US2007 levels, with this latter improvement attributed to the lack of active DPF regenerations in the 2010 engines (Khalek et al., 2015). On a custom 16-hour drive cycle, relative to 2007 engines, PM and PN emissions were reduced 71 percent, NO_x and NO_2 by 94 percent, hydrocarbons by 99+ percent, highly toxic dioxins and furans by 88 percent, and CO_2 by 3 percent. As reported in the final ACES health effects study (2015), health effects observed in rats after long-term exposure to diesel exhaust from new technology engines (compliant with 2007 regulations) were consistent with effects observed after exposure to NO_2 (NO_2 was reduced 94 percent in 2010 engines). Importantly, there was no increase in tumor formation over the background in the lung or any other organ compared to control animals; this was a major difference in long-term exposures to "traditional" diesel exhaust containing high levels of PM. Genotoxic endpoints showed no exposure-related changes. Some histopathologic changes observed in the gas-exchanging region of the lung were similar to changes after long-term exposures to oxidizing pollutant gases, such as NO_2 and ozone. There were few changes in respiratory function endpoints, which occurred more in females than males. Effects in the respiratory tract were mild and generally seen only at the highest exposure level. There were also few changes in inflammatory endpoints in blood, bronchoalveolar lavage, or lung tissue. Vascular endpoints were mostly unchanged, with a few scattered exposure-related changes (again mostly in females).

21CTP Response to Recommendations from NRC Phase 2 Review

The NRC Phase 2 review committee commented that significant progress was being achieved on emissions control understanding either through formal work in the program or through industry efforts. The only specific comments concerned researching CO_2 reduction pathways and characterizing particle number emissions to support future regulatory initiatives.

NRC Phase 2 Recommendations: Aftertreatment Program Activities

NRC Phase 2 Recommendation 3-7. The aftertreatment program within the 21CTP should be continued, and DOE should continue to support the activities of CLEERS that interface with the activities of the aftertreatment technical community at large.

21CTP Response: The Partnership agrees with this assessment to continue the aftertreatment programs. Combustion and aftertreatment activities are continuing under the SuperTruck projects, which are looking to achieve stretch efficiency goals while meeting current stringent emission standards: this produces a need for continuing aftertreatment research.

NRC Phase 2 Recommendation 3-8. In light of the progress being made with new combustion technologies, which show potential for very low cylinder-out NO_x and particulate emissions, the 21CTP should incorporate studies of particulate number emissions into their research portfolio.

21CTP Response: The Partnership is aware of the evolving interest in particulate number regulation (number of particles and size distribution), especially in Europe. We are currently measuring these parameters in several projects with the national laboratories, universities, and industry.

Committee Comment on 21CTP Responses

In general, the committee thought the progress was substantial and the effort should continue. The exception was a need for more work on characterizing PN emissions.

Findings and Recommendations: Aftertreatment Systems

Finding 3-6. The research agenda for 21CTP is focused on a wide diversity of heavy duty (HD) emissions control work. There are impressive fundamental studies on SCR catalysts, DPF fundamentals, low-temperature SCR and oxidation catalysts, passive NO_x adsorbers, multifunctional components, emissions measurement and modeling, system models, fuel effects, aging, and sensor development. Work is not continuing in 21CTP on lean-NO_x traps but has become part of the light-duty vehicle programs. These programs are delivering valuable results, but there are no program goals to guide future directions.

Recommendation 3-3. The Partnership should continue work on aftertreatment and emissions control, but the DOE should develop specific aftertreatment goals for the 21CTP. These goals will serve as a focal point for researchers to submit proposals and for the DOE to assess them.

Finding 3-7. Lacking are fundamental studies or even reported results on unregulated emissions, such as CH_4 and N_2O. Also lacking are projects or objectives aimed at post-2010 regulations, specifically supporting the CARB low-NO_x initiatives, in particular cold-start NO_x control using, for example, low-temperature NO_x adsorbers and SCR filters.

Recommendation 3-4. The Partnership should continue to fund work on improved SCR NO_x efficiency (mainly at low temperature, without compromising high-temperature efficiency) and aging and poisoning effects. California's and, potentially, EPA's move toward further HD NO_x reductions to meet National Ambient Air Quality Standards (NAAQS) for ozone will be critical. These new targets need to be set for the research efforts.

Recommendation 3-5. New fundamental emphasis on N_2O formation to support lower NO_x emissions should be added, as there is an apparent trade-off between low NO_x and higher N_2O caused by the need to inject more urea.

Finding 3-8. To achieve 50 percent BTE in the SuperTruck Program (Chapter 8), the engine compartment has limited space for the cooling system, the waste heat recovery system, and the aftertreatment system. The aftertreatment system volume, weight, and cost are important for the design of the engine compartment for trucks that are developed for 50-55 percent BTE.

Recommendation 3-6. Technologies such as an SCR catalyst on a DPF or others that have the potential to reduce the volume, weight, and cost of the aftertreatment system should be a part of the program to develop a 55 percent BTE engine.

Finding 3-9. OBD is a key industry need. It is a primary consideration in emissions control design and architecture and a major cost component. OBD technology is not available to meet the expected California low-NO_x regulations.

Recommendation 3-7. DOE should determine the gaps in OBD understanding and in sensor technology, especially to meet the tight California regulations, and should implement programs to help fill these gaps.

FUEL PROGRAMS

This review of fuel programs affecting the 21CTP begins where the NRC Phase 2 review finished (NRC, 2012). Some of the findings and recommendations regarding fuel technologies in the NRC Phase 2 report are still valid and will be briefly mentioned. New fuel issues, many reviewed in the 2013 21CTP Roadmap and Technical White Papers (21CTP, 2013), are described, and additional research is recommended.

The most significant change in the hydrocarbon petroleum resource pool since the NRC Phase 2 report has been the substantial increase in the supply of crude oil derived from directional drilling and hydraulic fracturing ("fracking") (Harvey and Loder, 2013). The properties and characteristics of hydraulically fractured crude oil samples from shale are generally different than those of crude samples pumped from large pools or deposits of oil. Crude oil from shale tends to be lighter than crude from many other sources (GAO, 2014). The blending of light crude oil derived from fracking operations with heavy crude oil streams (such as from tar sands deposits) creates an unusual blend of hydrocarbon components (sometimes referred to as a "dumbbell" oil blend) that can be challenging for some refineries trying to produce a complete slate of fuel products (Gonzalez, 2014). Although crude oil from shale is increasing the national energy supply, its characteristics are more suited for producing gasoline rather than middle distillates such as diesel fuel. This could impact the properties of diesel fuel in the future as well as force refiners to make capital upgrades to their refinery operations (Gonzalez, 2014).

Efforts to produce greater amounts of biodiesel (fatty acid esters) and to create renewable diesel derived from biomass have increased modestly since the NRC Phase 2 review.[7]

The use of natural gas as a fuel for medium- and heavy-duty trucks is increasing slowly based mostly on increased gas supplies generated from new extraction techniques and the resulting lower fuel costs. A discussion of the pros and cons of natural gas as a transportation fuel was recently published by the NRC (2014). Natural gas has been suggested as a method to reduce transportation sources of greenhouse gases (GHGs), although this conclusion requires further validation.

Advanced Petroleum Fuels

The DOE has active research programs that address diesel fuel properties, whether they are petroleum-derived or not. As explained earlier in the discussion on engine systems research, cooperative work between the DOE and the CRC (2012) has created a well-characterized, petroleum-based fuel matrix called FACE. This fuel matrix is specifically designed so that "researchers evaluating advanced combustion systems may compare results from different laboratories using the same set (or sets) of petroleum fuels for consistency" (CRC, 2005). For diesel fuel, the matrix consists of nine fuel blends varying in cetane number, aromatic content, and T_{90} values (temperature at which 90 percent of the fuel is distilled). A graphic depicting the matrix is shown in Figure 3-8.

The fuels described by this matrix are designed around the fuel properties and characteristics likely to affect advanced combustion system performance. They are not "an implied or explicit recommendation or endorsement for the adoption of any of these research fuels as implied or explicit fuel standards" (CRC, 2005). They do serve a valuable purpose, however, by providing a basis for comparison of test results from research programs intended to evaluate the performance of advanced combustion engine technology. The use of these fuels has already been described in some of the projects reviewed in the Engine Systems portion of this chapter and will be described further in some of the fuels projects.

Petroleum-based diesel fuel properties could become an issue in future years because of the production of greater amounts of light crude oils from fracking. As previously stated, crude oils from hydraulic fracturing processes contain significant amounts of highly volatile, low-molecular-weight hydrocarbons (GAO, 2014). When blended with other heavier crude sources either produced in or imported into the United States, the result is a crude oil blend with large amounts of light distillates and of heavy distillates but fewer components in the middle of the distillation range. Although

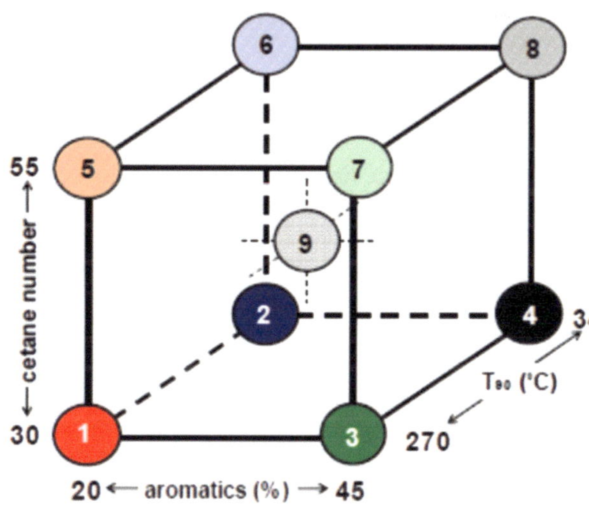

FIGURE 3-8 Fuels for advanced combustion engines (FACE) diesel fuel set. SOURCE: B. Zigler, "Fuels for Advanced Combustion Engines," DOE Annual Merit Review FJ002, May 15, 2012. National Renewable Energy Laboratory.

---

[7] See biodiesel production statistics at http://www.biodiesel.org/production/production-statistics.

such crude blends can be handled within modern refineries, a potential impact could be a decline in middle-distillate yields (Gonzalez, 2014).

The United States has for many years been a large importer of crude oil, mainly to meet the needs of transportation. Recent increases in domestic production have largely alleviated the concerns over crude supplies (GAO, 2014). If the demand for diesel fuel continues to increase and demand for gasoline for light-duty vehicles decreases due to fuel economy and emissions regulations, an adequate diesel fuel supply may become the most important future transportation energy concern. Research targeted at utilizing renewable fuels in spark-ignition engines may not be as critical as that targeted at diesel engine fuel efficiency strategies and the use of biodiesel or renewable diesel fuel blending components in advanced combustion engines.

The cost of producing petroleum-based fuels and the price of commercial fuels at the pump reached new highs during the last decade. Today, due to greater crude production, fuel costs have dropped substantially. This is important to OEMs in determining the cost of new technology they would be willing to invest in to improve efficiency and reduce fuel consumption. It is also important in determining the extent to which alternative fuels will be economically competitive with petroleum-based fuels. If the cost of petroleum-based fuels remains low, and if alternative or renewable fuels are desired for reduced GHG emissions, the use of such alternative fuels will be realized only through application of regulatory initiatives.

Biofuels

For many years, biofuels have been promoted by some government agencies for four reasons:

(1) To extend petroleum resources,
(2) To reduce petroleum imports,
(3) To reduce GHG emissions, and
(4) To increase domestic jobs.

Given the increase in domestic oil and gas supplies in recent years, the first two justifications for greater biofuel generation and sale are much less significant than at the beginning of the twenty-first century, when future oil supplies were thought to be limited.

The biofuels industry has continued to grow since the NRC Phase 2 review, but the use of biofuels in commercial fuel formulations is still limited. DOE has contributed to the development of technology and processes for producing cellulosic ethanol and biodiesel fuels, with the ultimate goal of commercialization. Congress established the Renewable Fuel Standard (RFS) in 2007, which set a goal of using 36 billion gallons of biofuels per year by 2022. Congress has provided tax credits and incentives for biofuels production, including that of renewable, ester-based diesel fuels. These credits and incentives generally remain in effect. An NAS-NAE-NRC (2009) report concluded that sufficient resources for biomass were available in the United States, and that substantial amounts of biofuels could be produced by 2020.

Despite the plans for increased production promoted by federal agencies, the only significant source of biofuels today is corn-based ethanol. This ethanol is added to gasoline in the United States, mostly at a concentration of 10 percent, although the EPA in 2010 allowed as much as 15 percent ethanol in gasoline for use in 2007 and later-model-year, light-duty vehicles. In early 2011, the EPA expanded its waiver to allow up to 15 percent ethanol in gasoline used to fuel 2001 through 2006 model-year, light-duty vehicles. The EPA cannot force fuel stations to provide gasoline blends containing 15 percent ethanol without the approval of Congress, which at this time it does not have. The use of 15 percent ethanol has been opposed by automotive OEMs due to concerns over durability in engines designed for 10 percent ethanol in gasoline (Shepardson, 2010), and to date only limited amounts of gasoline containing 15 percent ethanol have been sold at U.S. commercial fuel pumps.

The commercial production of cellulosic-derived ethanol is only now beginning to be realized. Three companies, Abengoa, Poet, and DuPont, announced plans to produce ethanol from cellulosic materials in either 2014 or 2015 (Abengoa, 2014; Poet, 2014; DuPont, 2015). The combined production of denatured ethanol from these three plants could reach 81.6 million barrels per year (ca. 0.61 percent of total denatured ethanol consumption in 2014). Commercial production of ethanol derived from cellulose has been supported by DOE (2014).

The production of biodiesel (essentially fatty acid methyl ester [FAME] and other esters) and renewable diesel (a pyrolyzed/hydrotreated, biomass-derived feedstock used in refineries for diesel fuel production) has been minimal but continues to increase.[8] In 2013, the EPA concluded that the industry was not going to be able to produce the amount of biofuels called for by the RFS (EPA, 2013). Thus, the total 2014 requirement for biofuel for use in transportation applications was temporarily set at 15.21 billion gallons per year (EPA, 2013). In 2013, ethanol production in the United States (virtually all corn-based) amounted to 13.31 billion gallons per year (RFA, 2014). Based on Energy Information Administration (EIA) estimates for U.S. gasoline consumption in 2013 of 135 billion gallons and distillate fuel consumption in 2013 of 58.7 billion gallons, this amount of ethanol is roughly equivalent to its use in gasoline at 10 percent (EIA, 2014b). By 2022, the RFS has a requirement for the use of 4 billion gallons of advanced biofuels, which can be just about any renewable fuel except corn-based ethanol. Even if all of this was biodiesel fuel, it would still meet only a relatively small percentage of diesel fuel demand.

[8] Ibid.

To meet future RFS requirements, advances in manufacturing processes and reductions in manufacturing costs are needed for ensuring growth in the use of biofuels. Looking ahead, biofuels for use in petroleum-based diesel fuel could be manufactured in one of two ways:

(1) Make a biodiesel fuel, such as FAME or another ester, from a specific feedstock, such as soybeans, and blend it into existing diesel fuel.
(2) Make a bio-based, renewable bio-crude oil by pyrolysis or hydrotreatment that can be used at a refinery in the production of conventional diesel fuel.

Much of the early effort to develop biofuels for blending with diesel fuel was for the development of ester-based fuels such as FAME, as described in the first option. The process for making simple, ester-based biofuels is well documented and has been in use for many years. In the biodiesel production process, triglyceride oils derived from biomass are reacted with methanol to produce a fatty acid alkyl ester and glycerin as a by-product. The quality of such biodiesel components is defined by ASTM standard D6751. These biofuels are now blended into diesel fuels in some regions of the United States and, to a greater extent, in Europe.

Significant commercial effort has been directed toward the production of renewable diesel fuels, as described by the second option, although such fuels are currently not much used in the United States (Peckham, 2014a). Renewable diesel fuel uses feedstocks from gasified biomass to generate a hydrocarbon stream that is processed at a refinery during the production of petroleum-based diesel fuel.

Renewable diesel fuel is viewed by the oil industry as a better option than biodiesel (Peckham, 2014a) for the following reasons:

(1) It is all hydrocarbon and is chemically more like diesel fuel;
(2) It is more compatible with diesel fuel infrastructure and engines than biodiesel and in many instances provides a fuel blend that meets ASTM D975 specifications; and
(3) It avoids unwanted effects associated with ester-based biodiesel fuels (e.g., FAME), such as lower volumetric energy content, instability, hygroscopicity, injector fouling, and low-temperature operability, among others.

Alternative Fuels

One of DOE's original 21CTP goals was to replace some of the petroleum fuels used for transportation with nonpetroleum-sourced alternatives. Other than ethanol from corn and natural gas, use of other alternative fuels has been limited. The status of ethanol use as an alternative fuel is discussed in the preceding section. A detailed review of the use of natural gas in medium- and heavy-duty trucks has been published by the NRC (2014). The reader is referred to this review for information on issues related to the use of natural gas.

In 2012 there were still only 127,000 natural gas vehicles (NGVs) of all classes in the United States, or 0.05 percent of the total vehicle population (NGV Global, 2012). Despite this small percentage, the EIA predicts that medium- and heavy-duty trucks will be the largest transportation consumer of natural gas by 2040. This natural gas consumption will most likely be divided between compressed natural gas (CNG)-fueled, medium-duty trucks and liquefied natural gas (LNG)-fueled, heavy-duty trucks. Even in 2040, however, the use of natural gas will represent only 3 percent of total transportation energy consumption (EIA, 2014b).

The NRC (2014) report provides a good explanation of the changes in the natural gas supply in the United States that are due to the development of new processes for fracturing shale deposits. Since the publication of that report, the recovery of natural gas has grown significantly, such that the United States is currently the world's leading gas producer. In addition to there being greater amounts of natural gas available, this increased production is also being driven by potential GHG regulations. It has been estimated in some well-to-wheel (WTW) energy and emissions analyses that, even taking into account increased methane emissions from natural gas vehicles, total GHG emissions will be lower than from pure gasoline- or diesel-fueled vehicles (NRC, 2014). However, lower GHGs with use of natural gas will have to be balanced against the greater energy consumption that occurs due to lower engine efficiencies when using natural gas as a fuel.

Additional calculations are needed to demonstrate the GHG emissions reduction benefits and the drawbacks of new manufacturing facilities and technologies for production of liquid hydrocarbon fuels from natural gas. It is well known that large-scale gas-to-liquid (GTL) manufacturing plants are planned for the United States (Berman, 2014). In addition, new technology has been proposed and developed that uses mini gas-to-liquid processing equipment (Peckham, 2014b). Such equipment could be employed at stranded gas deposits for which there is no connection to a gas pipeline. The liquid fuel produced would be transported by truck, eliminating the need for gas pipeline construction. The resulting GTL fuel would be subsequently blended with conventional petroleum fuels at a refinery, creating a blend that meets commercial fuel standards. If commercially profitable, these mini GTL facilities could further expand the use of natural gas for the production of synthetic hydrocarbon fuels.

As pointed out in the NRC (2014) report, there is a need for developing lower cost, smaller natural gas fuel storage systems. A Class 8 natural gas truck can cost \$50,000 to \$100,000 more than its diesel fuel counterpart. The cost of installing a refueling facility is extra (the cost will depend on the number of trucks that must be refueled in a given time period). Much of the increased truck cost is associated with the fuel storage system: high-pressure tanks for CNG or cryogenic tanks for LNG.

Biomass-derived dimethyl ether (DME) has received attention, especially in Europe, as a sulfur-free diesel fuel substitute because of its high cetane number (55) and very low emissions of PM, NO_x, and CO. This fuel would require minor engine and fuel system modifications, but would necessitate a dedicated infrastructure for production, distribution, and storage, which is expected to be a major hurdle for its commercialization in the United States.

Review of 21CTP Fuel Technology Objectives

In the NRC Phase 2 report, it was pointed out that DOE had established three different sets of goals for the fuels research program from 2008 to 2011 (NRC, 2012). It was further noted that changing goals during that period made it difficult to assess progress against those goals. In fairness, it must be said that significant fuel research has been conducted and much important information gained at the DOE laboratories and by academic institutions and industrial research facilities since 2010. It is not of value to recount past changes in goals for fuel research. Instead, it will be assumed that the goals identified in the 21CTP Roadmap and Technical White Papers published in 2013 accurately reflect the current technology goals for the program (21CTP, 2013).

In the 2013 Roadmap and Technical White Papers, three specific technology goals are listed under Engine Systems. Goals 1 and 2 have already been reviewed in the engine systems section of this chapter. Goal 3 (21CTP, 2013) is as follows:

> Through experiments and models with FACE fuels and other projects, determine the most essential fuel properties, including renewables, needed to achieve 55 percent engine brake efficiency. (2014)

Progress on fuel technology in order to meet Goal 3 is as follows: A series of FACE diesel fuels has been selected and defined in cooperation between DOE and the CRC. These fuels have a selection of specific physical and chemical properties (cetane number, aromatics content, and T_{90}) and are commercially available for laboratories to purchase for individual research projects. FACE fuels and other surrogate diesel fuels are being used in several DOE laboratory programs designed to quantify advanced combustion engine performance and efficiencies when using fuels having well-defined characteristics.

This most recent version of Goal 3 is unrealistic in its currently stated objective of identifying "essential" fuel properties given the ongoing research on a wide variety of different combustion strategies. The committee believes that identification of a range of fuel properties that could enhance the performance of advanced combustion modes, such as LTC, would be a more realistic objective. FACE provides researchers with the ability to perform experiments with fuels of known characteristics, having property ranges that are within a range of variations that might be seen in future fuels. This is superior to running specific blends of research-grade fuels that are not representative of what an engine will experience in the field.

Although not specified as goals for Engine Systems within the 21CTP Roadmap and Technical White Papers, it is important to note that the document also listed a number of needed fuel research efforts that would help the 21CTP meet its overall objectives. Some of these include the following (21CTP, 2013):

- Develop [a] fundamental understanding of fuel effects on in-cylinder combustion and emissions formation processes in advanced combustion regimes;
- Develop predictive tools that relate molecular structure to ignition behavior and heat release for fuels used in advanced combustion engines;
- Evaluate new fuels and fuel blends for efficiency, emissions, and operating stability with advanced combustion regimes;
- Evaluate the potential of reforming small amounts of fuel to generate additives that can be used to achieve fast control in LTC modes;
- Identify renewable and synthetic fuel blending components that provide enhanced efficiency, performance, and emissions characteristics; and
- Evaluate performance of traditional lubricant formulations in engines using advanced combustion regimes.

21CTP, DOE, and DOD Fuel Projects

The DOE provided the committee with a list of 22 DOE, DOD, and NSF research programs investigating advanced fuel technologies that the committee believes should be considered as part of the 21CTP. The fuel projects, which are listed in Table 3-6, amount to a total budget in 2014 of $7,669,000.

The fuel research projects identified as linked with the 21CTP span the gamut from fundamental research (analytic modeling and laboratory-scale experiments), to fired single-cylinder, to full-scale dynamometer engine tests. Many of the fuel projects have the objective of evaluating the effects of fuel composition (including both hydrocarbons and biofuels) on advanced combustion strategies and emissions control systems performance (McCormick and Ratcliff, 2014; Mueller, 2014; Kurtz, 2014; Szybist et al., 2014; Reitz, 2014; Toops et al., 2014b). For example, Mueller at SNL (Project FT004) has utilized FACE fuels, surrogate fuels, and other diesel fuels to improve understanding of fuel effects on advanced-mixing, controlled-combustion strategies such as leaner-lifted flame combustion (LLFC). Surrogate fuels are well-defined formulations of specific chemical compounds for which models can more easily be derived in order to

TABLE 3-6 Major 21CTP-Related Projects Funded in FY 2014 Addressing Advanced Fuels (federal dollars)

Public Review Project No. / Title	Recipient	2012 Funding	2013 Funding	2014 Funding	Note
FT001 Fuel and Lubricant Effects	ORNL	1,400,000	1,250,000	1,465,000	
FT002 Advanced Combustion and Fuels	NREL	935,000	822,000	697,000	
FT003 Performance of Biofuels and Biofuel Blends	NREL	800,000	700,000	400,000	
FT004 Fuel Effects on Mixing-Controlled Combustion Strategies for High-Efficiency Clean-Combustion Engines	SNL	800,000	800,000	800,000	
FT007 Fuel and Lubricant Effects on Emissions Control Technologies	ORNL	1,445,000	700,000	825,000	
FT008 Gasoline-Like Fuel Effects on Advanced Combustion Regimes	ORNL	615,000	400,000	450,000	
FT010 Chemical Kinetic Modeling of Non-Petroleum Based Fuels	LLNL	750,000	–	500,000	
FT011 Impact of Biodiesel Metals on Aftertreatment System Durability	NREL	400,000	–	–	Project ended in 2012
FT015 Demonstration/Development of RCCI Combustion for High Efficiency, Low Emissions Vehicle Applications	Univ. of Wisconsin-Madison	500,000	640,000	360,000	Project ends in 2015
FT016 High Compression Ratio Turbo Gasoline Engine Operation Using Alcohol Enhancement	MIT	408,000	235,000	320,000	
FT017 Fuel Properties to Enable Lifted Flame Combustion	Ford Motor Company	436,904	406,000	694,000	
FT022 CFD Simulations and Experiments to Determine the Feasibility of Various Alternate Fuels for Compression Ignition Engine Applications	ANL	150,000	150,000	–	
Natural Gas Engine Development with CEC and SCAQMD	NREL	–	–	–	No project number. Last funding for this project in 2010.
PNNL Unconventional Fuels	PNNL	450,000	–	220,000	No project number
WTW analysis, refinery modeling of high-octane, NG pathways analysis, FT fuels, XTL fuels pathways.	ANL	500,000	500,000	575,000	No project number
Validation of JP-8 Surrogates in an Optical Engine	TARDEC	–	–	123,000	No project number
Ignition Models for Heavy Hydrocarbons Fuels	TARDEC/SNL	200,000	–	–	No project number
Fuel Bulk Modulus	TARDEC	551,000	30,000	10,000	No project number
Bulk Modulus of Compressibility Measurements of Conventional and Alternative Military Fuels	TARDEC	–	–	110,000	No project number
Reaction Pathway and Elementary Ignition Behavior of Surrogates for JP-8 and Alternative JP-8 Fuels	TARDEC	–	–	120,000	No project number
Potential for Dimethyl Ether to Yield Low Gaseous Emissions and Improve Efficiency under Lean-Burn Conditions	Michigan Tech University	–	650,000	–	No project number
Sooting Characteristics of Surrogate Fuels	Yale University	–	600,000	–	No project number
Total				7,669,000	

NOTE: Acronyms are defined in Appendix E. Some of the projects included in Table 3-6 for the fuels budget apply to both light- and heavy-duty vehicles. Dash denotes no funding. SOURCE: U.S. Department of Energy, Office of Energy Efficiency and Renewable Energy.

describe combustion under different operating conditions. This work has led to the development of a new "soot diagnostic" (Vertical Sheet Laser Induced Incandescence of Soot, VLII) for measurement of fuel effects on in-cylinder soot distributions and for assessment of soot models (Mueller, 2014). In a related study, Kurtz at Ford (Project FT017) demonstrated that increasing oxygen content in the diesel fuel increased the flame lift-off length and led to less soot formation (Kurtz, 2014). McCormick and Ratcliff (2014) at NREL (Project FT003) are investigating the use of biofuels containing minimum amounts of oxygen as "drop-in," nonpetroleum fuel components. Eight different oxygenates have been evaluated. Aromatic oxygenates were found to lower cetane number and increase premixed burn fractions, while di-isoamyl ether was found to raise diesel cetane number and reduce ignition delay and premixed burn fractions (McCormick and Ratcliff, 2014). At ORNL, Szybist et al. (2014) (FT008) are evaluating different biofuel blends in conventional spark ignition (SI), dilute SI, homogeneous charge compression ignition (HCCI), and reactivity controlled combustion ignition (RCCI) engines. In a downsized, downspeeded SI engine, a renewable super premium (RSP or E30) gasoline demonstrated improved efficiency relative to regular gasoline over limited operating cycles (the efficiency improvement outpaced the energy density loss). The RCCI operating range was expanded to 75 percent of its theoretical maximum while maintaining low soot and NO_x emissions when using the combination of B20 (20 percent biodiesel in petroleum diesel) and gasoline (Szybist et al., 2014). A research project at the University of Wisconsin partially funded by the DOE (FT015) is developing prototype light-duty and heavy-duty vehicles using RCCI engine technology employing a combination of gasoline and diesel fuel (Reitz, 2014). Toops et al. (2014b) (Project FT007) have investigated the effects of fuel and lubricant formulations on both gasoline and diesel emissions control system components. As a follow on to the work of Szybist et al. (2014), Toops and his team at ORNL have demonstrated that E30 ethanol/gasoline blends produce particulates with a higher reactivity (oxidize at lower temperatures) than regular gasoline in gasoline direct-injection (GDI) engines. The ORNL team is also developing accelerated laboratory diesel emissions durability tests in order to identify the effects of metals in biodiesel formulations on catalyst degradation. Currently, the focus is on creating a correlation between emissions control components that have been subjected to commercial long-term, heavy-duty service and those that have been subjected to severe laboratory tests (Toops et al., 2014b).

In addition to these experimental research programs, efforts are being conducted to develop comprehensive chemical kinetic mechanisms for petroleum-based and bio-derived fuels (Zigler, 2014; Som et al., 2014). For example, Som et al. at ANL (Project FT022) is using a three-component diesel surrogate fuel to develop models for biodiesel that predict both in-nozzle flow and spray characteristics, as well as combustion kinetics used in CFD simulations (Som et al., 2014). Zigler at NREL (Project FT002) has combined laboratory experiments with modeling efforts to identify combustion characteristics of specific fuel components and blends. This work has developed an ignition quality test (IQT) that allows the calculation of a derived cetane number (DCN) using as little as 25 ml of fuel. The IQT also provides data for calculation of Arrhenius parameters, used in combustion kinetics modeling. Importantly, the IQT is capable of measuring ignition performance and providing kinetic data for fuels ranging from gasoline to jet fuel to diesel to associated biofuels.

Based on the fuels project portfolio provided by the DOE, there is only one project that involves research on the development of natural gas engine technology. This project, conducted in cooperation with the South Coast Air Quality Management District (SCAQMD) and the California Energy Commission (CEC), was last funded in 2010. The committee received no update on the results or status of this program during any of its meetings.

Although the fuels research portfolio covers critical issues related to improving engine efficiency needed to meet 21CTP goals, the organization of DOE fuels projects relative to each other appears only loosely coordinated. Many useful fuel experiments are being conducted at different DOE laboratories, but the research is being conducted on a variety of advanced combustion engines using different fuels. It is not clear what the downselect process for focusing future (beyond 2015) engine/fuel research will be. How will DOE identify the most promising engine/fuel combinations for improving engine efficiency or reducing GHG emissions? How will DOE identify the most promising projects for future funding? Admittedly, some projects need to be completed before their full value is determined, but at this point the research path forward is unclear. Today, fuel projects seem to be generated by a bottom-up process based on recommendations of individual researchers at different laboratories. A top-down process via DOE management or peer review that identifies good options for commercial success would help focus limited resources on potential best outcomes.

The military fuels program has had a consistent set of objectives throughout the life of the 21CTP. Those objectives are to (1) minimize the number of fuel types that the military must purchase and transport, (2) minimize the amount of fuel used in operation through engine and vehicle efficiency improvements, and (3) increase the amount of non-petroleum-based fuels used in both tactical and combat applications. The military would prefer to use the same fuel (the most desired fuel is JP-8) in all of its vehicles. As described in the NRC Phase 2 report (NRC, 2012), the Army, through TARDEC, has the unique role of qualifying alternative fuels for use in tactical and combat vehicles having diesel engines.

The Army would like to use fewer petroleum-derived fuels, although there is a realization that it will be difficult to do so. It is exploring biodiesel fuels, but the lower energy density of most biofuels is a drawback. The military has a rigorous procedure for qualifying alternative fuels. This procedure could prevent some bio-based fuels from being accepted based on technical specifications. However, it is possible that federal military fuel procurement regulations could be written to require the military to develop technologies that are compatible with bio-based fuels. For that reason, the military continues to conduct and support research projects (at national laboratories and universities) to develop advanced diesel engine technologies that deliver improved efficiency and vehicle range when using bio-based fuels.

The fuels portfolio in Table 3-6 includes five such research projects managed by TARDEC. Research topics include development of new methods for determining fuel bulk modulus, development of ignition models for heavy hydrocarbon fuels, and development of combustion models using surrogate fuels for alternative JP-8 fuel formulations. A new method for measuring bulk modulus is needed because this parameter can vary greatly between biofuel samples and hydrocarbon fuels, affecting high-pressure injector flows.

Response to Recommendations from the NRC Phase 2 Report

NRC Phase 2 Recommendation 3-4. The DOE should reinstate its program for advanced petroleum-fuels (they will be transportation's primary fuels for many years to come) with the objective of maximizing the efficiency of their use.

21CTP Response: The new consolidated line incorporates the activities of both previous lines. Advanced petroleum-based fuels are already the subject of a large portion of the projects supported under the new line.

Committee Comment on Response to 3-4

The committee is satisfied with the total research effort directed at understanding the effects of petroleum-based fuels on advanced combustion technologies. However, at this time, the different research laboratories do not appear to have a coordinated research plan on how to identify the fuel properties that are most appropriate for the varied combustion strategies being investigated. Current research directions and test fuel formulations seem to be selected by individual laboratories without coordination.

***

NRC Phase 2 Recommendation 3-6. The DOE fuels goals should be re-evaluated in line with the FY 2012 budget and the recommendations of this report. Specific plans for achieving these goals should be established.

21CTP Response: We are continually open to re-evaluation of our goals in light of budget changes. Recent budgets have been volatile, which complicates the effort—e.g., between the FY12 Omnibus appropriation and the FY13 marks there has been a greater-than-40%-cut—but we will continue to re-evaluate as appropriate.

Committee Comment on Response to 3-6

Other than the development of a FACE fuels set, the committee is not aware of any other specific 21CTP fuels goals. In view of laboratories seeming to select the fuels for their individual research projects, it would benefit DOE to improve the coordination of fuel sets used by different research groups. In addition, specific combustion objectives should be assigned to each fuel set in order to determine the benefits that are available from various fuel compositions.

Findings and Recommendations: Future Fuel Research

Finding 3-10. A series of fuels for advanced combustion engines (FACE) and surrogates have been identified in cooperation between the DOE and the Coordinating Research Council (CRC). These fuels have specific physical and chemical properties and are being used in several advanced combustion research programs, including the evaluation of various LTC concepts (e.g., LLFC, RCCI), the development of CFD models for in-nozzle flow, spray formation, and combustion, and the development of new analytical techniques (e.g., IQT, VLII).

Finding 3-11. Currently, fuel projects appear to be generated by a bottoms-up process based on recommendations of individual researchers at different laboratories without guidance from DOE management on the practical ramifications of specific fuel choices or on the chances of commercial success.

Recommendation 3-8. The DOE should continue to explore how the United States might use its abundant petroleum, natural gas, and biofuel resources in the most efficient manner. Studies, some of which are under way that contribute to this objective, should strive to answer the following questions:

(1) What fuel properties (e.g., ignition characteristics, volatility, composition) of diesel fuel and gasoline maximize the efficiency of various advanced combustion engines? FACE and a common set of surrogate fuels should be utilized by all DOE facilities involved in combustion research programs in order to provide consistent fuel characteristics when evaluating laboratory experiment and engine test results.
(2) Based on well-to-tank analyses, what fuel properties and processing procedures result in the lowest GHG emissions for hydrocarbon-based and bio-based fuel components?

Finding 3-12. Goal 3 in the Engine Systems chapter of the 21CTP 2013 Roadmap and Technical White Papers identifies the objective of determining the essential fuel properties required to enable advanced combustion systems that can achieve 55 percent BTE. This 2014 goal as currently stated seems unrealistic to the committee. It suggests that there are specific fuel properties and values that will expand the operating range of advanced combustion strategies, such as various LTC concepts. The committee feels that the importance of fuel research is much broader than this. The FACE fuels provide researchers with the ability to perform experiments with fuels of known characteristics, having property ranges that are within a range of variations that might be seen in future fuels. This is superior to running specific blends of research-grade fuels that are not representative of what an engine will experience in the field. Using FACE fuels also helps with the kinetic model development being pursued in the surrogate fuel simulation program. Researchers can now test their advanced kinetic models against realistic, but known, fuels in real engines, an important step in developing simulation capabilities for predictive behavior. The committee believes a more detailed understanding of the impact of fuel properties on engine operation and potential facilitation of advanced combustion operation will also facilitate the high-level objective of maximizing the utility of our fossil fuels, thus reducing their use.

Recommendation 3-9. The Partnership should consider revising Goal 3 in its 2013 White Papers to make it more consistent with what the committee observes it is doing within the current fuel research programs—for example, facilitating the development of kinetic models for realistic fuels that could embody a range of properties and understandings of how fuel characteristics either probably, or even just possibly, will impact or facilitate current and potential combustion strategies. Also it is suggested that consideration should be given to whether enhanced understanding of the interplay between fuel characteristics and engine performance can suggest powertrain–fuel system combinations that further reduce fossil fuel consumption.

Finding 3-13. Despite past efforts to increase the use of renewable fuels, petroleum will remain the primary source for light-duty and heavy-duty vehicle fuel for the foreseeable future. U.S. gasoline demand is expected to decrease during the next 25 years, while diesel fuel demand is expected to grow. If regulators continue in their efforts to meet the Renewable Fuel Standard goals, production of biodiesel and renewable diesel will need to increase.

LUBRICANT PROGRAMS

Over the last 35 years, low-viscosity, low-friction engine oils, transmission fluids, and axle lubricants have contributed to improvements in light-duty vehicle fuel economy (Swedberg, 2012). Owing to their higher loads and lower speeds, it is not clear whether similar efficiency gains attributable to lubricant formulations can be achieved in heavy-duty diesel operation, but it is important to conduct the research that will define the benefits of such lubricants. Although the DOE has sponsored research on new lubricant additive technologies for reducing friction and thus fuel consumption, most of this work has focused on light-duty vehicle applications.

Traditionally, lubricants for truck engines have been developed and qualified for commercial sale through the combined efforts of the American Petroleum Institute (API), the American Chemistry Council (ACC), the American Society for Testing and Materials (ASTM), and the Truck and Engine Manufacturers Association (EMA). These organizations have developed test methods and labelling that ensure that oils meeting engine manufacturers' needs for durability, low emissions, and fuel efficiency are available in the U.S. market. The DOE has recently become active in developing new lubricant additive and base stock chemistries that might help in meeting the goals of the 21CTP. For example, it participated in the Collaborative Lubricating Oil Study on Emissions (CLOSE) project. Of particular concern in the development of new lubricant additives is any effect of sulfur from the engine oil (and the fuel), and any effect of phosphorus and ash from the engine oil on emissions control system performance, especially in regard to PM and NO_x reduction.

Relative to light-duty vehicles, not as much progress has been made in reducing friction through the use of advanced lubricants in heavy-duty vehicles, although low-friction, low-viscosity oils from the petroleum industry are being tested in the SuperTruck Program (see Chapter 8). Most commercial heavy-duty engine oils continue to be SAE 15W-40 viscosity grades. The use of these higher viscosity grades has generally been justified by the claim that greater viscosity provides better film thicknesses in heavily-loaded contacts within high-powered diesel engines.

Early industry tests have focused on improving MHDV truck fuel efficiency through the implementation of new, high-temperature, high-shear (HTHS) viscosity recommendations. HTHS viscosity specifications were introduced into the SAE J300 Engine Oil Classification in the 1980s in order to ensure that engine oil was viscous enough to protect heavily-loaded, fluid-film bearings. Tests with reduced HTHS viscosity oils instead of SAE 15W-40 oils have shown modest reductions in fuel consumption while other traditional additive components continue to protect durability and performance. Fuel consumption results are shown to be duty-cycle specific (NRC, 2012).

Review of 21CTP Lubricant Technology Objectives

As with the other goals related to fuel technologies, goals for improved lubricant technologies for the benefit of the 21CTP have changed over the years. The NRC Phase 2 report (2012) stated the following:

> The DOE has a target objective of reducing parasitic losses in system efficiency by developing improved engine and transmission lubricants. The target benefits are as follows: for 2016, 10% engine/15% drivetrain friction reduction lead-

ing to approximately 1.5% fuel economy benefit; for 2030, 25% engine/35% drivetrain friction reduction leading to approximately 3 to 4% fuel economy benefit.

Recently, the 21CTP Roadmap and Technical White Papers (21CTP, 2013) stated in Goal 5 for Vehicle Power Demands that DOE's objective is to "develop and demonstrate parasitic friction reduction technologies that reduce driveline losses by 50%, thereby improving Class 8 fuel efficiencies by 3%." Based on experience in the light-duty vehicle industry beginning in the 1980s and until today, this is an ambitious but achievable target. If it is accepted that this statement of achievability for a lubricant goal reflects DOE's current objective for the 21CTP, progress toward meeting this goal is described below.

A research portfolio consisting of projects whose objectives are the development of new friction-reducing/antiwear additive, base oil, and viscosity index technologies has been created. The projects are being conducted at the national laboratories and by universities and private companies. Recent results have shown promise in reducing friction in heavily-loaded laboratory contacts and in improving fuel economy in limited vehicle tests, but almost all of the results have been collected in light-duty vehicles, or under laboratory conditions designed to mimic light-duty service. Collaborations have been established between OEMs and oil and additive companies that are participating in the "SuperTruck" program, but no data on the specific benefits of the advanced lubricants used in the program have been provided to the committee.

As previously mentioned in the discussion of fuel technology, it is important to note that the 2013 Roadmap and Technical White Papers also lists needed lubricant and tribology research efforts that would help the 21CTP meet its overall objectives. These include the following:

- Evaluate performance of traditional lubricant formulations in engines using advanced combustion regimes.
- Determine tribological limits of current materials and sensors.

DOE and DOD Lubricant Programs

The 21CTP provided the committee with a list of 10 DOE and DOD projects related to the development of advanced lubricant technologies. The lubricant projects are listed in Table 3-7 and amount to a total budget in 2014 of $3,653,000 (see Appendix D).

The lubricant research projects identified as linked with the 21CTP also range from fundamental research (analytic modeling and laboratory-scale experiments), to single-cylinder engine tests, to full-scale dynamometer engine tests.

At ORNL, Toops et al. (2014b) and Qu et al. (2014a) are studying the use of ashless, ionic fluid, friction modifiers/antiwear additives in engine oils (Projects FT001 and FT014). The ionic fluid additive has been shown to reduce catalyst poisoning slightly relative to the commonest phosphorous-containing antiwear additive. In addition, this additive has demonstrated a 2 percent improvement in fuel economy relative to a commercial SAE 5W-30 engine oil in light-duty vehicle service. There are no data at this time from tests in heavy-duty vehicle service. This research project includes representatives of a major additive company who could help identify other commercial opportunities for this technology, such as use in axle lubricants.

Work at ANL (Erdemir, 2014) has focused on research using nanoparticles suspended in engine oils to reduce friction (Project FT018). Boron-containing additives have been the most effective. In low-speed, high-load diesel engine screening tests, boron-containing lubricants improved fuel economy 1.5 to 2.5 percent. Emissions catalyst poisoning tests need to be conducted.

Additional research (Projects FT021, VSS058) on advanced lubricants includes the development of low-friction, hard-surface coatings (Qu et al., 2014b; Ajayi et al., 2014). Further research results (preferably in heavy-duty engines) are needed to evaluate these lubricants more thoroughly.

The Army keeps many of its engines for 40 to 50 years, and so there is concern about the compatibility of newer fuels and lubricants in these engines. For example, diesel engine fuel pump and injector life can be an issue with low-lubricity fuels. Research on new technologies such as ashless antiwear and lubricity additives could provide oil and fuel formulations that are compatible not only with advanced engines but also with existing engines in operation today. For this reason, the Army, in its facilities at TARDEC, manages and conducts fundamental studies of lubricant additive technology.

Response to Recommendation from the NRC Phase 2 Report

NRC Phase 2 Recommendation 3-5. The DOE must work closely with industry in exploring improved lubricants that reduce fuel consumption, especially with regard to using such lubricants in existing truck engines and transmissions.

21CTP Response: The lubricants activity is relatively new, but the DOE has always strived to work with vehicle and engine OEMs, as well as oil and additive companies. DOE is currently partnered directly with vehicle OEMs such as Ford and GM on projects looking at next generation oils. DOE also has partnerships on projects with engine manufacturers such as Cummins to look at advanced engine oil additives. The program also interacts with OEMs to develop lower-friction engine components through participation in the MIT Lubrication in Internal Combustion Engines which includes Daimler, Volkswagen, Volvo, Toyota, PSA, Renault, and Mahle. DOE intends to continue and expand these collaborations in the future. It is also important to note lubricants will likely never drive major decisions at either engine companies or oil companies; therefore a government

TABLE 3-7 Major 21CTP-Related Projects Funded in FY 2014 Addressing Advanced Lubricants (Federal Dollars)

Public Review Project Title	Recipient	2012 Funding	2013 Funding	2014 Funding	Note
FT012 Engine Friction Reduction Technologies	ANL	500,000	1,140,000	500,000	
FT 014 Ionic Liquids as Anti-Wear Additives for Next-Generation Low-Viscosity Fuel-Efficient Engine Lubricants	ORNL	400,000	400,000	400,000	
FT018 Advanced Nanolubricants for Improved Energy Efficiency and Reduced Emissions in Engines	ANL	–	268,000	267,000	
FT019 Lubricant Formulations to Enhance Engine Efficiency (LFEEE) in Modern Internal Combustion Engines	MIT	630,000	870,000	–	
FT021 Can Hard Coatings and Lubricant Anti-wear Additives Work Together?	ORNL	–	250,000	250,000	
NWU/ANL Novel Lube Formulations	ANL	–	–	286,000	No project number
Hyperbranched Polymers as Lubricants	PNNL	–	519,375	200,000	No project number
Lubricant Formulations to Enhance Fuel Efficiency	TARDEC	–	500,000	500,000	No project number
Advanced Lubricants	TARDEC	–	–	900,000	No project number
VSS058 Development of High Power Density Driveline for Vehicles	ANL	350,000	300,000	350,000	
Total				3,653,000	

NOTE: Acronyms are defined in Appendix E. Some of the projects included in Table 3-7 for the fuels budget are applicable to both light- and heavy-duty vehicles. Dash denotes no funding. SOURCE: U.S. Department of Energy, Office of Energy Efficiency and Renewable Energy.

role is essential in assuring this social good, i.e., a 2% increase in fuel economy.

Committee Comment on Response to 3-5

The committee supports DOE's basic research efforts related to development of low-friction additive technology and concepts as long as they are targeted at improved fuel efficiency in medium- and heavy-duty driveline applications and if they are carried out in cooperation with commercial oil and representatives from the additive companies.

Findings and Recommendations: Future Lubricant Research

Finding 3-14. There is a DOE 21CTP lubricant goal, but the current lubricant research portfolio appears to focus more on light-duty vehicle applications than on heavy-duty powertrain applications. Since the evaluation of advanced lubricant technology in Class 8 trucks is only occurring in the SuperTruck program in cooperation with private oil and additive companies, it is not possible to quantitatively separate out lubricant benefits and judge progress against the 21CTP goal.

Recommendation 3-10. A greater portion of the laboratory lubricant development projects should be redirected at meeting the requirements and test conditions associated with heavy-duty truck use. Tests in HD vehicles or in full powertrain dynamometer tests should be conducted in order to verify friction reduction benefits of advanced lubricant technologies relative to conventional lubricants and to judge progress against the 21CTP goal. This work should be conducted in close coordination and involvement with OEMs and with companies in the additive and petroleum industry.

PROPULSION MATERIALS–MATERIALS PROCESSING

Current heavy-duty engines have demonstrated the long-term reliability and durability required for use in Class 8 trucks. The next generation of HD engines will pose new materials challenges that are being addressed in the 21CTP. HD engine development has as its objective to improve performance and engine efficiency. Furthermore, these engines must be lightweight, cost-effective, and meet all emission requirements. The materials challenge includes the development of propulsion materials that can withstand the high temperatures and pressures found in the engine environment. Manufacturing and inspection methods are an integral part of materials development. The advanced materials must enable cost-effective fuel savings. Several DOE projects address the critical need for propulsion materials, motors, and components that meet the constraints for HD transportation applications. Given the long time needed for the identification, development, and implementation of new materials, continuity of R&D is essential. Work is under way at the national laboratories on new alloys and on overcoming the mechanical property limitations of materials at high cylinder pressures and temperatures (see Table 3.8).

High-temperature materials work at ORNL involves the development of exhaust valve materials with high fatigue life for use in advanced engines. Computational methods are being used to predict alloy compositions with desired properties, including oxidation resistance up to 900°C. Other work at ORNL in advanced materials development for advanced turbocharger designs is under way. These materials have the high temperature capability and strength needed for sustained operation at high operating temperatures. The modification of surfaces is aimed at reducing friction between contacting surfaces in the engine. This work addresses the goal of 50 to 55 percent BTE.

PNNL work on materials for advanced diesel engines aims to develop and deploy engineered surfaces with improved thermal and mechanical properties using friction stir processing. To date, fatigue life has been improved by a factor of two. Materials projects at PNNL also include manufacturing technologies for high-power induction and permanent magnet motors. The goal is lower manufacturing cost and lighter weight assemblies. In other work, new aluminum alloy compositions are being developed with high strength at elevated temperatures.

The National Energy Technology Laboratory (NETL), together with Caterpillar, is exploring HD-high performance cast steels for crankshafts. Microstructure and processing parameters are being explored in an investigation of durability requirements. A related project has as its objective to improve component strength of new ferrous materials by 25 percent.

In summary, these materials projects address the following:

- Alloys for engines with improved strength, reduced friction, and better thermal and mechanical properties,
- Improved high-temperature performance for exhaust valve materials,
- Turbocharger designs,
- New cast steels for crankshafts,
- Alloy development for extrusions and forgings, and
- Permanent magnet electric motors.

It is beyond the scope of this report to review all projects in detail. However a few of the materials projects related to propulsion are summarized below.

ORNL Project PM053 (High Temperature Materials for High Efficiency Engines) began in September 2013 and is scheduled to end in August 2016. This project is 100 percent funded by DOE, and ORNL leads the project. Anticipated funding for FY 2014 is $200,000. This project has as its objective to develop cost-effective exhaust valve materials for use in advanced engines operating at temperatures up to 950°C. Computational methods are being used to predict new alloy compositions with needed oxidation resistance, fatigue properties, and stability. Recently, the effect of composition on oxidation resistance at 800°C was addressed and work is on track for evaluation of selected alloys up to 900°C. Alloying elements that enable the desired microstructural characteristics are being identified for new Ni-based alloys. Current low Ni alloys do not have good strength at 950°C, and alloys used in aerospace applications are expensive owing to both high Ni content and the use of expensive alloying elements. The goal of this work is to provide high cycle fatigue life comparable to that with the high Ni alloys but at lower Ni levels.

PNNL Project PM004 (Tailored Materials for Advanced CIDI Engines [through FY 2013]). This project, which began in FY 2008, is a CRADA with Caterpillar and PNNL. Its objective is to develop and deploy engineered surfaces via friction stir processing (FSP) in traditional engine materials and to develop FSP in aluminum. Treated engine materials exhibited better thermal and mechanical properties. Friction stir processing can selectively modify an area of a part for better properties. Project milestones reached include the demonstration of a twofold improvement in fatigue life and reduction in thermal crack initiation and growth. Results have been documented. Process parameters, prototype parts, and knowledge have been transferred to Caterpillar.

PNNL Project PM004 (Novel Manufacturing Technologies for High Power Induction and Permanent Magnet Electric Motors [FY 2014]). This project, which began in FY 2011, was scheduled to have ended in September 2014. It is a CRADA with General Motors and PNNL. The objective of the project is to develop and deploy high-power induction rotors and stators that are lightweight and less expensive to manufacture than current assemblies. The approach is to apply friction stir welding as a low-cost method to join the bars to the end caps. Process parameters are being developed. The microstructure and mechanical properties of Cu/Cu joints was examined. A welding fixture was developed for friction stir welding of Al and Cu rotor parts.

PNNL Project PM044 (High Temperature Aluminum Alloys). This project, which also began in 2011, is a CRADA between Cummins and PNNL. Its objective is to develop aluminum alloy compositions with high strength at elevated temperatures (300 MPa tensile strength at 300°C) using a

TABLE 3-8 21CTP Projects Related to Propulsion Materials and Materials Processing and Federal Budgets (dollars)

Public Review Project Title	Proj. No.	Recipient	2012 Funding	2013 Funding	2014 Funding	Note
Tailored Materials for Advanced CIDI Engines (through FY 13)/ Novel Manufacturing Technologies for High Power Induction and Permanent Magnet Electric Motors (FY 14)	PM004	PNNL	350,000	300,000	225,000	
Friction and Wear Enhancement of Titanium Alloy Engine Components	PM007	ORNL	125,000	–	–	Project ended FY 2012
HD-Cast Fe Alloys for High PCP Engines	N/A	NETL	–	3,477,000	–	Fully funded FY 2013, 3 yr project
Materials for HCCI Engines	PM018	ORNL	225,000	–	–	Project ended FY 2012
Materials for Advanced Turbocharger Designs	PM038	ORNL	300,000	–	250,000	
High-Temperature Aluminum Alloys	PM044	PNNL	395,000	300,000	125,000	
Design-Optimization of Piezoceramic Multilayer Actuators for Heavy-Duty Diesel Engine Fuel Injector	PM051	ORNL	300,000	190,000	175,000	
Friction Reduction through Surface Modification	PM052	ORNL	–	260,000	150,000	Project end FY 2014
High-Temperature Materials for High Efficiency Engines	PM053	ORNL	–	200,000	–	
Applied ICME for New Propulsion Materials	PM057	ORNL	–	68,711	825,176	
HD, High-Performance Cast Steels for Crankshafts (CAT/GM)	PM058 (ANL)/ PM059 (Cat)	NETL	–	2,100,000	–	Fully funded in FY 2013, 3 yr project

NOTE: CAT, Caterpillar; CIDI, compression-ignition direct injection. Dash denotes no funding.

melt spinning process to produce flakes, which are then consolidated by extrusion. Laboratory-scale extrusion tooling was developed for use in consolidation and extrusion. The mechanical properties of the extrusions and forgings were evaluated. Test results on three alloy compositions showed that two compositions, (AFCT [Al-Fe-Cr/Ti] and AFM-11 [Al-Fe-Mn]) had higher tensile strengths at 300°C than Al-8.5 Fe alloy. The AFM-11 alloy had a tensile strength exceeding 250 MPa. Full-scale components will be tested in the future.

ORNL Project PM057 (Applied ICME for New Propulsion Materials). This project, led by ORNL, began in FY 2013 and will run through FY 2017. The project makes use of integrated computational materials engineering (ICME) to address the need for more efficient, faster, and less expensive materials development for propulsion applications. The project is totally funded by the DOE. Funding received in FY 2013 was $70,000, and the FY 2014 budget is $580,000. Specific applications addressed are the development of ceramic perovskites composed of lead-zirconium-titanate (PZT). High-ZT thermoelectric materials are of interest for waste heat recovery and climate control; piezoelectrics for high-performance fuel injection; low-cost permanent magnets, eliminating rare earth elements for electric drive systems; and durable low-temperature catalysts for exhaust emission control that operate near 150°C. Progress reported to date includes the prediction of high-p-type thermoelectric performance in PbSe. Work is currently under way on first-principle exploration of alloys near PZT, and two alloys have been selected for further development. $Hf_2Co_{11}B$ and Fe_5PB_2 were identified to be promising materials for permanent magnet applications. Work is progressing on a hydrothermally stable CuFe-SSZ-13 catalyst composition with good low temperature activity for NH_3-SCR.

NETL Project PM058 (HD-High Performance Cast Steels for Crankshafts). This project, led by Caterpillar, started in March 2014, and the DOE budget for FYs 2014-2017 is $300,000. Partners include General Motors, ANL, Northwestern University, and the University of Iowa. The objective of the project is to develop cast steel alloys and processing techniques for high-performance crankshafts with as-cast properties of 800 MPa ultimate tensile strength and 615 MPa yield strength. ANL will apply high-energy x-ray imaging and diffraction techniques to correlate microstructure with processing parameters. Fatigue tests will be used to establish durability requirements.

NETL Project PM059 (Development of Advanced High Strength Cast Alloys for Heavy-Duty Engines). This project, led by Caterpillar, was started in December 2012 and is scheduled to end December 2016. Total project funding is $5.08 million, with a DOE share of $3.48 million and a contractor share of $1.6 million. The objective of this project is new high-strength ferrous materials with at least 25 percent improvement in component strength relative to components made with A842. At the same time, targets are set for cost that will speed the adoption of new materials. As of June 2014, 16 prototype casting samples had been designed and produced. Materials properties of prototype castings alloys are being evaluated. X-ray tomography was found to be capable of identifying graphite structures in the iron matrix, and fluorescence analysis provided chemical information. During FYs 2013-2014, work progressed on identifying and modeling critical mechanisms that govern microstructure development during cast iron solidification.

ORNL Project PM038 (Materials for Advanced Turbocharger Designs). This project began in September 2009 and was scheduled to end September 2014. The budget for FY 2014 was $150,000. This project is a CRADA with 50/50 cost sharing by DOE and Honeywell. The project supports the Advanced Combustion Engine goal for the 2015 commercial engine with a 20 percent improvement in efficiency over the 2009 baseline efficiency. Turbocharging improves fuel efficiency, but the higher temperatures (>750°C, diesel; >950°C, gasoline) exceed the strength and temperature capability of current materials. Turbocharger housing and other components with more temperature capability and strength are needed for higher sustained operating temperatures. The alloy being investigated is CF8C-Plus cast stainless steel, which has more strength than HK30Nb stainless alloy at 750°C and is 33 percent less expensive. This alloy was commercialized by Caterpillar in 2006 for its Cat Regeneration System (CRS), used to regenerate the diesel particulate filter. Recent progress on this CRADA includes diesel engine exhaust testing of CF8C-Plus steel at 800°C and evaluation of oxidation resistance of CF8C-Plus in diesel exhaust. The CRADA was not extended by DOE and Honeywell continued on its own.

ORNL Project PM007 (Friction and Wear Enhancement of Titanium Alloy Engine Components). This project started in October 2009 and had a project end date of September 2011. The project was funded by DOE at $350,000 per year for FYs 2010-2012. Informal collaborators on this project were Cummins, Greenleaf Corporation, and NASA Glenn Research Center. This project addressed the goal of 50 percent improvement in freight efficiency by substituting strong, durable corrosion-resistant alloys for steel components. Specifically, the goal was to increase the use of titanium alloys in friction-and-wear critical engine components such as connecting rods, valves and valve guides, pistons, movable vanes in turbochargers, and bushings in EGR systems. Initially, a test method was selected for baseline friction-and-wear tests, and reciprocating pin-on-flat tests were conducted on materials, coatings, and surface treatments in order to select materials/treatments for the second phase of this project. No information was available to the committee beyond the 2011 project review.

ORNL Project PM052 (Friction Reduction through Surface Modification). This project started in October 2010 and was scheduled to end September 2014. The total project funding was $1,135,000. The objective of the project was to improve the fuel efficiency of HD diesel-powered vehicles by reducing the friction between contacting surfaces of the engine. It is estimated that in an HD engine, 10-15 percent of energy is lost to parasitic friction and that a 20-40 percent friction reduction would improve fuel efficiency by 2-6 percent. The target components include piston rings, connecting rod ends bearings/bushings, and cam followers. The method for reducing friction was a combination of surface texturing and coating technology. Milestones reached are (1) a report was produced describing the durability test procedure to be used for textured surfaces, (2) studies were completed on the effects of texturing on friction in a reciprocating piston ring/liner configuration, (3) wear-resistant thin coatings for textured bearing surfaces were selected, and (4) friction test specimens of textured and coated specimens were obtained.

NETL Project PM059 (HD-Cast Fe Alloys for High PCP Engines). The goal of this project is to develop new high-strength ferrous alloys for enabling increased peak cylinder pressures for improved performance and efficiency of heavy-duty engines. The project uses an integrated computational materials engineering (ICME) approach to computationally engineer new material compositions and manufacturing processes to achieve improved material performance, with a goal of a 25 percent improvement in component strength relative to A842 compacted graphite iron. At the time of the committee's meetings, the project, conducted with Caterpillar, had been under way for about a year, so it was not presented at the DOE 2014 AMR. It was presented at the DOE 2015 AMR, however. Accomplishments during the first year focused on establishing baselines for properties, structure, and machinability.

Response to Recommendation from the NRC Phase 2 Report

NRC Phase 2 Recommendation 3-10. The DOE should fund programs in the areas outlined in its "21CTP White Paper on Engines and Fuels" (February 25, 2011) in the section "Approach to Reaching Goals" covering materials R&D for valve trains, major engine components, air-handling systems (turbochargers and EGR systems), and exhaust manifold sealing materials.

21CTP Response: The Partnership agrees with this finding. The U.S. Department of Energy continues to fund research in materials that will enable improved efficiency in HD engines and after treatment devices.

Committee Comment on Response to 3-10

The committee is satisfied with this response.

High Temperature Materials Laboratory

The High Temperature Materials Laboratory (HTML) at ORNL is no longer operating as a national user facility as a result of federal budget reductions. The facility remains available, however, for project work under a CRADA that provides cost-recovery. The NRC Phase 2 report noted that this laboratory is a valuable resource for materials research for 21CTP in view of its specialized instrumentation and professional expertise.

Response to Recommendation from the NRC Phase 2 Report

NRC Phase 2 Recommendation 3-11. The DOE should continue to provide 21CTP researchers and other potential users access to HTML, and it should make every effort to maintain support for HTML and to maintain the cutting edge capability of the facility. Moreover, the DOE should provide sufficient funding for HTML, and for the research specialists who oversee and operate the facility, to enable continued research collaboration with the academic community, other government laboratories, and industry. In particular, HTML support should not be reduced to a level that allows only maintenance of the equipment for paying users.

21CTP Response: The Partnership agrees with this finding stating that the HTML is a valuable resource to 21CTP researchers. The prioritization of funding for DOE programs resides with Congressional budget authority and is beyond the scope of the 21CTP.

Committee Comment on Response to 3-11

The committee agrees the HTML is a valuable resource for industry collaboration. For future collaboration, the DOE needs to maintain the facility and associated expertise and to review whether the budget changes have affected the state-of-the-art of the HTML capabilities, including staff expertise.

REFERENCES

21CTP (21st Century Truck Partnership). 2013. Roadmap and Technical White Papers. U.S. Department of Energy Office of Energy Efficiency and Renewable Energy. http://energy.gov/sites/prod/files/2014/02/f8/21ctp_roadmap_white_papers_2013.pdf. Accessed March 14, 2015.

Abengoa. 2014. Abengoa Celebrates Grand Opening of Its First Commercial-Scale Next Generation Biofuels Plant. Abengoa Bioenergy. http://www.abengoabioenergy.com/web/en/prensa/noticias/historico/2014/abg_20141017.html?q=when%20did%20the%20hugoton%20plant%20start%20production/. Accessed August 20, 2015.

Advanced Collaborative Emissions Study (ACES). 2015. Advanced Collaborative Emissions Study (ACES): Lifetime Cancer and Non-Cancer Assessment in Rats Exposed to New-Technology Diesel Exhaust. Research Report 184. Boston, Mass.: Health Effects Institute.

Ajayi, O., C. Martin, A. Greco, and G. Fenske. 2014. Development of High Power Density Driveline for Vehicles (Developing Enabling Tribological Technologies). Argonne National Laboratory. DOE Annual Merit Review VSS058, Washington, D.C., June 19. http://energy.gov/sites/prod/files/2014/07/f17/vss058_ajayi_2014_o.pdf. Accessed March 14, 2015.

Amar, P. 2014. Development and Demonstration of a Fuel-Efficient Class 8 Highway Vehicle. Volvo Technology of America. DOE Annual Merit Review VSS081, Washington, D.C., June 19. http://energy.gov/sites/prod/files/2014/07/f17/vss081_amar_2014_o.pdf. Accessed March 7, 2015.

Bartley, G.J., C.J. Chadwell, T.W. Kostek, and R. Zhan. 2012. SCR Deactivation Kinetics for Model-Based Control and Accelerated Aging Applications. SAE Technical Paper 2012-01-1077. doi:10.4271/2012-01-1077.

Bergmann, D. 2013. Developing the Technology Innovation Process for Further Emissions Reduction. Presentation at Integer Emission Conference, Atlanta, Ga., October 22.

Berman, D. 2014. As fracking keeps pumping, A Qatar grows on the bayou. *Wall Street Journal*, May 28.

CARB (California Air Resources Board). 2015. DRAFT Heavy-Duty Technology and Fuels Assessment: Overview. www.arb.ca.gov/msprog/tech/report.htm. Accessed September 16, 2015.

Cargnello, M., J.J. Delgado Jaén, J.C. Hernández Garrido, K. Bakhmutsky, T. Montini, J.J. Calvino Gámez, R.J. Gorte et al. 2012. Exceptional activity for methane combustion over modular Pd - CeO_2 subunits on functionalized Al_2O_3. *Science* 337(713). doi: 10.1126/science.1222887.

Carrington, D. 2014. 2014 KIVA Development. Los Alamos National Laboratory. DOE Annual Merit Review ACE014, June 17. http://energy.gov/eere/downloads/vehicle-technologies-office-merit-review-2014-2014-kiva-development. Accessed March 13, 2015.

Caton, J. 2000. A Review of Investigations Using the Second Law of Thermodynamics to Study Internal-Combustion Engines. SAE Technical Paper 2000-01-1081. doi:10.4271/2000-01-1081.

Charlton, S.J. 2010. Meeting the US Heavy-Duty EPA 2010 Standards and Providing Increased Value for the Customer. 6th AVL International Exhaust Gas and Particulate Emissions Forum. Ludwigsburg, Germany, March 9-10.

Chen, X., N. Currier, A. Yezerets, and K. Kamasamudram. 2013. Mitigation of platinum poisoning of Cu-zeolite SCR catalysts. *SAE Int. J. Engines* 6(2). doi:10.4271/2013-01-1065.

CRC (Coordinating Research Council). 2005. Fuels for Advanced Combustion Engines (FACE) Working Group Mission Statement, November 16. http://www.crcao.org/publications/advancedVehiclesFuelsLubricants/FACE/. Accessed February 28, 2015.

CRC. 2012. Methodology for formulating diesel surrogate fuels with accurate compositional, ignition-quality, and volatility characteristics. American Chemical Society. *Energy & Fuels* 26: 3284-3303. http://www.crcao.org/reports/recentstudies2012/AVFL-18/AVFL-18%20Final%20Report%20[E&F%20Journal%20Article].pdf. Accessed March 14, 2015.

Daw, S., J. Szybist, J. Pihl, D. Splitter, V. Kalaskar, C. Xie, G. Fisher, and C. Chiangmai. 2014. Stretch Efficiency for Combustion Engines: Exploiting New Combustion Regimes. Oak Ridge National Laboratory and University of Michigan. DOE Annual Merit Review ACE015, June 18. http://energy.gov/eere/vehicles/downloads/vehicle-technologies-office-merit-review-2014-stretch-efficiency-combustion. Accessed March 13, 2015.

Dec, J. 2014. Low-Temperature Gasoline Combustion (LTGC) Engine Research – Previously known as HCCI/SCCI. Jeremie Dernotte and Chunsheng Ji, Sandia National Laboratories. DOE Annual Merit Review ACE004, June 17. http://energy.gov/eere/vehicles/downloads/vehicle-technologies-office-merit-review-2014-low-temperature-gasoline. Accessed March 13, 2015.

Delgado, O., and N. Lutsey. 2014. The U.S. Super Truck Program Expediting the Development of Advanced Heavy-Duty Vehicle Efficiency Technologies. ICCT White Paper, June.

DuPont. 2015. The DuPont Cellulosic Ethanol Facility in Nevada, Iowa: Leading the Way for Commercialization. DuPont.com. http://www.dupont.com/products-and-services/industrial-biotechnology/advanced-biofuels/cellulosic-ethanol/nevada-iowa-cellulosic-ethanol-plant.html. Accessed August 20, 2015.

EIA (Energy Information Administration). 2014a. Short-Term Energy Outlook (STEO), September. http://www.eia.gov/forecasts/steo/archives/Sep14.pdf. Accessed February 28, 2015.

EIA. 2014b. Annual Energy Outlook with projections to 2040. U.S. Department of Energy. http://www.eia.gov/forecasts/aeo/pdf/0383(2014).pdf. Accessed March 14, 2015.

EPA (Environmental Protection Agency). 2000. Heavy-Duty Engine and Vehicle Standards and Highway Diesel Fuel Sulfur Control Requirements. EPA420-F-00-057, December. Washington, D.C. http://www.epa.gov/otaq/highway-diesel/regs/f00057.pdf.

EPA. 2013. EPA Proposes 2014 Renewable Fuel Standards, 2015 Biomass-Based Diesel Volume. Regulatory Announcement, November. http://www.epa.gov/otaq/fuels/renewablefuels/documents/420f13048.pdf. Accessed March 14, 2015.

Erdemir, A. 2014. Advanced Nanolubricants for Improved Energy Efficiency and Reduced Emissions in Engines. Argonne National Laboratory. DOE Annual Merit Review FT018, Washington, D.C., June 19. http://energy.gov/eere/vehicles/downloads/vehicle-technologies-office-merit-review-2014-advanced-nanolubricants. Accessed March 14, 2015.

GAO (Government Accounting Office). 2014. Changing crude oil markets. http://www.gao.gov/assets/670/666274.pdf. Accessed February 28, 2015.

Goldsborough, S., M. Johnson, and D. Banyon. 2014. Collaborative Combustion Research with BES. Argonne National Laboratory. DOE Annual Merit Review ACE054, June 18. http://energy.gov/eere/vehicles/downloads/vehicle-technologies-office-merit-review-2014-collaborative-combustion. Accessed March 13, 2015.

Gonzalez, R. 2014. Blending the Optimal Cocktail. *FUEL*, September: 20-22.

Harvey, C., and A. Loder. 2013. Fracking Boom Pushes U.S. Oil Output to 25-Year High. *Bloomberg Business*, December 11. http://www.bloomberg.com/news/2013-12-11/fracking-boom-pushes-u-s-oil-output-to-25-year-high.html. Accessed March 14, 2015.

Hohl, Y. 2014. SCR on Filter-The Future for Construction Machinery. 10th International CTI Conference, SCR Systems, Stuttgart, July 8.

Ishizaki, K., N. Mitsuda, N. Ohya, H. Ohno, T. Naka, A. Abe, et al. 2012. A Study of PGM-Free Oxidation Catalyst $YMnO_3$ for Diesel Exhaust Aftertreatment. SAE Paper 2012-01-0365.

Kamp, C., A. Sappok, and V. Wong. 2012. Soot and ash deposition characteristics at the catalyst-substrate interface and intra-layer interactions in aged diesel particulate filters illustrated using focused ion beam (FIB) milling. *SAE Int. J. Fuels Lubr.* 5(2). doi:10.4271/2012-01-0836.

Khalek, I.A., M.G. Blanks, P.M. Merritt and B. Zielinska. 2015. Regulated and unregulated emissions from modern 2010 emissions-compliant heavy-duty on-highway diesel engines. *Journal of the Air and Waste Management Association* 65(8): 987-1001.

Kim, J., E. Kim, J. Han, and H. Han. 2013. Pt/Pd Bimetallic Catalyst with Improved Activity and Durability for Lean-Burn CNG Engines. *SAE Int. J. Fuels Lubr.* 6(3), doi:10.4271/2013-01-2591.

Kim, Y.J., H.J. Kwon, I. Heo, I-S. Nam, B.K. Cho, et al. 2012. Mn–Fe/ZSM5 as a low-temperature SCR catalyst to remove NO_x from diesel engine exhaust. *Applied Catalysis B: Environmental* 126 (2012) 9-21.

Koeberlein, D. 2014. Cummins SuperTruck Program: Technology and System Level Demonstration of Highly Efficient and Clean, Diesel Powered Class 8 Trucks. DOE Annual Merit Review ACE057, June 20.

Kumar, A., K. Kamasamudram, and A. Yezerets. 2013. Hydrocarbon storage on small-pore Cu-zeolite SCR catalyst. *SAE Int. J. Engines* 6(2). doi:10.4271/2013-01-0508.

Kurtz, E. 2014. Fuel Properties to Enable Lifted Flame Combustion. Ford Motor Company. DOE Annual Merit Review FT017, June 19. http://energy.gov/sites/prod/files/2014/07/f17/ft017_kurtz_2014_o.pdf. Accessed April 6, 2015.

Laroo, C., C. Schenk, J. Sanchez, J. McDonald, et al. 2013. Emissions of PCDD/Fs, PCBs, and PAHs from a modern diesel engine equipped with selective catalytic reduction filters. *SAE Int. J. Engines* 6(2). doi:10.4271/2013-01-1778.

McCormick, R., and M. Ratcliff. 2014. Performance of Biofuels and Biofuel Blends. National Renewable Energy Laboratory. DOE Annual Merit Review FT003, June 19. http://energy.gov/sites/prod/files/2014/07/f17/ft003_mccormick_2014_o.pdf. Accessed March 14, 2015.

McNenly, M.J., S.M. Aceves, D.L. Flowers, N.J. Killingsworth, G.M. Oxberry, G. Petitpas, and R.A. Whitesides. 2014. Improved Solvers for Advanced Engine Combustion Simulation. Lawrence Livermore National Laboratory. DOE Annual Merit Review ACE076, June 17. http://energy.gov/eere/vehicles/downloads/vehicle-technologies-office-merit-review-2014-improved-solvers-advanced. Accessed March 13, 2015.

Mueller, C. 2014. Fuel Effects on Mixing-Controlled Combustion Strategies for High-Efficiency Clean-Combustion Engines. Sandia National Laboratories. DOE Annual Merit Review FT004, June 19. http://energy.gov/sites/prod/files/2014/07/f17/ft004_mueller_2014_o.pdf. Accessed March 14, 2015.

Musculus, M. 2014. ACE001: Heavy-Duty Low-Temperature and Diesel Combustion & Heavy-Duty Combustion Modeling. Sandia National Laboratories. DOE Annual Merit Review ACE001, June 17. http://energy.gov/eere/vehicles/downloads/vehicle-technologies-office-merit-review-2014-heavy-duty-low-temperature-and. Accessed March 13, 2015.

Narula, C.K., M. Moses-DeBusk, X. Yang, L.F. Allard, X. Chen, and M.G. Stocks. 2010. Catalysts via First Principles. DOE Annual Merit Review PM011, June 10. http://energy.gov/sites/prod/files/2014/03/f12/pm011_narula_2010_o.pdf. Accessed August 19, 2015.

NAS-NAE-NRC (National Academy of Sciences-National Academy of Engineering-National Research Council). 2009. *Liquid Transportation Fuels from Coal and Biomass: Technical Status, Costs, and Environmental Impacts.* Washington, D.C.: The National Academies Press.

NGV (Natural Gas Vehicle) Global. 2012. http://www.iangv.org/current-ngv-stats/. Accessed March 14, 2015.

NRC (National Research Council). 2012. *Review of the 21st Century Truck Partnership – Second Report.* Washington, D.C.: The National Academies Press.

NRC. 2014. *Reducing the Fuel Consumption and Greenhouse Gas Emissions of Medium- and Heavy-Duty Vehicles, Phase Two: First Report.* Washington, D.C.: The National Academies Press.

Oefelein, J., G. Lacaze, and L. Hakim. 2014. Large Eddy Simulation (LES) Applied to Advanced Engine Combustion Research. With contributions from R.N. Dahms and A. Ruiz. Sandia National Laboratories. DOE Annual Merit Review ACE007. http://energy.gov/eere/vehicles/downloads/vehicle-technologies-office-merit-review-2014-large-eddy-simulation-les. Accessed March 13, 2015.

Partridge, W.P., M.-Y. Kim, J.A. Pihl, R.M. Connatser, J.-S. Choi, N. Currier, A. Yezerets, and K. Kamasamudram. 2014. Cummins-ORNL\FEERC Emissions CRADA: NO_x Control and Measurement Technology for Heavy-Duty Diesel Engines, Self-Diagnosing SmartCatalyst Systems. DOE Annual Merit Review ACE032, June 19. http://energy.gov/eere/vehicles/downloads/vehicle-technologies-office-merit-review-2014-cummins-ornlfeerc-emissions. Accessed March 13, 2015.

Patel, B. 2012. Innovative NO_x Abatement Alternatives to Liquid Urea SCR–A Technical Update on Hydrocarbon LNC. Integer Diesel Emissions Conference and ARLA32 Forum, Brazil, June.

Peckham, J. 2014a. Rivals to biofuels emerge. *FUEL*, June: 36-40.

Peckham, J. 2014b. Gas-to-liquids breakthrough. *FUEL*, June: 54-56.

Picket, L., and S. Skeen. 2014. Spray Combustion Cross-Cut Engine Research. Sandia National Laboratories. DOE Annual Merit Review ACE005, June 17. http://energy.gov/eere/vehicles/downloads/vehicle-technologies-office-merit-review-2014-spray-combustion-cross-cut. Accessed March 13, 2015.

Pitz, W., M. Mehl, and C.K. Westbrook. 2014. Chemical Kinetic Models for Advanced Engine Combustion. Lawrence Livermore National Laboratory. DOE Annual Merit Review ACE013, June 17. http://energy.gov/eere/vehicles/downloads/vehicle-technologies-office-merit-review-2014-chemical-kinetic-models. Accessed March 13, 2015.

Poet. 2014. The Project: POET's Project Liberty. http://www.renewableenergyworld.com/articles/print/volume-17/issue-6/bioenergy/the-project-poets-project-liberty.html. Accessed August 20, 2015.

Powell, C. 2014. Fuel Injection and Spray Research Using X-Ray Diagnostics. Argonne National Laboratory. DOE Annual Merit Review ACE10, June 18. http://energy.gov/eere/vehicles/downloads/vehicle-technologies-office-merit-review-2014-fuel-injection-and-spray. Accessed March 13, 2015.

Qu, J., P. Blau, H. Luo, S. Dai, T. Toops, B. West, B. Bunting, M. Viola, G. Mordukhovich, and D. Smolenski. 2014a. Ionic liquids as anti-wear additives for next-generation low-viscosity fuel-efficient engine lubricants. http://energy.gov/eere/vehicles/downloads/vehicle-technologies-office-merit-review-2014-ionic-liquids-anti-wear. Accessed March 14, 2015.

Qu, J., H. Meyer, Y. Zhou, Z. Cai, C. Ma, M. Chi, and H. Luo. 2014b. http://energy.gov/sites/prod/files/2014/07/f17/ft021_qu_2014_o.pdf. Accessed March 14, 2015.

Reith, C., F. Dornhaus, and A. Geisselmann. 2013. System Optimization and Integration for Heavy Duty Emissions Control Systems. FAD Emissions Conference, Dresden.

Reitz, R. 2014. Demonstration/Development of Reactivity Controlled Compression Ignition (RCCI) Combustion for High Efficiency, Low Emissions Vehicle Applications. Wisconsin Engine Research Consultants. DOE Annual Merit Review FT015, June 16-20. http://energy.gov/sites/prod/files/2014/07/f17/ft015_reitz_2014_o.pdf. Accessed March 14, 2015.

RFA (Renewable Fuels Association). 2014. Monthly U.S. Fuel Ethanol Production/Demand. http://ethanolrfa.org/pages/monthly-fuel-ethanol-production-demand. Accessed March 14, 2015.

Rohe, R., R. Marques, D. Harris, and C. Jones. 2012. New Acidic Zirconia Mixed Oxides for NH_3-SCR Catalysts for Better Passenger Cars Integration. Presented at IAV MinNO_x Conference, Berlin.

Sappok, A., I. Govani, C. Kamp, Y. Wang, and V. Wong. 2013. In-situ optical analysis of ash formation and transport in diesel particulate filters during active and passive DPF regeneration processes. *SAE Int. J. Fuels Lubr.* 6(2). doi:10.4271/2013-01-0519.

Schmeisser, V., M. Weibel, L. Sebastian Hernando, I. Nova, E. Tronconi, and M.P. Ruggeri. 2013. Cold start effect phenomena over zeolite SCR catalysts for exhaust gas aftertreatment. *SAE Int. J. Commer. Veh.* 6(1). doi:10.4271/2013-01-1064.

Shenk, M. 2014. U.S. Oil Output Surges to Highest Since 1986 on Shale. Bloomberg, September 17. http://www.bloomberg.com/news/print/2014-09-17/u-s-oil-output-surges-to-highest-since-1986-on-shale.html. Accessed March 14, 2015.

Shepardson, D. 2010. Higher ethanol blend gets OK, *The Detroit News*, October 14.

Singh, S. 2014. SuperTruck Program: Engine Project Review. Recovery Act–Class 8 Truck Freight Efficiency Improvement Project. Detroit Diesel Corporation, Project ID ACE058. http://energy.gov/sites/prod/files/2014/07/f17/ace058_singh_2014_o.pdf. Accessed March 7, 2015.

Sjoberg, M. 2014. Advanced Lean-Burn DI Spark Ignition Fuels Research. Sandia National Laboratories Combustion Research Facility. DOE Annual Merit Review FT006, Washington, D.C., June 19. http://energy.gov/sites/prod/files/2014/07/f17/ft006_sjoberg_2014_o.pdf. Accessed March 7, 2015.

Som, S., D. Longman, A. Ramirez, and Z. Wang. 2014. CFD Simulations and Experiments to Determine the Feasibility of Various Alternate Fuels for Compression Ignition Engine Applications. Argonne National Laboratory. DOE Annual Merit Review FT022, Washington, D.C., June 19. http://energy.gov/sites/prod/files/2014/07/f17/ft022_som_2014_p.pdf. Accessed March 14, 2015.

Som, S., Q. Xue, M. Battistoni, Y. Pei, J. Kodavasal, and D. Longman. 2014. Advancement in Fuel Spray and Combustion Modeling for Compression Ignition Engine Applications. Argonne National Laboratory. DOE Annual Merit Review ACE075, June 18. http://energy.gov/eere/vehicles/downloads/vehicle-technologies-office-merit-review-2014-advancement-fuel-spray-and. Accessed March 13, 2015.

Song, X., J.H. Johnson, and J.D. Naber. 2014. A review of SCR catalysts integrated into diesel particulate filters. *Int. J. of Engine Research.*

Stanton, D. 2013. Systematic development of highly efficient and clean engines to meet future commercial vehicle GHG regulations. *SAE Int. J. Engines* 6(3): 1395-1480.

Swedberg, S. 2012. The ups and (mostly) downs of viscosity. *Lubes 'n' Greases*, April: 6-11.

Strots, V., G. Nicol, M. Weibel, E. Tronconi, et al. 2014. Advanced NO_x Exhaust Aftertreatment Development for Long-Distance Transport. Presentation at IAV MinNO_x Conference, June 2014, Berlin.

Szybist, J., S. Curran, D. Splitter, V. Kalaskar, A. Dempsey, S. Sluder, B. West, and R. Wagner. 2014. Gasoline-Like Fuel Effects on Advanced Combustion Regimes. Oak Ridge National Laboratory. DOE Annual Merit Review FT008, Washington, D.C., June 19. http://energy.gov/sites/prod/files/2014/07/f17/ft008_szybist_2014_o.pdf. Accessed March 14, 2015.

Tang, W., D. Youngren, M. SantaMaria, and S. Kumar. 2013. On-engine investigation of SCR on filters (SCRoF) for HDD passive applications. *SAE Int. J. Engines* 6(2). doi:10.4271/2013-01-1066.

Theis, J., and C. Lambert. 2014. Assessment of Passive NO_x Adsorbers for Diesel Applications. 8th International Conference on Environmental Catalysis, August 24-27, Asheville, N.C.

Toops, T., C.E.A. Finney, and E. Nafziger. 2014a. Neutron Imaging of Advanced Transportation Technologies. Oak Ridge National Laboratory. DOE Annual Merit Review ACE052, June 18. http://energy.gov/eere/vehicles/downloads/vehicle-technologies-office-merit-review-2014-neutron-imaging-advanced. Accessed March 13, 2015.

Toops, T., J. Storey, M. Eibl, S. Lewis, J. Pihl, C. Xie, M. Debusk, V. Prikhodko, J. Parks II, and M. Lance. 2014b. Fuel and Lubricant Effects on Emissions Control Technologies. Oak Ridge National Laboratory. DOE Annual Merit Review FT007, Washington, D.C., June 19. http://energy.gov/sites/prod/files/2014/07/f17/ft007_toops_2014_o.pdf. Accessed March 14, 2015.

Viswanathan, S., N. Rakovec, and D.E. Foster. 2012. Microscale Study of Ash Accumulation Process in the DPF Walls Using the Diesel Exhaust Filtration Analysis (DEFA) System. Proceedings of the ASME 2012 Internal Combustion Engine Division Fall Conference, Paper ICEF2012-92104, September, Vancouver.

Walker, A. 2012. Current and Future Trends in Catalyst-Based Emission Control System Design. Presentation at the SAE Heavy-Duty Diesel Emission Control Symposium, September 2012, Gothenburg.

Wang, W., G. McCool, N. Kapur, G. Yuan, B. Shan, M. Nguyen, U.M. Graham, B.H. Davis, G. Jacobs, K. Cho, and X. Hao. 2012. Mixed-phase oxide catalyst based on Mn-mullite (Sm, Gd)Mn_2O_5 for NO oxidation in diesel exhaust. *Science* 337: 832. doi: 10.1126/science.122509.

Whitesides, R., N. Killingsworth, G. Petitpas, M. McNenly, and D. Flowers. 2014. Model Development and Analysis of Clean & Efficient Engine Combustion. Lawrence Livermore National Laboratory. DOE Annual Merit Review ACE012, June 17. http://energy.gov/eere/vehicles/downloads/vehicle-technologies-office-merit-review-2014-model-development-and-analysis. Accessed March 13, 2015.

Yezerets, A., N. Currier, K. Kamasamudram, and J. Li. 2014. Understanding Various Deactivation Mechanisms of Cu-Zeolite SCR Catalysts. Presentation at IAV MinNO_x Conference, June, Berlin.

Zigler, B. 2012. Fuels for Advanced Combustion Engines. National Renewable Energy Laboratory, Project ID FJ002, May 15.

Zigler, B. 2014. Advanced Combustion and Fuels. National Renewable Energy Laboratory, Project ID FT002, June 19. http://energy.gov/sites/prod/files/2014/07/f17/ft002_zigler_2014_o.pdf. Accessed August 20, 2015.

4

Hybrid Vehicles

INTRODUCTION

The hybridization of medium- and heavy-duty vehicles (MHDVs) is given a high priority among the technology objectives of the 21st Century Truck Partnership (21CTP). The roadmap and technical white papers document issued by 21CTP in 2013 (21CTP, 2013) states that "Hybrid electric vehicle (HEV) technology is a key enabler that will help 21CTP achieve its goals" by allowing MHDV manufacturers "to simultaneously improve fuel economy, emissions, and performance." The objective of this chapter is to (1) review the progress that the 21CTP has made toward accomplishing its ambitious technology objectives in the hybridization of medium- and heavy-duty vehicles; (2) identify the areas where its efforts to date have fallen short of these objectives; and (3) to provide recommendations for actions that should be taken to enhance the effectiveness of its coordinated project efforts in this area.

Hybrid propulsion systems for MHDVs exhibit both similarities and differences when compared to the more mature versions of this hybrid technology that are available in a growing number of hybrid electric light-duty vehicle (LDV) passenger models. Broadly speaking, the overall objectives of the hybrid drive propulsion equipment are the same in both vehicle classes. That is, hybrid drives are supplementary propulsion drives that augment the core internal combustion engine (ICE) powertrain in order to accomplish one or more of the following objectives:

- Recover braking energy that would otherwise be dissipated as heat and convert it instead to either electrical or hydraulic energy that can be stored and then used later for propulsion or vehicle auxiliary purposes. In addition to reducing fuel consumption and emissions, this approach can dramatically reduce brake wear in some MHDV classes such as urban buses, yielding significant maintenance savings.
- Provide additional acceleration that is added to the baseline ICE acceleration, making it possible to downsize the engine in some cases, resulting in lower fuel consumption and emissions.
- In some hybrid configurations, allow the ICE to spend much more of its operating time in its sweet spot range of torque and speed for maximum efficiency, thereby reducing its fuel consumption and emissions.
- Create opportunities for MHDVs to operate with their engines off (i.e., zero emissions operation) either for operating over some distance in an all-electric mode with zero emissions or for stop/start-mode operation in heavy traffic to eliminate unnecessary fuel consumption and emissions when the vehicle is temporarily stopped.
- Provide a source of stored auxiliary electric or hydraulic power that can be valuable to MHDV operators for a variety of purposes that include the powering of tools and lifts in vocational trucks or the powering of electric generators in long-haul Class 8 trucks in order to achieve idle reduction objectives (see Chapter 6).

Although this list is not exhaustive, it captures the most important operational objectives for incorporating hybrid drives into MHDVs. In many of the demonstration versions of hybridized MHDV trucks that have been designed and built to date, a conscious effort has been made to incorporate as many of the special functions listed above as possible in order to maximize the value extracted from the hybrid propulsion equipment.

Since the details of many alternative hybrid propulsion drive configurations developed for MHDVs have been provided in another NRC report (NRC, 2010), no attempt will be made to reproduce this valuable tutorial information here. However, it does deserve to be pointed out that two major types of hybrid propulsion drives are being developed that on the one hand are complementary and on the other, competitive. The most widely adopted hybrid drive architecture is based on electric power, using a combination of generators and electric motors and electrical energy storage devices (typically batteries) as the basic building blocks. Since all hybridized passenger LDVs now in production use some version of hybrid electric propulsion drives, this approach is becoming well established in the vehicle manufacturing industry. However, a second type of hybrid propulsion drive has been developed based on hydraulic power, using a combination of hydraulic pumps, motors, and hydraulic energy storage in pressurized accumulators. Although the hybrid hydraulic vehicular drive technology is less mature than its hybrid electric counterpart, the hydraulic hybrid drive exhibits some noteworthy characteristics, including the ability to store short bursts of energy more economically than batteries in electric-based drives. This feature makes the hybrid hydraulic drive appealing for medium-duty delivery and refuse trucks that undergo large numbers of start/stop cycles during a typical day, with short total travel distances.

CURRENT STATUS AND CHALLENGES FOR MHDVS

The commercialization of hybrid propulsion drives for MHDVs has been focused on a few specific segments of the MHDV market that are particularly well-suited to benefit from the performance advantages of the hybrid powertrains. These include

- Class 2b pickup trucks and vans,
- Classes 4-6 box-and-bucket trucks,
- Class 8 refuse trucks, and
- Class 8 urban transit buses (see Annex at the end of this chapter for discussion)

One of the key features that nearly all of these MHDVs share is a typical duty-cycle that includes frequent stops and starts that are well-suited for braking energy recovery using a hybrid powertrain. As a result, each of these vehicles can achieve significant improvements in fuel consumption from hybridization. Several manufacturers have been actively participating in the commercialization of hybrid powertrains for these vehicles, including 21CTP partners Allison, BAE Systems, and Eaton.

Despite the progress, some serious obstacles have impeded the growth of these markets, leading to notable setbacks. In particular, the lower cost of gasoline, diesel fuel, and natural gas and the disappearance of federal financial incentives have made it more difficult for MHDVs to succeed in competitive markets such as Class 4-6 delivery trucks. One notable recent example has been Eaton's decision to discontinue sales of its diesel-electric hybrid drive system in North America, announced in September 2014 (Eaton, 2014). A year earlier, Eaton had decided to discontinue its mild hydraulic drive system aimed at refuse trucks (Eaton, 2013), and there are indications that other hybrid hydraulic powertrains are encountering similar cost competition problems in the marketplace (21CTP-1, 2014). Other major manufacturers have also encountered serious problems with commercialization of hybrid drive equipment for medium-duty trucks in North America, including Allison Transmission (see section titled "Overview of Medium-Duty Hybrid Vehicles in Classes 2b to 6" later in this chapter for more details).

These setbacks are particularly disappointing since studies and field data confirm that hybrid propulsion systems can make significant contributions to reducing both the fuel consumption and greenhouse gas (GHG) emissions of several different types of heavy-duty trucks (NRC, 2012). This takes on increased importance since Phase I of new federal standards limiting the maximum fuel consumption and GHG emissions of MHDVs became active for the first time in model year (MY) 2014 (EPA, 2011, 2013), and Phase II limits are being formulated by the Environmental Protection Agency (EPA) and the Department of Transportation (DOT).[1] Although these standards do not require manufacturers to adopt hybrid drives or any other specific technology, the substantial improvements made possible by hybridization of some classes of MHDVs are expected to become increasingly valuable in the future. As a result, there is a risk that decreases in fuel prices that are unlikely to last forever will retard the development of technology for achieving long-term improvements in the fuel consumption and GHG emissions of MHDVs that are 21CTP objectives.

The current marketplace challenges for hybrid power trains can be largely attributed to the basic problem of payback periods that are too long for MHDV customers to accept. Even though the trucks are expected to have lifetimes longer than 10 years, the payback periods that are acceptable to truck purchasers are often 2 years or less (CALSTART, 2010), making it difficult for hybrid technology to compete for new business. This same CALSTART study includes analysis showing that the shorter the time that a hybrid truck is owned, the more difficult it is for an owner to recoup any premium for the hybrid propulsion equipment before the vehicle is sold.

The payback period barrier to market acceptance of hybrid trucks is recognized by the 21CTP leadership, who have expressed an interest in exploring alternatives, including "regulatory pull, robust incentives, and, perhaps, alternative business models (e.g., battery leasing)," to make up for the cost disadvantage of current hybrid powertrains (21CTP-1, 2014). For example, one approach to addressing the payback period problem would be to extend prorated purchase incentives to the second and subsequent generations of MHDV hybrid truck purchasers, making it easier for vehicle sellers to recoup their investments in this fuel- and emissions-reduction technology. Finally, if the long-term goal of the 21CTP program—to significantly drive down

[1] See the National Highway Traffic Safety Administration (NHTSA)/EPA Phase 2 Notice of Proposed Rulemaking for medium- and heavy-duty vehicle fuel efficiency and GHG standards, June 19, 2015, www.nhtsa.gov/fueleconomy.

the cost of MHDV hybrid equipment—is achieved, MHDV hybrid trucks will progressively compete more effectively in the international truck marketplace as the technology improves and matures.

Although the term "MHDV" covers a wide range of truck vehicles from Class 2b to Class 8 with a wide range of applications and duty cycles, the majority of the development effort for hybridized MHDVs has been focused on four areas: (1) medium-duty Class 3 to Class 6 delivery and electric utility trucks; (2) long-haul Class 8 trucks; (3) urban passenger transit buses; and (4) Class 8 vocational trucks, most notably refuse trucks. The first two of these are addressed in the body of this report in the following two subsections since they focus on truck classes that have been the target of multiple 21CTP R&D projects since the beginning of the Partnership. A discussion of hybrid buses is provided in an Annex to this chapter.

Overview of Medium-Duty Hybrid Vehicles in Classes 2b to 6

While heavy-duty long-haul tractor-trailers may travel 80,000 to more than 225,000 miles per year, medium-duty (MD) trucks in Class 6 are much more varied in configuration and often referred to as "work trucks" or, more specifically, as "vocational trucks" in regulations (see, e.g., Figure 4-1). In many cases, the specialized equipment on the truck (such as the bucket for an electric-utility vehicle or the suction pumps on specialized vehicles for cleaning sewers) can greatly exceed the cost of the truck itself. Since these vehicles are intended to perform work, often in stationary positions, their annual mileage may be as low as 20,000 miles (NRC, 2010). There are several vocational truck applications for which the same electric power converter equipment originally developed for use in hybrid power trains has been adapted to power the apparatus mounted on the vehicle for use in stationary operation.

FIGURE 4-1 Odyne Class 6 plug-in hybrid utility lift truck. SOURCE: Green Fleet Magazine (2008). Courtesy of Odyne Systems, LLC.

In addition to vehicles in the heavier classes, there is activity in battery-electric and hybrid vehicles (both electric and hydraulic) for package delivery vehicles and small commercial vehicles. These are typically categorized in Classes 2b to 5 based on their weights. The CALSTART organization has published a useful collection of photos illustrating the many different types of MD hybrid vehicles (HTUF, 2014).

Similar to the case of heavy-duty (HD) hybrid trucks, the cost of the constituent powertrain parts and the resulting payback period have slowed the market acceptance of MD hybrid vehicles. Since many of these vehicles approach the size and weight of passenger vehicles, including sport utility vehicles and vans, the idea of leveraging technology developed for passenger cars becomes increasingly reasonable as the vehicle class number drops. However, the special needs of commercial vehicle operation, including vibration, higher lifetime miles traveled, longer hours of operation, and temperature, need to be specifically factored into research and development activities aimed at these vehicles.

Previous studies have shown that hybrid powertrains in MD vocational trucks can provide as much as a 30 percent improvement in fuel consumption (NRC, 2010). Much of that savings comes from not idling the engine while operating the equipment on the vehicle. However, this drop in idling time can also be achieved in MD trucks with conventional ICE powertrains if the apparatus mounted on the vehicle does not depend on a running engine for operation. For example, the bucket of an electric-utility vehicle can be designed for operation without a running engine using power electronics, motors, and batteries that share a common heritage with the electric traction drive equipment that would be used in a hybrid-electric drivetrain. This type of MD vocational truck helps to reduce fuel consumption for the fleet but would not be classified as a hybrid vehicle, even though it benefits from the significant work done on components and subsystems developed for hybrid-electric powertrains.

There are other benefits of applying this electric drive technology to truck-mounted apparatus, including quieter operation in residential areas. Effort has also been devoted to implementing remote control of the engine so that the vehicle can make limited movements under the control of the person in the raised bucket. This feature improves the efficiency of the vehicle's operation, providing a tangible financial benefit that helps to pay for the technology.

Since many of the MD vehicles are operated in urban environments, there is an opportunity for turning off the engine whenever the vehicle is stopped in traffic, a feature commonly referred to as stop/start operation. Since the electric drive equipment in a stop/start system is used primarily to start the internal combustion engine and to provide modest support for acceleration and regenerative braking, the power ratings and cost of the hybrid-electric equipment can be reduced from a "full" hybrid to a "mild" or "micro" hybrid configuration. For example, UPS (United Parcel Service) previously worked with Eaton and Daimler to enable its package delivery vehicles to stop their engines in traffic whenever the vehicle's brake is depressed and the vehicle is stopped for more than a preset short time. The system restarts the engine as soon as the driver releases the brake pedal in order to avoid any perceptible delay in accelerating the vehicle due to restarting the engine. This stop/start feature is commonplace in European passenger cars and is in the process of becoming available in passenger vehicles in the United States. This passenger car technology is expected to find its way into medium-duty and some heavy-duty commercial vehicles.

One of the keys to applying hybrid drive technology in these MD vocational truck applications and to meeting the required commercial payback threshold is understanding how the vehicle is used. This understanding makes it possible for the hybrid equipment to be tailored to optimize the cost/benefit ratio. The SuperTruck program that has been supported by the 21CTP for HD trucks helped to better understand fuel use in HD Class 8 long-haul trucks by standardizing based on a 24-hour duty cycle. Using this approach, fuel use required during off-hours to keep the driver comfortable and entertained is included in the fuel consumption calculations. (Interested readers are encouraged to read Chapter 8, which is devoted to the SuperTruck program as well as the "Class 8 Long-Haul Trucks" section in this chapter for more details.)

This fuel use issue is even more important for vocational trucks because of their complicated and varied duty cycles, which include long periods of stationary operation. In November 2013,[2] Allison Transmission presented to the National Research Council (NRC) an analysis of the duty cycles of vocational trucks. Southwest Research Institute presented the results of its analysis of various technologies for saving fuel in both MD and HD trucks to the NRC in April 2014, followed by a workshop with EPA in December 2014.[3] The duty cycles used for the various analyses included several that are tailored for Class 6 vocational and delivery trucks, such as Parcel Cycle High-Efficiency Truck Users Forum (HTUF) Class 6 and the California Air Resources Board (CARB) truck urban cycle (Heavy-Heavy Duty Diesel Truck [HHDDT] Phase 1 GHG).

Several funded projects are collecting real-world data on duty cycles for MD vocational and delivery trucks. Examples include monitoring programs for UPS hydraulic package-delivery vehicles, FritoLay battery-electric package-delivery vehicles, Smith Electric vehicles, and Odyne plug-in hybrid electric vehicles. A significant effort is the Fleet DNA project led by NREL,[4] which is collecting information on hundreds of vocational vehicles in multiple locations around the country. This project is expected to deliver valuable information for understanding real-world operation of these MD vehicles, as well as to provide guidance for future hybrid system development and fuel consumption regulations.

A case history that illustrates both the technology opportunities and market barriers encountered by manufacturers of hybrid drivetrain equipment for trucks is provided by the H 3000 hybrid transmission developed for commercialization by Allison Transmission (Allison, 2015). Allison is one of the major commercial developers of hybrid transmissions systems for installation in buses and medium-duty trucks (H 40/50 EP series), with over 6,600 hybrid systems currently operating in transit buses. These transmissions are designed to recover braking energy from vehicles that have frequent stops and use this energy to reduce the vehicle's fuel consumption, power accessory equipment, and reduce brake wear (see Figure 4-2). Target applications for the H 3000 transmission are MD vocational trucks for applications including pickup and delivery, shuttle buses, utility service, and small refuse trucks in the Class 5 to light Class 8 weight ranges. Fuel savings are projected to be up to 25 percent depending on the truck duty cycle. The U.S. Department of Energy (DOE) has made an investment of $68 million in this program using American Recovery and Reinvestment Act (ARRA) stimulus funds. This support has been used to help develop this technology and to help build a new factory capable of building as many as 20,000 units of the H 3000 hybrid transmission annually. Although field demonstrations have been successful, the H 3000 hybrid transmission has not yet officially been brought to market, reflecting the market barriers to commercial acceptance currently being encountered by hybrid drive equipment manufacturers such as Allison. Since the ARRA-funded H 3000 development was not carried out as a project in the 21CTP portfolio of funded projects, no further evaluation of this project is included in this report.

Despite the commercial barriers to greater use of MD battery-electric and hybrid vehicles of all types in Classes 2b to 6, there continues to be interest in the industry and incentives to drive it forward. California provides a Hybrid and Zero-Emission Truck and Bus Voucher Incentive Project (HVIP) incentive and lists a number of MD vehicles available for this compensation through its website (HVIP, n.d.). New York also provides incentives; eligible vehicles are listed in NY Truck VIP (2015). The city of Chicago has provided incentives for battery-electric vehicles and buses, and new regulations are expected to extend the incentives to include MD hybrid trucks (Drive Clean Chicago, 2015). DOE main-

[2] M. Howenstein, Allison Transmission, "Transmission Technology and Fuel Consumption." Presentation to NRC Committee on Assessment of Technologies and Approaches for Reducing the Fuel Consumption of Medium- and Heavy-Duty Vehicles, Phase 2, on November 21, 2013.

[3] T. Reinhart, Southwest Research Institute, "Technologies for MD/HD GHG and Fuel Efficiency," Presentation to NRC Committee on Assessment of Technologies and Approaches for Reducing the Fuel Consumption of Medium- and Heavy-Duty Vehicles, Phase 2, on April 29, 2014.

[4] D. Anderson, "Vehicle Systems Simulation and Testing," Presentation to the committee on September 3, 2014.

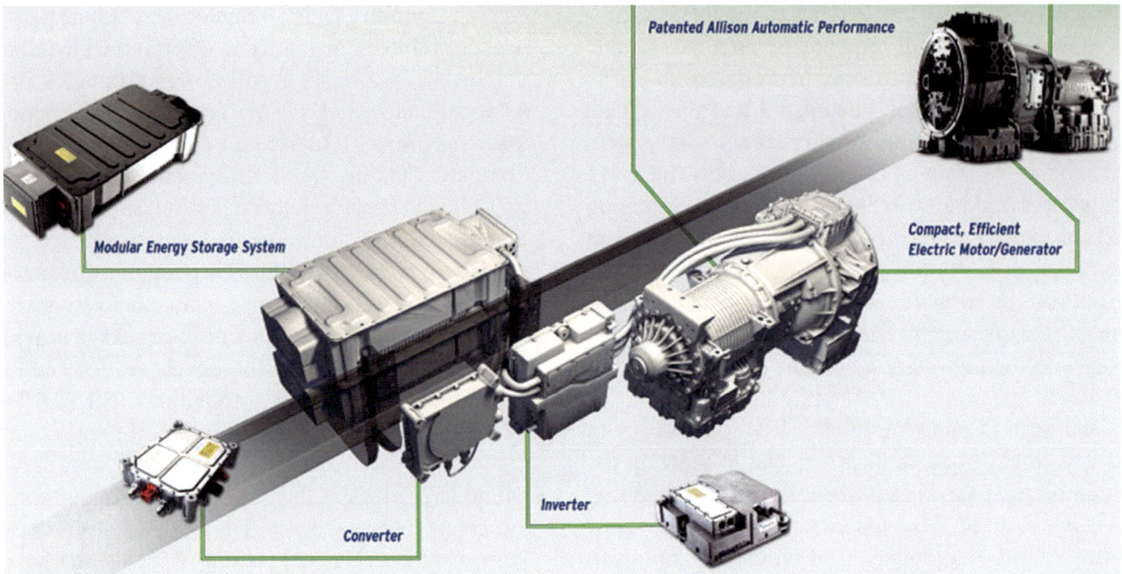

FIGURE 4-2 Allison H 3000 hybrid transmission major components. SOURCE: Allison (2015). © Allison Transmission 2011. All rights reserved.

tains a list of available vehicles in the Alternative Fuels Data Center (DOE, 2014). Some additional products can be found by looking at the member list for HTUF (HTUF, 2014).

HTUF, founded several years ago by CALSTART (CALSTART, n.d.), is a valuable source of information about the commercialization of MD hybrid vehicles. Bill Van Amburg, the senior vice president of CALSTART, has provided the committee with a list of 29 manufacturers from around the world that are currently actively involved in the development and production of MD hybrid and battery-electric trucks[5] Although the list is not reproduced here, the significant number of manufacturers reflects the continuing international interest in the development and commercialization of MD hybrid trucks. It should be noted that many of the manufacturers on this list are currently focused on research or development activities and are not yet offering their vehicles for sale, often attributable to the very soft market conditions that currently prevail for hybrid MDHVs in North America. This reflects the immature state of the MD hybrid technology in this field as well as the challenging market conditions noted earlier.

Class 8 Long-Haul Trucks

Heavy-duty Class 8 long-haul trucks present a challenging target for hybrid powertrain equipment. On the one hand, they do not fit the standard pattern for the other popular HD vehicle targets for hybrid drives. That is, the large majority of Class 8 long-haul trucks travel long distances without stopping, preventing these vehicles from taking advantage of the hybrid drive's special ability to recover energy from frequent stops and starts. Studies have indicated that the potential fuel consumption reduction benefits from hybridization fall into a range of 2 to 8 percent depending on assumptions about the duty cycle and the frequency and amplitude of elevation changes (i.e., hilliness) along the truck route (NRC, 2010). These reductions are much smaller than comparable numbers for other MHDV candidates such as Class 6 delivery trucks. Nevertheless, the fact that the total fuel consumption of Class 8 long-haul trucks is higher than that of any other MHDV type means that even small percentage reductions in fuel consumption can yield impressively large total reductions in fuel use and emissions.

The Daimler SuperTruck team chose to incorporate a hybrid-electric drive into its vehicle and is close to completing its evaluation. The Navistar SuperTruck team dropped its preliminary plans to include a full hybrid drive in its truck and is now evaluating microhybrid units for possible inclusion in its demonstrator truck (see Chapter 8 for more details). Despite the success of the Daimler SuperTruck demonstrator vehicle in achieving its goal of a 50 percent increase in freight efficiency (freight ton-miles per gallon), corresponding to a 33 percent reduction in its fuel consumption per ton-mile, the net contribution of the hybrid drive to this reduction is modest, making it difficult to justify the cost of the hybrid drive in the Daimler demonstrator truck. As discussed earlier in this chapter, the payback period for the hybrid drive equipment is typically much longer than the short periods of 2 years or less demanded by vehicle owners for justifying their investments. As a result of experiences such as this one in the Daimler Supertruck project, the future of the hybrid electric drive for Class 8 long-haul trucks is cloudy.

[5] Van Amburg, personal communication, 2014.

One of the questions left unanswered by the Daimler SuperTruck program is whether the trade-off between vehicle performance (measured in terms of fuel consumption, GHG emissions, and other metrics) and cost could be improved significantly by changing the ratings of the battery, electric motor, or power electronics. This is a serious question that applies not only to hybrid drive systems in Class 8 long-haul trucks, but also to a much wider range of vehicle sizes and applications, including vocational trucks and buses described in the preceding sections. In some cases, the adoption of mild hybrid systems having smaller motors and batteries, also referred to as microhybrids, can provide better performance vs. cost trade-offs than larger full hybrid drives. Since work on the simulation of MHDV hybrid trucks has already been carried out using federal funding at Argonne National Laboratory (ANL) (ANL, 2013) and the National Renewable Energy Laboratory (NREL) (NREL, 2014), the prospect of combining these simulations with real-world data collected from hybrid truck programs such as the Daimler SuperTruck demonstrator vehicle suggests some opportunities for investigating hybrid drive optimization.

An additional factor for consideration is the development by Daimler engineers of a vehicle speed control algorithm called eCoast, which modulates the truck speed on grades so that the vehicle mass becomes the energy storage means (via potential energy) for reducing the vehicle's fuel consumption without the need for expensive batteries or hybrid drive equipment (NACFE, 2014). This eCoast software algorithm is capable of delivering much of the value of the hybrid drive equipment at a fraction of the cost. On the other hand, the hybrid electric drive is tightly integrated into the overall vehicle propulsion drive, allowing it to play a role in enabling several advanced vehicle features. This fact complicates the engineering matter of deciding whether a hybrid drive can be justified or not. This example illustrates the challenges faced by truck vehicle designers in choosing which technologies to add or remove as they design new truck propulsion systems that can meet complex combinations of performance criteria in addition to achieving the demanded reductions in fuel consumption and emissions.

REVIEW OF THE 21CTP HYBRID VEHICLE TECHNOLOGY PROGRAM

Overview

In its 2013 roadmap document (21CTP, 2013), the 21CTP states its strategic approach in five thrusts, summarized as follows: (1) Develop hybrid propulsion systems for MHDVs; (2) overcome the technical barriers that inhibit the technologies; (3) educate interested parties on the importance of MHDV hybrid systems; (4) stimulate market demand for MHDV hybrid products; and (5) establish confidence in MHDV hybrid technologies by providing unbiased testing and evaluation of hybrid MHDVs.

The material presented to the committee by 21CTP indicates that most of the Partnership's effort has been focused on the first, second, and fifth thrust, with much less evidence of initiatives focused on either hybrid MHDV education or market stimulation.

In this same document, the top priority R&D areas that require government funding to meet 21CTP's hybrid vehicle technology goals are identified as (1) drive unit optimization; (2) drive unit cost; (3) energy storage system reliability; and (4) energy storage system cost. It is notable that all four of these identified R&D areas focus on components or subsystems in a hybrid propulsion unit and not on the integrated hybrid system. The implications of this intentional focus on component and subsystem R&D rather than integrated hybrid drive systems will be addressed in more detail later in this chapter.

Other key 21CTP initiatives that have a significant impact on MHDV hybrid drive systems include the following:

- The SuperTruck program for long-haul Class 8 trucks. The Daimler project includes a full hybrid drive system and the Navistar team is investigating mild hybrid options. This is discussed in more detail in Chapter 8.
- Participation in development of test procedures for MHDV powertrains that include both conventional and hybrid powertrain configurations, using the Vehicle Systems Integration (VSI) Laboratory at Oak Ridge National Laboratory (ORNL) (Smith et al., 2014).

Review of 21CTP Hybrid Vehicle Technology Goals

In 2006 the 21CTP established three hybrid technology goals for 2012. These goals established design life and cost targets for the hybrid drive unit and the energy storage system, as well as fuel economy and emission improvement targets for a heavy hybrid propulsion system operating on an urban driving cycle. Progress toward achieving these goals was reviewed in the NRC Phase 2 report (NRC, 2012), and this discussion will not be repeated or updated here since the target date for those three goals has passed.

Instead, this report will focus its discussion on six stretch goals for MHDV hybrid propulsion technology that were defined by the 21CTP in February 2011 (DOE, 2011). The target areas for these stretch goals are generally the same as those targeted by the three 2006 goals, but the six stretch goals are updated and made more specific. The reason these goals were explicitly designated as "stretch" goals when they were created is that these goals "can only be accomplished with increased funding through the 21st Century Truck Partnership" (21CTP, 2013). Since little funding has been allocated to MHDV hybrid component R&D since FY 2007 (NRC, 2012, Table 4-2), these stretch goals remain unfulfilled from the standpoint of 21CTP. As a result, a thorough discussion of each of the six stretch goals would not be meaningful under the circumstances.

However, hybrid drive and energy storage R&D, primarily for light-duty passenger vehicles, are being supported by DOE. The objectives of this R&D program overlap the needs of the MHDVs to some degree. As stated in the 21CTP roadmap document, "There is a common perception that investments in passenger car (LDVs) technology benefit HD trucks. This is not entirely true" (21CTP, 2013). In light of this observation, it makes sense to broadly address the relevance and impact of this DOE-sponsored R&D on the MHDV stretch goals. These six goals will be reviewed in sequence. (The goal text is summarized; the complete version is available in the hybrid propulsion white paper dated February 28, 2011 (DOE, 2011)).

- *Goal 1—Electric Machines.* Develop advanced motor technology that will deliver electric machines with improved durability, lower cost, better power density, and alternatives to rare earth permanent magnets.
  —Greater than 1 million miles (Class 8 line-haul application) or 15 years of life (vocational applications).
  —Power density for some motor designs today is at approximately 0.5 kW/kg. The objective is to nearly double the power density to approximately 1 kW/kg. A cost target of $50/kW by 2016 has been established.
  —Motors and generators have efficiencies typically at approximately 94 percent today. The objective is 96 to 97 percent by 2016.
  —Demonstrate a nonpermanent magnet motor technology in a commercial vehicle application that would equal or meet current hybrid system requirements by 2013.

- *Goal 2—Inverter Design/Power Electronics.* Develop technologies that will improve the cycle life of critical components within the inverter and other power electronics within the hybrid system.
  —Develop an improved switching device (insulated gate bipolar transistors [IGBTs] or other) that has a broader operating temperature range and a top temperature higher than today's 50°C and offers improved system life and durability. Develop this improved switching system and demonstrate benefits by 2016.
  —By 2016, reduce the overall weight of inverter designs by 20 percent through more efficient switching devices with higher operating temperatures and potential integration with engine cooling systems.

Goals 1 and 2 focus on achieving significant improvements in the performance, lifetime, and cost of the electric machine and power electronic inverter, respectively. The good news is that technology trends in these areas, supported in part by the DOE investments in R&D aimed at LDVs, are yielding improvements, in particular in the performance and power density of the power electronics. However, the likelihood of developing electric machines that do not contain high-grade rare earth magnets that exceed the performance and power density of the existing machines that use these magnets, as called for in the first goal, is low. Nevertheless, this R&D effort is spurring the development of so-called non-rare-earth traction machines that are likely to be attractive for some hybrid applications that do not need the weight and efficiency advantages provided by rare-earth magnets.

Progress toward achieving the cost targets and lifetimes called for in these first two goals is less clear and more difficult to evaluate. Project principal investigators employed by established suppliers of this equipment make positive claims about progress toward these objectives, but lifetime and cost are notoriously difficult to evaluate quantitatively. Indications are that the trends are in the right direction, but significant work still needs to be done to reach the aggressive targets. It should be noted that the lifetime targets for MHDVs are more challenging than those for passenger vehicles, both because of the length of the MHDV lifetime goal (15 years, compared to <10 years for passenger vehicles), and the more challenging physical environment (e.g., temperature, vibration, corrosive agents) experienced by hybrid drive equipment in MHDV applications.

- *Goal 3—Energy Storage Systems.* Develop an energy storage system with 15 years of design life, a broader allowable temperature operating range, improved power density and energy density, and significantly lower cost.
  —Develop a system that can provide a cycle life of 5,000 full cycles, which should achieve the target of 1 million miles (on the highway) or 15 years (vocational). Current state-of-the art energy storage systems are typically rated for 8 years of life.
  —By 2017, extend the acceptable operating temperature range for lithium ion batteries, currently at 0°C, to 55°C.
  —Develop battery technologies that will significantly increase power and energy densities.
  —Proposed cost targets:
    - $45/kW and/or $500/kWh for an energy battery by 2017;
    - $40/kW and/or $300/kWh for a power battery by 2020; and
    - By 2016, the cost of the overall battery pack should not exceed the cost of the cells themselves by more than 20 percent.
  —Establish an "end-of-life" strategy for advanced batteries and provide the necessary funding related to either the remanufacturing or recycling of batteries by 2017.

The status of the third goal, which targets power density, cost, and lifetime targets, but for energy storage equipment (primarily batteries) rather than electric machines and power electronics. This has been the single largest area for R&D investment of any of the five stretch goals during the past 5 years, although the focus has been on energy storage components for LDVs, not MHDVs. Steady progress is being reported on increasing both the energy and power density of new batteries for propulsion applications, but their ability to meet the challenging lifetime and cost targets for hybrid MHDVs, as stated in the goals, is much less certain. The specialized aspects of the hybrid MHDV application that distinguish them from the LDV hybrid system, together with the much lower volumes associated with the MHDV market, combine to make these targets particularly challenging for commercial suppliers to meet.

- *Goal 4—Hybrid System Optimization, Medium Duty.* To develop and demonstrate medium-duty hybrid system technology that can deliver substantial increases in fuel economy, beyond what is available with today's systems:
  —Potential applications for demonstration include MD shuttle buses, vocational trucks, and on/off highway MD work trucks.
  —A vehicle demonstration program that provides a platform for developing these medium-duty technologies (similar to the SuperTruck program for heavy-duty technologies) is one potential approach, with development and demonstrations to be completed by 2017.

- *Goal 5—Hybrid System Optimization, Heavy Duty.* An overarching goal is to develop and demonstrate HD hybrid system technology that can deliver substantial increases in fuel economy.
  —For urban, heavy start-and-stop driving cycles, a stretch goal of 60 percent (38 percent reduction in fuel consumption) has been identified.
  —For regional haul and line-haul applications, the percentage improvements would be more modest, with a stretch goal of 25 percent (20 percent reduction in fuel consumption).
  —Additional review and development need to be considered for those vehicles that would possess alternative anti-idling devices that could be provided without additional infrastructure changes.

Stretch goals 4 and 5 are different from the first three goals by virtue of targeting hybrid system optimization objectives for MD and HD vehicles, respectively. The intent of these two stretch goals to focus R&D effort on *system* design issues rather than *components* is highly commended because the value proposition for hybrid systems in MHDVs can be significantly improved by carefully integrating such systems into the MHDV systems, where it can contribute to the implementation of several valuable and innovative features that support 21CTP objectives. As noted earlier, the Daimler SuperTruck team[6] adopted this approach in its demonstrator truck. The specific targets included in these two goals were purposely chosen to extend well beyond the objectives of the current SuperTruck projects, and the cited target numbers (e.g., 20 percent fuel consumption reduction for line-haul trucks) are very ambitious, exceeding anything that has been demonstrated to date. As of this time, no funding has been allocated to achieving these goals. There is little sign that MHDV equipment or vehicle manufacturers will make these R&D investments on their own. The 21CTP program is in the process of revising its goals in this area, as discussed later in this chapter in the section "21CTP Hybrid Team Restructuring," and the committee supports these efforts, as reflected in its Recommendation 4-1.

- *Goal 6—Electrified Power Accessories.* Develop robust, durable, efficient electric power accessories for use with medium- and heavy-duty hybrid systems:
  —Electrifying accessories such as power steering, air compressors, and air-conditioning compressors can significantly reduce parasitic losses by powering them on-demand.
  —Target date for availability of such improved accessories: 2016.

The final stretch goal, Goal 6, focuses on the development of electrified power accessories such as power steering and air-conditioning compressors that are "robust, durable, [and] efficient." Unlike the preceding five goals, there are no quantitative targets established for this goal because of its breadth. This goal is closely related to the preceding two because the MHDV system integration and optimization efforts associated with Goals 4 and 5 can be expected to extend into the areas of electric accessories, as was the case in the Daimler SuperTruck project.[7] Although there has been no R&D funding invested in the development of electric accessories specifically for MHDVs, some of the results of R&D efforts that have been supported to advance accessory electrification technology in LDVs will broadly benefit MHDVs as well. However, as noted previously, the specialized nature of the MHDV specifications, the challenging physical environment for this equipment, and the longer lifetime targets make it difficult to directly apply the results from passenger vehicle R&D projects without significant additional effort tailored to MHDVs.

Before closing this section, it should be said that the 21CTP leadership, in response to a committee question, answered that it had "not conducted a full planning effort

[6] D. Kayes, D. Rotz, and S. Singh, Daimler Truck North America LLC, "SuperTruck Team," Presentation to the committee on May 15, 2014.

[7] Ibid.

for reviewing these goals and defining the specific budget required to meet each one. . ." (21CTP-1, 2014) even though this was a specific recommendation of the NRC Phase 2 report. (See discussion on H-6 Revised Hybrid Goals, Recommendation 4-5, in the section "21CTP Response to Recommendations in NRC 2nd Review Report" near the end of this chapter for more details.) In this same response, the 21CTP leadership said "the Hybrid team has been revisiting the goals in light of the current and near-term future outlook for hybrid technology in commercial trucks" (21CTP-1, 2014). The latest information from the 21CTP leadership indicates that the hybrid team is in the process of undergoing reorganization, and it is expected that this process, when completed, will have a significant impact on the future of these goals. More details about this reorganization can be found in the section "21CTP Hybrid Team Restructuring" later in this chapter.

Assessment of Progress and Key Accomplishments

The committee has reviewed available information about several projects funded by federal agencies in the Partnership that relate to battery-electric vehicles, hybrid-electric vehicles, and hybrid-hydraulic vehicles. Information provided by 21CTP leadership indicates that total expenditures by three of the 21CTP-affiliated federal agencies (DOE, DOD, and DOT) on hybrid electric drive technology for both LDVs and MHDVs during the 2012-2014 totaled $63.4 million. Table 4-1 lists all of the electric drive technologies projects identified by the 21CTP leadership as part of 21CTP project portfolio. This funding can be broken into two main portions, one consisting of projects that are specifically focused on MHDVs with hybrid and battery-electric drives. The other portion of the projects claimed by 21CTP is focused on developing electric drive technology for LDVs, accompanied by the claim that the technology targeted by these projects is applicable to larger MHDVs.

Closer examination of the projects that make up this inventory reveals that the dominant portion of this project funding (>75%) is directed at vehicle demonstration projects, and much of the funding was supplied by the ARRA stimulus program. Although such demonstration (pilot) programs are valuable, it should be noted that less that 20 percent of this funding is associated projects that can be objectively categorized as R&D.

Attention will first be addressed to projects that specifically address MHDVs. A review of information available to the committee reveals that these 21CTP-affiliated MHDV projects, with only a couple exceptions, can be placed in one of four categories:

TABLE 4-1 Electric Drive Technologies Projects Identified as Part of 21CTP Inventory

Agency	Subgrouping	Internal Project Title	Recipient	2012 Funding	2013 Funding	2014 Funding
DOD		Non-Rare-Earth Materials for Motors	N/A	1,500,000	500,000	500,000
DOD		Modeling and Optimization of Electrified Propulsion Systems	N/A			98,000
DOD		High Energy Density Asymmetric Capacitors	N/A			99,000
DOD		Powertrain Thermal Management - Integration and Control of a Hybrid Electric Vehicle Battery Pack, E-Motor Drive, and Internal Combustion Engine Multiple Loop Cooling System	N/A			50,000
DOD		Advanced Models for Electric Machines	N/A			78,000
DOE	Advanced Electric Drive Technologies R&D	Various (see Annual Report, http://energy.gov/sites/prod/files/2014/04/f15/2013_apeem_report.pdf)	3 National labs (ORNL, NREL, Ames), universities, and industry partners	13,266,940	11,053,327	9,704,789
DOE	Electric Drive Technologies	Various (see Annual Report, http://energy.gov/sites/prod/files/2014/04/f15/2013_apeem_report.pdf)	6 Industry (GE, UQM, GM, Sigma, APEI, Synthesis Partners) and 1 lab (ANL)	5,500,000	9,587,357	11,500,000
Annual totals				20,266,940	21,140,684	22,029,789
Total for 2012-2014						63,437,413

Category 1. Projects that support specific truck manufacturers to develop new MHDV hybrid or battery-electric trucks or improve existing models and then collect data about their performance during field tests. In some cases, the project involves building significant numbers of the vehicles. Examples of these projects include ARRA VT072, "Smith Electric Vehicles: Advanced Vehicle Electrification & Transportation Sector Electrification" (Mackie, 2014) and ARRA VT083, "Plug-In Hybrid Electric Commercial Fleet Demonstration and Evaluation" (Cox, 2014). These projects are by far the largest of all the projects associated with hybrid drive technology that are part of the 21CTP portfolio, representing a total federal funding commitment of $77 million over 5 years for these two named projects. They were both made possible using funding provided to DOE under the 2009 ARRA, and they are the only current 21CTP projects that use federal funding to design and construct new MD hybrid trucks (780 total vehicles for the two projects) that are then placed in the field for testing and data collection. (One of the ARRA-funded SuperTruck projects also includes a hybrid drive as part of its drivetrain; see Chapter 8.) Both projects have suffered setbacks that have delayed their completion, but progress is being made by each toward meeting their truck delivery targets during 2015. Preliminary field test results look promising for achieving reduced fuel consumption and emissions, but more field testing is necessary to provide a more complete evaluation.

Category 2. Projects that support work at the national laboratories or third-party organizations to evaluate the performance of several different types and models of MHDV hybrid or battery-electric trucks (or buses) without requiring government funding to directly support the development and manufacturing of those vehicles. Examples of these projects include NREL VSS001, "MHDV Field Evaluations" (Walkowicz, 2014), and the project led by the South Coast Air Quality Management District (SCAQMD), VSS115, "Zero Emissions HD Drayage Truck Demonstration") (Choe, 2014). Both projects involve agreements with multiple truck manufacturers who are responsible for developing and building the hybrid trucks, which are field-tested by an independent third-party organization, which collects extensive data. Federal funding for projects in this group is typically $600,000 or less per year, approximately ten times smaller than annual federal expenditures for projects in the first group. The skills of the funded researchers are applied primarily to directing the data collection, followed by rigorous evaluation. Results from these projects range from very promising in the case of VSS001, cited above, to disappointing in the case of another project, VSS116, "Hydrogen Fuel Cell Electric Hybrid Truck and Zero-Emission Delivery Vehicle Deployment" (Carr and Williams, 2014), which has been significantly delayed by problems with the manufacturers that were originally selected to provide the vehicles. It should be noted that this last-mentioned project included funding to provide financial incentives for customers to purchase the fuel-cell-powered vehicles for field testing, contributing to its larger total DOE budget of $2.4 million over 3 years.

Category 3. Projects that involve a closer relationship between one or more national laboratories and a manufacturer of hybrid or battery-electric MHDVs to pursue the development of improved vehicle technology involving the architecture, subsystems, or control of the drivetrain. Examples of this group of projects include VSS133, "Cummins MD & HD Accessory Hybridization CRADA," which combines the efforts of Cummins and ORNL (Deter, 2014), and VSS134: "Vehicle Thermal System Modeling in Simulink," which represents a collaboration between NREL and three truck industry partners (Lustbader and Kiss, 2014). Typically, these projects are organized so that the national laboratory focuses its efforts on modeling, analyzing, and, in some cases, testing the new technology while the industry partner takes responsibility for implementing the new technology in hardware and/or software, as appropriate. The key difference between this project group and the preceding one is that, in this group, national laboratory researchers are more directly involved in technical activities such as modeling, analysis, and laboratory testing that directly influence the development of the new technology that is being conducted under the leadership of the industry partners. Annual funding levels for these projects are typically $600,000 or less per year, again far less than the annual budgets for projects in the first group. These projects are particularly appealing since they provide opportunities to harness the special technical skills of researchers at the national laboratories to provide valuable assistance to the manufacturers in areas that complement the skills of their in-house technical staffs. Of the two projects, the first one, involving the Cummins-ORNL cooperative research and development agreement (CRADA), is further along due to its earlier start date, and the technical results to date from the modeling and simulation work appear to be very promising.

The plans to test prototype hardware and software developed during the course of the Cummins-ORNL project on dynamometers in the new Vehicle Systems Integration (VSI) Power Train Test Laboratory at ORNL will provide a welcome opportunity to demonstrate the unique strengths and features of this impressive new facility (Smith et al., 2014). The facility includes two high-power (500 kW), high-performance dynamometers that can be used to test large truck engines and hybrid drivetrains using sophisticated controls and instrumentation that can apply any desired drive cycle to the engine system while a thorough set of measurements are being made, including fuel usage and exhaust emissions (see Figure 4-3). The VSI Laboratory also includes high-speed computers and high-quality real-time control equipment that make it possible to conduct tests that combine actual equipment with simulated components or controllers in the same experiment. There are very few test facilities like this in the world today specifically designed for testing MHDV truck engines and powertrain equipment.

Category 4. Projects that address issues associated with the introduction of MHDV hybrid and battery-electric trucks from a higher level by developing models to analyze and project the performance, cost, market opportunities, and the resulting reductions in fuel consumption and emissions of these vehicles for several years in the future on a regional and national scale.

Examples of this last group of projects include VAN001, "Impact Analysis: VTO Baseline and Scenario (BaSce) Activities," led by ANL (Stephens, 2014), and VAN012, "Modeling for Market Analysis: HTEB, TRUCK, and LVChoice," led by an ANL contractor, TA Engineering, Inc. (Birky, 2014). Although the number of projects in this group and the total annual funding (<$700,000 for the two projects listed above) are lower than for the other three groups, these modeling studies can play an important role in helping 21CTP leadership to evaluate the areas in which their future project funding can have the largest positive impact for achieving the Partnership objectives of reducing fuel consumption and emissions. This effort takes on special importance because of the challenging conundrums associated with evaluating the trade-offs between the national benefits of hybridization in MD trucks (higher percentage benefits per truck but lower total fleet fuel savings) and HD trucks (lower percentage benefits per truck but higher total fleet fuel savings). Unfortunately, the validity and credibility of these models is heavily dependent on the quality of the input data which, in some cases, requires detailed cost and sales data that manufacturers are traditionally hesitant to provide. Regardless of the inevitable debates about the accuracy of their future projections, the models developed as a result of these projects play a useful role by encouraging leaders in both government and industry to consider the high-level impact issues when deciding on future technology investments.

FIGURE 4-3 500 kW dynamometer with diesel engine-under-test in ORNL VSI Laboratory with truck outline overlay. SOURCE: Smith et al. (2014).

In addition to the projects that are specifically focused on MD and HD trucks discussed above, a number of other projects are being funded by DOE and DOD to develop technology for automobiles and combat vehicles that may be applicable to hybrid and battery-electric MHDVs in the future. Although the specific objectives and status of these projects will not be discussed in detail here, it should be stated that none of the DOE-funded projects are defined to directly address the technical demands that distinguish light-duty automobiles from medium- and heavy-duty trucks. However, there are some projects that address these issues indirectly. For example, there are a few projects, such as APE063, "Performance and Reliability of Bonded Interfaces for High-Temperature Packaging" (DeVoto, 2014), and APE061, "Cost-Effective Fabrication of High-Temperature Ceramic Capacitors for Power Inverters" (Balachandran, 2014), that address important technical challenges associated with achieving rugged and reliable power electronics in hostile high-temperature environments with semiconductor junction temperatures up to 200°C, conditions that are relevant to the longer lifetimes expected for power electronics operating in future hybrid/electrified trucks. Several of the other projects, such as those focused on developing traction motors that do not use expensive rare-earth magnetic material, could eventually lead to scaled-up versions that would be useful in hybrid drives for trucks, but when this might happen is uncertain.

There are also a few projects identified by DOD that could yield electric drive technology that matches or exceeds the power ratings and ruggedness requirements for electric drives in commercial MHDVs. One example is a project funded by the U.S. Army Tank Automotive Research, Development and Engineering Center (TARDEC) named "Integrated Starter Generator (ISG)" (DOD, 2014) that is developing high-performance electric machines and inverters with ratings of 120 kW and 160 kW for use in future combat vehicles—power ratings that fall within the range required for hybrid drive systems in HD trucks. This is a multiyear project led by General Dynamics Land Systems (GDLS) that began in FY2013 with a planned duration of 6 years and a total budget of $26.9 million. The project is planned to include significant efforts to build prototype ISG hardware that will be tested in both the laboratory and actual combat vehicles (the Army's Stryker). Although the technical details provided by DOD about this project are quite limited, it is interesting to note that the project leaders specifically call out "Intelligent Engine Start/Stop" as one of the target operating modes, highlighting its potential relevance to MHDV hybrid trucks.

Before closing this section, it is worth noting that the ARRA-funded projects that comprise the dominant share of the 21CTP projects in the hybrid electric drives area are due to end in FY2015. As a result, the annual expenditures in the future 21CTP budgets associated with hybrid electric drive technology are likely to drop significantly for FY2016 and beyond unless hybrid-related projects emerge to take the place of the ARRA-funded projects. The committee supports investments by 21CTP in longer-term R&D projects focused on promising approaches that might lead to significant improvements in hybrid-related technology, as reflected in Recommendation 4-2.

Summary of Key Barriers and Future Opportunities

The environment for the hybridization of MHDVs has changed significantly during the past 5 years. Some of the factors influencing these changes include (1) experience is being gained with hybrid drives in MHDVs for a variety of vehicle classes and applications that has been clarifying the real-world fuel consumption improvements that can be achieved and (2) natural gas and petroleum prices have dropped considerably, reducing the economic attractiveness of commercially offered hybrid drivetrains. Against this backdrop, the key barriers and issues that are slowing the commercial acceptance of hybrid trucks include the following:

- While the prices of key components—including the power electronics, motors, and batteries—in hybrid drivetrains for light-duty passenger cars are decreasing, the rate of decrease is not been sufficiently fast to allow hybrid drivetrains to meet payback period criteria set by the truck purchasers under current conditions of falling fuel prices and the absence of any direct price for emissions other than regulatory limits. The cost of batteries or alternative energy storage components has been particularly troublesome despite continuing progress on reducing battery costs.
- Despite claims that the significant investment by the federal government in hybrid drive technology R&D for LDVs is directly applicable to MHDVs, there are substantive differences between the requirements of LDVs and MHDVs that create important gaps in the technology readiness of commercial hybrid drive components for truck drivetrain applications.
- The truck manufacturing industry is characterized by a significant number of small- to medium-size firms compared to the smaller number of much bigger passenger vehicle manufacturers. This difference makes it much less likely that truck manufacturers will invest in the long-term R&D needed to develop mature and cost-effective hybrid drivetrain technology for their future truck products.
- One of the biggest conundrums is that while hybrid technology is most beneficial for MD trucks that experience large numbers of start/stop cycles, collectively, these Classes 3 to 6 trucks combined consume less than half of the total fuel consumed by long-haul Class 8 trucks, which do not benefit as much from the

introduction of hybrid drivetrains (NRC, 2010). This has made it more difficult for the 21CTP to justify R&D investments in hybrid technology for any of the MD truck classes.
- Federal test procedures for evaluating fuel consumption and emissions are still incomplete for hybrid truck drivetrains, complicating the process of quantitatively evaluating their performance for the purpose of qualifying for regulatory credits or state tax incentives. Historically, these fuel consumption and emissions tests for conventional MHDVs have always been conducted using the diesel engine alone. More details are provided in the subsection "Hybrid Vehicle Fuel Consumption and Emissions Certification," later in this chapter.
- The early years of hybrid truck manufacturing, the past 10-15 years, have been marked by a number of immature products that resulted in poor vehicle experiences for the buyers, as well as orphaned products caused by manufacturers who prematurely left the market or went out of business. This checkered history has hurt resale values and discouraged fleets from adopting hybrids.

The net impact of these barriers on the current market for hybrid trucks has been summarized by Deborah Gordon, executive director of Regulatory Issues and Hybrid Programs at Allison, as follows:

> When considering volume production over the next 10 years for commercial truck hybrids, Allison believes that fleet operators and vehicle manufacturers currently lack a viable business case to support widespread deployment. At least a few considerable barriers remain; the technology is considered somewhat unproven in terms of 'real world' reliability in truck vocations, a lack of significant financial incentives to offset costs, and the current impact of low fuel prices (Gordon, personal communication, 2015).

Despite this discouraging assessment of the hybrid truck market in North America, it should be noted that the market for hybrid trucks is much stronger in other parts of the world, including China and Europe, where concern about air pollution is high, particularly in densely populated urban environments. For example, a recently completed study by Frost & Sullivan predicted that global sales of MD hybrid and electric trucks will grow from 2,200 annually in 2013 to 84,000 annually in 2022, corresponding to a compound annual growth rate (CAGR) of nearly 50 percent (Frost & Sullivan, 2014). The projections of that same study for HD hybrid and electric trucks are even more attention-grabbing, predicting global annual sales growth from 300 in 2013 to 51,000 in 2022, corresponding to a CAGR value of 77 percent. The study predicts that the largest purchaser of these hybrid and electric trucks will be China, with sales driven primarily by government mandates. Consistent with these predicted trends, Dr. Mihai Dorobantu, the Eaton representative on the 21CTP hybrid team, has noted that Eaton's sales of hybrid truck drivetrain equipment are very strong in China for MHDV applications such as city buses (see Figure 4-4), despite the fact that Eaton's U.S. sales dropped to the point of causing it to suspend market activities in North America, as noted earlier in this chapter (Dorobantu, personal communication, 2015).

These sharply countervailing trends suggest that, while the technology and market barriers to hybrid MHDVs are currently high, particularly in North America, the long-term international market opportunities are substantial. In this situation, the role of the 21CTP in applying its R&D resources to achieve positive long-term objectives could have a significant impact on the prospects for U.S.-based manufacturers to succeed in the future domestic and international markets for hybrid MHDVs.

21CTP Hybrid Team Restructuring

The 21CTP organization includes a hybrid team that consists of representatives of 21CTP industry partner companies that have a commercial interest in hybrid drive components, subsystems, or complete vehicles. This hybrid team provides advice to the 21CTP executive committee and to the leadership of the overall 21CTP program inside DOE.

During its recent meetings in July and November 2014, members of the hybrid team reviewed the special challenges associated with the commercialization of MHDV hybrid drive equipment and the vehicles in which they are installed. An important topic of discussion during those meetings was that, beyond hybrid drives, "there are opportunities for efficiency gains in the remainder of the drivetrain (conventional, automatic, or AMT transmissions, axles, etc.)" (21CTP-1, 2014). Members of the hybrid team are now in the midst of reorganizing the group to broaden its focus to include advanced drivetrains. To this end, the currently proposed name of this reorganized team is the Hybrid and Drivetrain

FIGURE 4-4 HD hybrid city buses in Jining City, China, built with Eaton hybrid drives. SOURCE: Eaton Corporation, "Eaton drops hybrids in North America," *Fleets & Fuels*, June 26, 2013. http://www.fleetsandfuels.com/fuels/hybrids/2014/06/eaton-drops-hybrids-in-north-america/.

Working Group. Although the changes are not official, it is expected that the terms of the team restructuring will be worked out with the 21CTP leadership and completed by midyear in 2015.

Some valuable insights into the objectives and future directions of this reorganized working group have been provided by Mihai Dorobantu, director of Technology Planning and Government Affairs in the Eaton Vehicle Group, who is a member of the current hybrid team (Dorobantu, 2014, personal communication). He has been actively involved in the team restructuring process and is well versed in the current status of hybrid MHDV technology and markets. According to Dr. Dorobantu, the current hybrid team members collectively have a much better understanding of the capabilities of today's hybrid technology in MHDVs as well as the commercial obstacles that it currently faces. At the same time, there is a much better understanding of the need to take a more integrated systems view of future MHDVs, including the role of the powertrain beyond the engine as well as the broader electrification trends that are affecting all forms of land transportation, including LDV automobiles. Armed with their years of collective experience in the hybrid MHDV field, the team members are proposing that the scope of the new working group should be broadened to include a wider variety of technologies that can provide cost-effective reductions in truck fuel consumption and GHG emissions.

Questioned about the role of hybrids in the scope of the new working group, Dr. Dorobantu indicated that it was not the intention of the team members to eliminate hybrid technology from the 21CTP program since there are specific market segments (e.g., urban buses) and market locations (e.g., China) where hybrid technology is experiencing considerable market success. Instead, a goal of the new working group will be to determine how hybrid technology can be utilized in innovative ways as part of a broader powertrain electrification process to achieve critical technology breakthroughs that will be commercially successful in a wider range of MHDV classes and application categories. For example, Dr. Dorobantu expressed optimism about the longer-term market opportunities for advanced power train technology in HD Class 8 long-haul trucks because of emerging market trends toward regional-haul trucks with smaller engines that spend considerably more of their operating time in congested urban environments requiring frequent stops and starts.

Based on this discussion and other information provided by the 21CTP leadership cited above, it is apparent that one of the important achievements of this hybrid team restructuring, when completed, will be a greater focus on the interactions and potential integration of several key subsystems in the vehicle, starting with the engine. This proposed system focus contrasts with the current 21CTP strategy of focusing on the development of component technology, a topic that will be addressed in more detail later in this chapter.

Hybrid Vehicle Fuel Consumption and Emissions Certification

The fuel consumption and emissions certification of a conventional diesel-powered MHDV is currently accomplished by running the engine over a combination of transient and steady-state operating conditions on an engine dynamometer. The certification process is considerably more complicated for hybrid trucks since the emissions cannot be accurately determined unless the complete hybrid powertrain, including both the engine and the electric machine(s), is tested as an integrated unit. Progress has been made by the National Highway Traffic Safety Administration (NHTSA) and EPA in recent years toward defining alternative approaches to evaluating the fuel consumption and emissions of hybrid MHDVs using either simulation, dynamometer testing of hybrid drivetrain power packs without the rest of the vehicles, or chassis dynamometer testing of the complete hybrid truck. However, these evaluation procedures are marginal for the Phase I MHDV fuel consumption and emissions regulations issued in 2011 (EPA, 2011), hindering their ability to accurately evaluate the fuel consumption and emissions characteristics of production hybrid MHDVs.

Improved validation techniques will give the industry and the regulators the information needed to assure compliance with new fuel consumption and GHG emissions standards in a way that reflects the benefits that hybridization provides to vehicle performance. The need for these MHDV test procedures has been apparent for several years, providing the basis for one of the recommendations (Recommendation 4-8) in the NRC Phase 1 report (NRC, 2008). The current committee received promising reports about the recent completion of the Vehicle Systems Integration Laboratory at ORNL,[8] including reports that this facility will be used in the development of MHDV hybrid power pack test procedures in collaboration with EPA and other agencies. However, no specific date has been set for the completion and release of the new test procedures.

Role of the Federal Government and the States

Hybrid Truck Incentives and Tax Credits

Incentives were established to help accelerate the development and implementation of high-efficiency HD vehicles, taking the form of fuel consumption credits. More specifically, manufacturers earn credits for HD vehicles and engines they produce that exceed the fuel consumption standards. Credits are calculated at the end of each model year based on

[8] D. Smith, Vehicle Systems Integration (VSI) Laboratory, Oak Ridge National Laboratory, Presentation to the committee on November 18, 2014.

the fleet average fuel consumption. A manufacturer is permitted to average, bank, or trade credits that it accumulates by complying with the standards. Carbon dioxide (CO_2) credits can be used to offset compliance with the nitrous oxide (N_2O) and methane (CH_4) vehicle standards (i.e., 0.10 g/bhp-hr N_2O and 0.10 g/bhp-hr CH_4 for engine testing of long-haul tractors and vocational vehicles in 2014 and beyond for compression ignition [CI] engines, and 2016 and beyond for spark ignition [SI] engines).

Several states have adopted incentives (NCSL, 2014) for the purchase of hybrid vehicles or for conversions. These state incentives take a variety of forms, including grants, rebates/vouchers, loans, tax credits, or tax exemptions. California has a hybrid truck and bus voucher program. New York has an alternative fuel vehicle voucher/incentive program. Colorado offers tax credits for either the purchase or lease of qualified vehicles or for qualified conversions. These Colorado credits apply to battery-electric and plug-in hybrid-electric vehicles, hydraulic hybrid trailers, and to alternative fuel vehicles (AFV), including liquefied natural gas, compressed natural gas, liquefied petroleum gas or hydrogen. The Texas Clean Fleet Program (TCFP) offers grants to replace HD on-road diesel vehicles with alternative fuel and hybrid vehicles.

The Federal Transit Administration (FTA) provides incentives for mass transit buses, including those powered by conventional diesel engines (alone), hybrid powertrains, and by other types of nondiesel engines. The federal incentive is 80 percent of the purchase price for buses with conventional diesel engines, plus 90 percent of the differential price for a bus equipped with a hybrid powertrain. If the bus is powered by a nondiesel (e.g., natural gas) engine, 82 percent of the differential price is covered by the federal incentives.

Fuel Efficiency Standards

In 2011 the EPA and NHTSA announced the Heavy-Duty National Program, establishing standards that reduce GHG emissions and improve the fuel efficiency for medium- and heavy-duty engines and vehicles. These standards require that the HD tractors used in tractor-trailer combinations for long-haul service reduce fuel consumption and GHG emissions by 9 to 23 percent compared to their 2010 baseline values, starting in model year (MY) 2017 (EPA, 2011). The standards apply to HD long-haul trucks (Class 7 or 8), large pickup trucks (Class 2b), and vocational vehicles (Classes 2b to 8). The rules cover both engines and vehicles.

Two approaches for phase-in of the 2018 standard have been established to provide manufacturers with some flexibility on how they comply with the new standards. One of these is based on an engine averaging, banking, and trading (ABT) program and the other is based on a vehicle ABT program. Both programs are designed to apply with increasing stringency during the 5-year period from MY 2014 to MY 2018.

The EPA and NHTSA together with CARB plan to extend the HD program beyond 2018 to achieve further reductions in fuel consumption and GHG emissions. In establishing these standards, the EPA will likely have to consider new technologies that are not currently in production such as advanced forms of hybridization.

RESPONSE TO RECOMMENDATIONS FROM THE NRC PHASE 2 REPORT

H-2 Hybrid Goals. NRC Phase 2 Recommendation 4-1. The DOE should provide an up-to-date status with respect to the heavy-duty hybrid goals. The DOE should partition the available hybrid funds between heavy-duty and light-duty hybrid R&D technology to promote the R&D required for the development of heavy-duty hybrid technologies, since heavy-duty hybrid requirements are significantly different from light-duty requirements.

21CTP Response: DOE does not have specific hybrid goals for light-duty hybrids. Research and Development (R&D) and corresponding goals are for component technologies (e.g. batteries, electric motors, etc.). These technologies and the R&D advances should be scalable across vehicle weight classes in many cases.

Committee Comment on Response to 4-1

In the DOE 21CTP roadmap and technical white papers (21CTP, 2013), Section 2/10.1 summarizes six stretch goals for the MHDV Hybrid Group; these were originally formulated and published in 2011 (DOE, 2011). The goals address motor technology, power electronics, energy storage, system optimization for MHDVs, and electrified power accessories. In response to a question posed by the committee to the 21CTP management about the status of work on these stretch goals, the following response was received:

> Although we have not conducted a full planning effort for reviewing these goals and defining the specific budget required to meet each one, the Hybrid team has been revisiting the goals in light of the current and near-term future outlook for hybrid technology in commercial trucks (21CTP-1, 2014).

No further information has been received from the 21CTP management team regarding the results of the hybrid team's efforts to revisit the goals, suggesting that a significant amount of uncertainty still exists in this area. It is important for these goals to be clarified as soon as possible so that definite plans can be made for accomplishing those goals.

***

H-3 Hybrid Goals. NRC Phase 2 Recommendation 4-2. The DOE should determine what is needed for the battery cells and other electric drive components in the ARRA-Transportation Electrification programs aimed at development and manufacturing in the United States, as specified in the objectives of these programs.

Partnership Response: The objective of the ARRA Transportation Electrification grants are to demonstrate, collect data, and evaluate potential grid impacts of electric-drive vehicles that are ultimately produced in the United States. While DOE encourages domestic sourcing of components used in the vehicles, there is no requirement that the components be manufactured in the United States.

Committee Comment on Response to 4-2

From the context of this recommendation and its associated finding (Finding 4-1) in the NRC Phase 2 report, it appears that the Phase 2 report was recommending that DOE make an effort to deliver information about its battery and motor drive component development programs to the major industry recipients of ARRA-funded awards who were using the federal funding to develop both battery-electric and hybrid trucks and buses. No specific action was apparently taken by DOE in response to this recommendation, and the time that has now elapsed since the ARRA awards were made makes the recommendation moot at this point.

***

H-4 Hybrid Emissions Certification. NRC Phase 2 Recommendation 4-3. As partners of the 21CTP, EPA and DOT's NHTSA should work with CARB to develop test procedures for the certification process for criteria emissions so that the emissions benefits of hybridization will be recognized, allowing the reduction in size or simplification of the emission control system of hybrid heavy-duty vehicles to be realized.

Partnership Response: DOE agrees that the proposed test procedure development should be performed by EPA and DOT's NHTSA.

Committee Comment on Response to 4-3

The importance of the development of these test procedures has been discussed earlier in this chapter and will not be repeated here. Recommendations to accelerate the development of these test procedures date back to the NRC Phase 1 report in 2008 and an updated version of this recommendation is included in this report in order to spur further development of these certification procedures.

***

H-5 Hybrid Business Case/Break-even Time. NRC Phase 2 Recommendation 4-4. Dual paths should be pursued to achieve a break-even time of 5 years for heavy-duty hybrid vehicles. First, the DOE should use its vehicle simulation tools to determine the advanced technologies needed to meet the goal of 60 percent improvement in fuel economy (38 percent reduction in fuel consumption), from the current status of 20 to 40 percent improvement (17 to 29 percent reduction in fuel consumption) and initiate R&D programs to develop these technologies. Second, manufacturers should be encouraged to explore modular, flexible designs, which could yield higher production volumes and thus achieve significant reductions in capital costs of hybrid systems.

Partnership Response: DOE is prepared to assist industry in these types of studies. DOE does not plan to conduct or initiate hybrid centric R&D programs. DOE's focus is on electric-drive component R&D to develop technologies that can be integrated by manufacturers into advanced technology vehicles.

Committee Comment on Response to 4-4

There is significant value in taking a systems approach to designing hybrid drive systems for trucks, as was demonstrated in the SuperTruck program results. At least two of the six stretch goals set by 21CTP for its hybrid truck program in 2011 are specifically focused on system optimization of MD and HD hybrid trucks (Goals 4 and 5), and the available evidence indicates that R&D efforts on systems-based objectives is justified. For example, the 21CTP federal agencies, including DOE and EPA, are encouraged to make their vehicle simulation tools available to their industry partners to evaluate vehicle-level approaches to maximizing the benefits of hybrid drive systems for a variety of truck applications, in keeping with the 21CTP goals. Despite the ambitious quantitative objectives set in the 21CTP stretch goals, opportunities for applying smaller microhybrid units in MDHVs to achieve more modest fuel consumption and emissions reductions should be included in these evaluations if they can achieve significant improvements in the value proposition for new mild hybrid truck configurations.

***

H-6 Revised Hybrid Goals. NRC Phase 2 Recommendation 4-5. The 21CTP should establish plans and develop realistic budgets for accomplishing the six new stretch goals for heavy-duty hybrid vehicles in accordance with the committee's findings, explain the rationale behind the new goals, and provide the current status of the applicable technology for each of the goals so that the magnitude of the tasks for each can be assessed.

21CTP Response: The Partnership concurs that planning for these updated goals is critical: the Partnership industry and government members will be working as a team to conduct these planning efforts and identify the appropriate parameters for successful achievement of the goals, subject to available funding. Ongoing research results will inform goal revisions. Two of the SuperTruck teams are developing and integrating full hybrid systems into Class 8 vehicles. In addition, ORNL will be installing and testing a full heavy-duty hybrid system in a dedicated test cell. 21CTP will use these project findings to revise goals as appropriate.

Committee Comment on Response to 4-5

This recommendation is closely associated with Recommendation 4-1. As noted in the Comment on Response 4-1, the 21CTP hybrid team has not developed a plan for addressing the six stretch goals and is now in the process of revisiting the goals in conjunction with the restructuring of the team.

FINDINGS AND RECOMMENDATIONS

The MHDV hybrid propulsion initiative currently comprises a major thrust within the strategic approach defined for the 21CTP technology program. Despite some significant technical accomplishments that have been cited in this chapter, the MHDV hybridization program finds itself at a crossroads with countervailing forces and some internal inconsistencies that the 21CTP leadership must acknowledge and resolve in order to set a clear direction for moving forward. A couple of striking examples include these:

- The 2013 21CTP roadmap continues to place a high priority on hybridization in its technology plan for achieving major reductions in fuel consumption and emissions, yet no funding has been requested or allocated to explicit MHDV hybrid R&D projects for the past 7 fiscal years. Although claims are made by the 21CTP leadership that DOE expenditures on component-oriented R&D for LDV hybrid drives is sufficient to meet the requirements for MHDV hybridization, the 2013 21CTP roadmap document makes a special point of emphasizing the differences between the technology needs for LDV and MHDV hybrid systems (21CTP, 2013).
- Several MHDV hybridization projects funded by federal agencies and private industry have clearly demonstrated that these hybrid systems can successfully deliver major reductions in fuel consumption and emissions, but the cost of the hybrid drive equipment does not meet typical payback period requirements set by the MHDV truck purchasers. In some of these cases, the breakeven period for the hybrid drive equipment falls well within the expected lifetime of the MHDV and its propulsion drive, even without applying any price on the carbon that is released. However, the tight payback period, 2 years or less, that is typically required by industry (CALSTART, 2010) significantly impedes the ability of MHDV hybrid drive equipment to succeed in the marketplace. Reasons cited by truck purchasers for insisting on such short payback periods include the high cost of capital and typically short new truck ownership periods of 3 to 5 years, with little confidence that the new buyers will compensate them for any premium-cost features that reduce fuel consumption.

The demonstrated long-term benefits of hybridization for reducing fuel consumption and emissions in several types of MHDVs are too large to ignore despite the cost barriers faced by many of today's hybrid drive manufacturers for their commercial offerings. More stringent fuel consumption and GHG emissions standards for MD and HD trucks are now being developed by the federal government for the years beyond MY 2018, making it important for 21CTP to support the development of advanced technology, such as battery-electric and hybrid drives, that will help to meet those goals. Unfortunately, the cost of hybrid drive equipment is not likely to fall sufficiently fast to meet payback requirements in the near future. The following findings and recommendations have been formulated with the objective of learning as much as possible from past and current 21CTP projects focused on hybrid MHDV equipment, and applying these lessons to the future development of more cost-effective hybrid systems that can overcome the current market barriers.

Finding 4-1. The 21CTP is considering a proposal to restructure its hybrid team so that it can work on other drivetrain efficiency improvements, including other types of system integration opportunities that incorporate hybrid drive equipment.

Recommendation 4-1. The 21CTP hybrid team should use this opportunity to redefine its mission in a manner that will lead to vehicle efficiency and emissions reduction improvements via a range of technology options, including promising opportunities for electrification and other types of innovative drivetrain improvements. During the course of this restructuring, the six R&D stretch goals developed in 2011 for the MHDV hybridization program should be redefined as part of the development of strategic objectives of the restructured advanced drivetrain initiative. At the conclusion of this process, the 21CTP leadership, working together with DOE and the other 21CTP partner federal agencies, should make a serious effort to secure funding to pursue whatever goals emerge so that they have a realistic chance of being achieved.

Finding 4-2. Several manufacturers have commercialized MD hybrid trucks during the past several years and successfully demonstrated their ability to significantly reduce fuel consumption and emissions, particularly in vocational and delivery truck applications. Despite this progress, the high cost of the hybrid drive train equipment and batteries, combined with dropping prices for natural gas and oil, have significantly retarded their market penetration in the United States. This has caused economic hardships for many hybrid truck manufacturers, causing a widespread reevaluation of the current hybrid truck business viability, at least in North America. At the same time, there is evidence that business opportunities for MHDV hybrid equipment are growing in other parts of the world, particularly in China, where government mandates are having a major impact.

Recommendation 4-2. Recognizing the advantages that hybridization can offer in trucks, 21CTP should support the development of new technology that offers promise for significantly improving the performance and cost-effectiveness of hybrid truck technology in the longer term. Project opportunities should be pursued to evaluate cost-effective vehicle electrification configurations for trucks, including hybrid drives with optimized component ratings to minimize their payback periods in different vehicle classes and applications. This future work should take advantage of technology advances originally made and commercialized for light-duty vehicles, including new battery technologies as well as opportunities for integrated microelectrification of truck functions such as start/stop operation, idle reduction, waste heat recovery, engine starting, and accessory electrification.

Finding 4-3. Although EPA and NHTSA have made considerable progress toward defining the certification procedures for fuel consumption and emissions in hybrid MHDVs, these procedures are still incomplete and imprecise in some important areas, particularly with regard to chassis dynamometer testing of complete hybrid MHDVs, and dynamometer testing of hybrid drivetrain power packs to determine their emissions and fuel consumption performance.

Recommendation 4-3. 21CTP should make it a priority to encourage EPA and NHTSA to accelerate their efforts to strengthen and finalize procedures for certifying the fuel consumption and emissions of hybrid MHDVs, including procedures for chassis dynamomenter testing of complete hybrid vehicles and dynamometer testing of hybrid propulsion drivetrains alone. The 21CTP leadership is encouraged to work together with EPA and NHTSA to inform and educate the 21CTP stakeholders and the broader MHDV manufacturing community about the details of these procedures when they become available.

Finding 4-4. The 21CTP has articulated its strategy of depending on investments by DOE in the development of key components—that is, batteries, motors, and power electronics—in light-duty hybrid vehicles, based on the argument that this technology will be applicable to hybrid MHDV drivetrains as well. However, statements in the 2013 21CTP roadmap and technical white papers document pointedly note the limitations of this approach because of key differences between the performance and lifetime requirements for the two types of vehicles, as well as differences in their operating environments.

Recommendation 4-4. The 21CTP should acknowledge that there are substantive differences between the hybrid drive requirements of LDVs and MHDVs, making it sensible to focus future hybrid MHDV investments on those components and subsystems where these differences exist and have the highest impact on the performance and cost of hybrid drives in MHDVs. This strategy should be combined whenever possible with efforts to design the hybridization equipment to accomplish multiple functions in the integrated drivetrains of future hybrid MHDVs.

REFERENCES

21CTP (21st Century Truck Partnership). 2013. *Roadmap and Technical White Papers*. Office of Energy Efficiency and Renewable Energy, U.S. Department of Energy. https://www1.eere.energy.gov/vehiclesandfuels/pdfs/program/21ctp_roadmap_white_papers_2013.pdf.

21CTP-1. 2014. Questions from NAS Review Committee (June 2014), Responses as of September 29, 2014.

ANL (Argonne National Laboratory). 2013. Autonomie web page. http://www.autonomie.net.

Balachandran, U. 2014. *Cost-Effective Fabrication of High-Temperature Ceramic Capacitors for Power Inverters*. Argonne National Laboratory Team: B. Ma, T.H. Lee, J.E. Emerson, and S.E. Dorris. DOE Annual Merit Review APE061, Washington, D.C.

Birky, A.K. 2014. *Modeling for Market Analysis: HTEB, TRUCK, and LVChoice*. TA Engineering, Inc. DOE Annual Merit Review VAN012, Washington, D.C., June 18.

CALSTART. n.d. HTUF: High-Efficiency Truck Users Forum. http://www.calstart.org/Projects/htuf.aspx.

CALSTART. 2010. *Saving Fuel, Saving Money: An Assessment of Fleet Cost Savings from High Efficiency Trucks*, Pasadena, Cal. http://www.calstart.org/Libraries/Publications/Saving_Fuel_Saving_Money_-_Final_Fleet_Report.sflb.ashx.

Carr, A., and N. Williams. 2014. *Hydrogen Fuel-Cell Electric Hybrid Truck and Zero Emission Delivery Vehicle Deployment*. Houston-Galveston Area Council. DOE Annual Merit Review VSS116, Washington, D.C., June 19.

Choe, B. 2014. *Zero-Emission Heavy-Duty Drayage Truck Demonstration*. Principal Investigator: M. Miyasato, South Coast Air Quality Management District. DOE Annual Merit Review VSS115, Washington, D.C., June 19.

Cox, J. 2014. *Plug-in Hybrid Electric Commercial Fleet Demonstration and Evaluation*. Principle Investigator: M. Miyasato. South Coast Air Quality Management District. DOE Annual Merit Review ARRAVT083, Washington, D.C., June 19.

Deter, D. 2014. *Cummins MD & HD Accessory Hybridization CRADA*. Center for Transportation Analysis and ORNL. DOE Annual Merit Review VSS133, Washington, D.C., June 16-20.

DeVoto, D. 2014. *Performance and Reliability of Bonded Interfaces for High-Temperature Packaging*. National Renewable Energy Laboratory. DOE Annual Merit Review APE063, Washington, D.C., June 17.

DOD (U.S. Department of Defense). 2014. Response to TARDEC-Specific Questions from the National Academies Committee on Review of the 21st Century Truck Partnership, Phase 3, Advanced Starter Generator (DOD006), November.

DOE (U.S. Department of Energy). 2011. *Hybrid Propulsion White Paper*, Submitted by the DOE Office of Vehicle Technologies to the NRC Committee for Review of the 21st Century Truck Partnership, Phase 2, February 28.

DOE. 2014. *Alternative Fuel and Advanced Vehicle Search: Hybrid Systems*. Alternative Fuels Data Center, DOE Clean Cities program. http://www.afdc.energy.gov/vehicles/search/hybrid/?&current=true&display_length=50&all_fuels=y&all_manufacturers=y.

Drive Clean Chicago. 2014. *Drive Clean Truck–Eligible Vehicles*. Drive Clean Chicago, LLC. http://www.drivecleanchicago.com/HowToParticipate/EligibleVehicles.aspx.

Eaton. 2013. *Eaton Drops Hydraulic Hybrid System*, HDT Truckinginfo, September 3. http://www.truckinginfo.com/news/story/2013/09/eaton-drops-hydraulic-hybrid-system.aspx.

Eaton. 2014. Eaton drops hybrids in North America, *Fleets & Fuels*, June 26. http://www.fleetsandfuels.com/fuels/hybrids/2014/06/eaton-drops-hybrids-in-north-america/.

EPA (U.S. Environmental Protection Agency). 2011. *EPA and NHTSA Adopt First-Ever Program to Reduce Greenhouse Gas Emissions and Improve Fuel Efficiency of Medium- and Heavy-Duty Vehicles.* Office of Transportation and Air Quality. EPA-420-F-11-031. http://www.epa.gov/otaq/////climate/documents/420f11031.pdf.

EPA. 2013. Heavy-Duty Engine and Vehicle, and Nonroad Technical Amendments. *Federal Register*, vol. 78, no. 116, June 17.

EPA. 2014. Fact Sheet: Opportunity for All: Improving the Fuel Efficiency of American Trucks – Bolstering Energy Security, Cutting Carbon Pollution, Saving Money and Supporting Manufacturing Innovation, Press Release, The White House, February 18, 2014. https://www.whitehouse.gov/the-press-office/2014/02/18/fact-sheet-opportunity-all-improving-fuel-efficiency-american-trucks-bol.

Eudy, L. 2014. *Technology Validation: Fuel Cell Bus Evaluations*. National Renewable Energy Laboratory. DOE Annual Merit Review TV008, Washington, D.C., June 19.

Frost & Sullivan. 2014. *Strategic Analysis of the Global Hybrid and Electric Medium- and Heavy-Duty Truck Market*, Brief summary of full study, NE0C-18, December 2014.

Green Fleet Magazine. 2008. Hybrid lift truck helps utilities clean the air. *Green Fleet Magazine*, January.

HTUF (High-Efficiency Truck Users Forum). 2014. HTUF National Meeting. Argonne National Laboratory, September 22-24. http://www.slideshare.net/CALSTART/htuf2014-recap.

HVIP (California Hybrid and Zero-Emission Truck and Bus Voucher Incentive Project). n.d. About the Hybrid Truck and Bus Voucher Incentive Project (HVIP). https://www.californiahvip.org.

Lustbader, J., and T. Kiss. 2014. *Vehicle Thermal System Modeling in Simulink.* National Renewable Energy Laboratory, Annual Merit Review VSS134, Washington, D.C., June 16.

Mackie, R. 2014. *Smith Electric Vehicles: Advanced Vehicle Electrification and Transportation Sector Electrification*. Smith Electric Vehicles Corporation. DOE Annual Merit Review ARRAVT072, Washington, D.C., June 19.

NACFE (North American Council for Freight Efficiency). 2014. *Tire Pressure Systems–Confidence Report.* North American Council for Freight Efficiency on the Confidence of Adopting Tire Pressure Systems. http://nacfe.org/wp-content/uploads/2014/01/TPS-Detailed-Confidence-Report1.pdf.

NCSL (National Conference of State Legislatures). 2014. *State Efforts Promote Hybrid and Electric Vehicles.* http://www.ncsl.org/research/energy/state-electric-vehicle-incentives-state-chart.aspx.

NRC (National Research Council). 2008. *Review of the 21st Century Truck Partnership, First Report.* Washington, D.C.: The National Academies Press.

NRC. 2010. *Technologies and Approaches to Reducing the Fuel Consumption of Medium- and Heavy-Duty Vehicles*. Washington, D.C.: The National Academies Press.

NRC. 2012. *Review of the 21st Century Truck Partnership, Second Report.* Washington, D.C.: The National Academies Press.

NREL (National Renewable Energy Laboratory). 2014. Vehicle Technology Simulation and Analysis Tools, NREL website, December 31. www.nrel.gov/transportation/systems_analysis_tools.html.

NY Truck VIP (New York Truck Voucher Incentive Program). 2015. NYNYCAFV-VIF "Alternative Fuel" Vehicle Eligibility List. New York Truck Voucher Incentive Program. https://truck-vip.ny.gov/NYCAFV-VIF-vehicle-list.php.

Smith, D., S. Curran and Z. Gao. 2014. *Powertrain Controls Optimization for Heavy Duty Line Haul Trucks*. Center for Transportation Analysis and Oak Ridge National Laboratory. DOE Annual Merit Review VSS141, Washington, D.C., June 16-20.

Stephens, T. 2014. *Impact Analysis: VTO Baseline and Scenario (BaSce) Activities*. Argonne National Laboratory. DOE Annual Merit Review VAN001, Washington, D.C., June 18.

Walkowicz, K. 2014. *Medium and Heavy Duty Vehicle Field Evaluations*. National Renewable Energy Laboratory. Team: M. Lammert, A. Ragatz, A. Duran, and M. Simpson. Annual Merit Review VSS001, Washington, D.C., June 17.

ANNEX

Hybrid Buses

Heavy-duty Class 8 transit buses are particularly good candidates for hybrid technology owing to their frequent starts and stops for passenger pickups and drop-offs, combined with occasional longer-distance trips to return to a terminal. In addition, their typical use in densely urban areas that often fail to comply with air quality standards increases their attractiveness because of the emissions reductions they can offer.

Since the transit bus application is so well suited to hybridization, it has received considerable attention and growth. The 21CTP roadmap and white papers (21CTP, 2013) provide details of some of the prior accomplishments and support for this hybrid vehicle application area. Important past programs that have provided financial support in recent years include the National Fuel Cell Bus Program, from 2006 to 2010, the Transit Investments for Greenhouse Gas and Energy Reduction (TIGGER) in fiscal year 2011,[9] and emissions certification support for hybrid buses. For the latter effort, a final report was issued by the Federal Transit Administration in August 2013 on the emissions testing done at West Virginia University's Center for Alternative Fuels, Engines and Emissions (Wayne, 2013). Data gathered during this program on emissions and fuel economy of hybrid transit buses up to model year 2009 are available at the Integrated Bus Information Systems (IBIS) website.[10]

Unlike the case for many of the other medium- and heavy-duty truck classes, the technology of hybrid transit buses has matured to the point that significant numbers of buses with hybrid propulsion systems have been manufactured by several companies and are now in daily use in many urban bus fleets in the United States and around the world. For example, the total number of BAE Systems transit bus hybrid electric propulsion systems manufactured and installed globally to date exceeds 4,500 units.[11] The corresponding total number of hybrid electric propulsion systems manufactured for transit buses by Allison Transmission over the past 11 years is greater than 6,500 (Allison, 2015).

However, there continue to be significant rebates and incentives available to purchasers of HD hybrid transit buses even today. The cost of a transit bus is paid 80 percent by the Federal Transit Authority in the United States. If the vehicle is a hybrid, the amount is 82 percent, or 90 percent of the differential, as stated by Bart Mancini, senior principal systems engineer at BAE Systems, in his presentation to the committee on December 4, 2014. Other countries are known to offer incentives as well. Yan Zhou and Thomas Stephens reported in the 2014 Annual Merit Review for DOE project VAN011 that China provides financial incentives of 420,000 to 500,000 yuan for hybrid and battery-electric buses longer than 10 meters. They also reported significant volumes for hybrid and battery-electric bus production of 4,000 units in 2010, growing to >10,000 units in 2013.

In November of 2014, Allison Transmission made announcements about its further development of hybrid bus products. These announcements included news of CARB's approval for the pairing of the Allison H40/50 EP hybrid transmission with either the Cummins ISB6.7 or the Cummins ISL9 engines (Allison Transmission Holdings Inc., 2014a) and a total electrification option for powering air conditioning, air compressors, and power steering (Allison Transmission Holdings, Inc., 2014b).

In October 2014, BAE Systems announced that it would install four hybrid drive systems in buses for the City of Honolulu, where there are already 80 hybrid buses in the fleet of 525 (15 percent of the fleet). The buses are being paid for using American Recovery and Reinvestment Act funds (Cresenzo, 2014). The same announcement indicates that Hawaii is pushing to move more of its bus fleet to low-emissions energy sources, including hydrogen fuel cells and batteries.

The third manufacturer in the IBIS list that has a history of producing hybrid transit bus transmissions, ISE, filed for bankruptcy in 2010 and then sold its assets to Bluways USA, a subsidiary of Bluways International in Belgium. ISE had sold 300 hybrid systems. No information on Bluways USA products is available online.[12]

Some of the technical challenges that must be overcome in order for hybrid and battery-electric drives to achieve greater market penetration have much in common with those that face hybrid drives in other MD and HD truck applications. The most important of these is the challenge of building hybrid drives that cost much less than they do today in order to make them more economically attractive to cities and municipalities that face tightening budgets for future bus purchases. For battery-electric drives, improvements in the battery energy and power density characteristics are critical in order to extend the buses' all-electric driving range and to improve their ability to absorb high peak regenerative braking power pulses without needing to use their mechanical brakes. Achieving these major battery performance improvements without increasing their cost (or, better yet, while decreasing their cost) is one of the biggest challenges facing hybrid- and battery-electric drive systems for virtually all truck, bus, and passenger vehicle applications.

---

[9] See TIGGER Program Overview at http://www.fta.dot.gov/12351_11424.html.

[10] See Integrated Bus Information System (IBIS) at http://ibis.wvu.edu.

[11] B. Mancini, BAE Systems, "Electric-Hybrid Powertrains: Past, Present, and Future," presentation to the NRC Committee on Assessment of Technologies and Approaches for Reducing the Fuel Consumption of Medium- and Heavy-Duty Vehicles, Phase 2, December 4, 2014.

[12] See www.bluways.com.

References

21CTP (21st Century Truck Partnership). 2013. *Roadmap and Technical White Papers*. U.S. Department of Energy, Office of Energy Efficiency and Renewable Energy. https://www1.eere.energy.gov/vehiclesandfuels/pdfs/program/21ctp_roadmap_white_papers_2013.pdf.

Allison Transmission, Inc. 2011. *Allison Hybrid H 40 EP/H 50 EP*. http://www.allisontransmission.com/docs/default-source/marketing-materials/sa5983en-h40-50-ep1BCB31AC06C2F2B94ACCEED0.pdf?sfvrsn=4. Accessed April 23, 2015.

Allison Transmission, Inc. 2015. H 3000TM Introduction and Hybrid Market Update for MD- and HD-Vehicles. Report provided to the committee by Allison Transmission, February.

Allison Transmission Holdings Inc. 2014a. Allison Transmission and Cummins receive ARB approval in California for the H 40/50 EP system paired with ISB6.7 and ISL9 engines. *PR Newswire*, November 13. http://www.prnewswire.com/news-releases/allison-transmission-and-cummins-receive-arb-approval-in-california-for-the-h-4050-ep-system-paired-with-isb67-and-isl9-engines-282555441.html. Accessed April 23, 2015.

Allison Transmission Holdings Inc. 2014b. Allison Transmission announces total electrification option for components used with its H 40/50 EP™ system. *PR Newswire*, November 7. http://www.prnewswire.com/news-releases/allison-transmission-announces-total-electrification-option-for-components-used-with-its-h-4050-ep-system-281910791.html. Accessed April 23, 2015.

Cresenzo, B. 2014. BAE Systems to equip new Honolulu buses with hybrid system. *Pacific Business News,* October 16. http://www.bizjournals.com/pacific/news/2014/10/16/bae-systems-to-equip-new-honolulu-buses-with.html. Accessed April 23, 2015.

Wayne, W.S. 2013. *Transit Vehicle Emissions Program Final Report*. Federal Transit Administration Report No. 0048. U.S. Department of Transportation.

5

Vehicle Power Demands

INTRODUCTION

Vehicle power demands are weight sensitive and encompass the engine power used to overcome inertia, rolling resistance, aerodynamic drag, drivetrain losses, and the power used for auxiliary loads. Figure 5-1 has often been used to summarize the vehicle power demands for a tractor-trailer with a gross vehicle weight (GVW) of 80,000 lb operating on flat terrain for 1 hour; the original paper used energy in units of kilowatt-hours (kWh) for each area rather than percentages (Woodrooffe and Vachon, 2000).

More recently, Figure 3-1 in Chapter 3 has been used to describe where the fuel energy goes. Beginning with 100 percent of the energy in the left hand bar, it moves to show the energy lost in the cylinder to heat and out through the exhaust. The middle column shows the energy lost to friction, pumps, and other accessories on the engine needed to meet emissions certification. The column titled "accessories" shows energy used for auxiliary loads such as air conditioning. The fourth column shows the energy lost through inefficiencies in the driveline, including the clutch, transmission, and axle. The fourth column shows the energy at the wheels and how it is used to propel the vehicle down the road or up a grade or to stop it.

Vehicle power demands are discussed in the NRC Phase 2 report (NRC, 2012). The present report provides an update[1] and delineates the 21st Century Truck Partnership (21CTP) technology goals for each of the vehicle power demand areas addressed by the Partnership; the present report then addresses each of these areas in separate sections: (1) aerodynamics, (2) tire rolling resistance, (3) auxiliary loads, (4) weight reduction, (5) thermal management, and (6) friction and wear. It should be noted that there are a limited number of activities related to vehicle power demands under the umbrella of 21CTP since many of these areas are being addressed by the SuperTruck projects. Additional material related to vehicle power demands is found in Chapter 4 on hybrid vehicles and Chapter 8 on the SuperTruck program.

The previous report (NRC, 2012) and discussions in the industry focus on vehicle power demands from the net output of the engine while driving at highway speeds on level ground with a fixed load. The various uses of the engine power (overcoming rolling resistance, overcoming aerodynamic drag, power train losses, auxiliary loads) were categorized and sized along with details of their efficiency. As regulations for emissions and fuel consumption have advanced, the effects of energy density of the fuel used and the specifics of the drive cycle have become important. Fleets

[1] For this section of the report, the following presentations to the committee were the source of information: D. Anderson, "Vehicle and Systems Simulation and Testing," September 2014; J. Gibbs, DOE Office of Vehicles Technologies, "NAS Review of VTO Materials Program in Support of 21CTP,"on September 3, 2014; A. Greszler, Vehicle Systems. Volvo Group Truck Technology, "SuperTruck: Development and Demonstration of a Fuel-Efficient Class 8 Highway Vehicle," on May 15, 2014; D. Kayes, D. Rotz, and S. Singh, SuperTruck Team, Daimler Trucks North America (DTNA), on May 14, 2014; G. Fadler, Navistar, "Navistar Fuel Economy and Emissions," presentation to the NRC Committee on Assessment of Technologies and Approaches for Reducing the Fuel Consumption of Medium- and Heavy-Duty Vehicles, Phase 2, 2013; K. Howden, DOE VTO, "Overview and Update of 21CTP Responses to NRC Recommendations," May 14, 2014; G. Keller, ANL, "Update on Idling Reduction Activities," September 4, 2014; E. Koeberlein, Cummins/Peterbilt, "Cummins SuperTruck Program: Technology and System Level Demonstration of Highly Efficient and Clean, Diesel Powered Class 8 Trucks," May 14, 2014; T. Reinhart, Southwest Research Institute, "Technologies for MD/HD GHG and Fuel Efficiency," November 18, 2014; K. Stork, DOE Vehicle Technologies Office, "Overview of the DOE Fuel and Lubricant Technologies R&D," September 3, 2014; Spears, Environmental Protection Agency and National Highway Traffic Safety Administration, "Looking Ahead to the Next Phase of Heavy-Duty Greenhouse Gas and Fuel Efficiency Standards," September 3, 2014; V. Sujan, Cummins, Inc., "SuperTruck: Vehicle Modeling, Optimization and Cycle Management," August 28, 2014; R. Zukouski, Navistar, "SuperTruck–Development and Demonstration of a Fuel-Efficient Class 8 Tractor and Trailer," November 18, 2014. The following Annual Merit Reviews were also the source of information: Ajayi et al., 2014; Benedict, 2014; Birky, 2014; Cox, 2014; Deter, 2014; Fenske et al., 2014; Gonder, 2014; Kambiz, 2014; Karbowski et al., 2014; Kim and Rousseau, 2014; Lustbader, 2014; Lustbader and Kiss, 2014; Martin et al., 2014; Meyer, 2014; Rask, 2014; Rugh, 2014; Stephens, 2014; Walkowicz, 2014a, b.

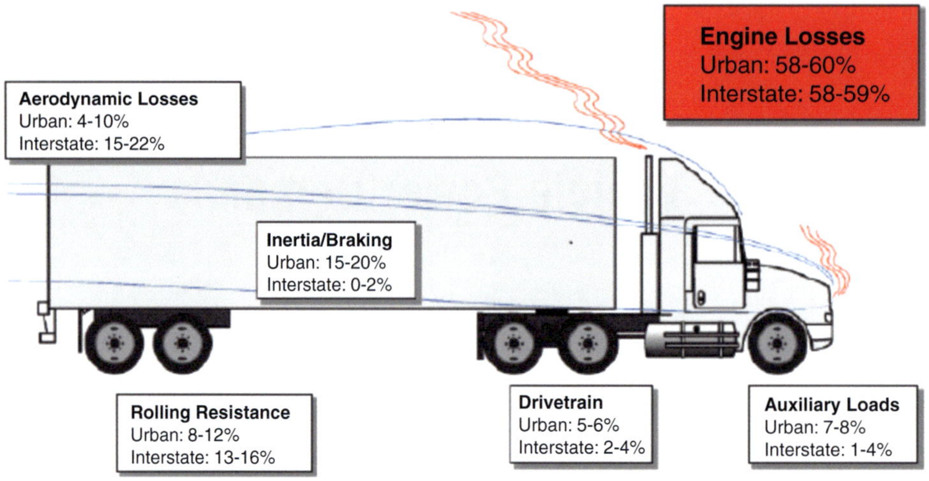

FIGURE 5-1 Energy "loss" range of vehicle attributes as impacted by duty cycle, on a level road. SOURCE: NRC (2010), Figure 5-8.

continue to work to reduce fuel costs over their drive cycles using a variety of component technologies, including fleet management systems and systems to monitor and control driver behavior (Technology and Maintenance Council, 2014). There is a need for a reevaluation of the overall energy use for a vehicle. The committee believes it is more appropriately called the Vehicle Energy Demand Over a Drive Cycle.

Recognizing the changes that are occurring, the Partnership updated the goals for fuel consumption in the 2013 publication of its *Roadmap and Technical White Papers* (2013). The roadmap provides a description of a baseline for "average power use inventory" and forward- looking goals in Figures 5, 6, 7, and 8 of that document. These enhanced goals take into account the varying applications of Class 6 through Class 8 vehicles; are very specific in terms of the assumptions for weight; include idling fuel used; and use specific fuel consumption in energy units. However, the specification of drive cycles, including speed, terrain, and 24-hour operation, is less precise. Also, the goals do not include the effect of energy density of the fuel used and do not allow for other energy sources such as solar power or off-board electrical connections. This restatement of the goals in the white papers is a good step. In particular, it is the first attempt to specify load-specific fuel consumption and to take account of the current Environmental Protection Agency/National Highway Traffic Safety Administration (EPA/NHTSA) regulations on fuel consumption and greenhouse gas (GHG) emissions for medium- and heavy-duty vehicles (MHDVs) through model year 2018. Further work is needed to ensure research is done in the right areas to achieve real-world savings (NRC, 2014).

Finding 5-1. Current regulations for MHDVs on fuel consumption and load-specific fuel consumption are in place through model year 2018. The continuing improvement in reducing fuel consumption calls for a new baseline of vehicle energy demands over a drive cycle based on real-world operation. This new baseline would take into account such factors as load, grade, speed, torque, distance-based target schedules, and drive cycles over an extended period of time as well as rest periods required by law. The 21st Century Truck Partnership has taken a step forward in redefining a new baseline with the proposed average power use inventory in the 2013 roadmap and technical white papers.

Recommendation 5-1. The Partnership, through DOE and NHTSA, should work with the California Air Resources Board (CARB) and the trucking industry to work out a comprehensive new baseline for vehicle power demands (in kilowatt-hours) of a circa 2020 vehicle that include an extended period of operation.

GOALS AND SELECTED RELEVANT PROJECTS

The 21CTP *Roadmap and Technical White Papers* (2013) offer five technology goals for vehicle power demands.

> **Technology Goal 1:** Develop and demonstrate advanced technology concepts that reduce the aerodynamic drag of a Class 8 highway tractor-trailer combination by 20% (from a drag coefficient of 0.69 to 0.55). Evaluate a stretch goal of 30% reduction in aerodynamic drag (from C_d=0.69 to C_d=0.48). The baseline for this goal is the proposed EPA/NHTSA baseline of C_d=0.69 with 9.2 m^2 frontal area for a conventional Class 8 tractor with high roof sleeper.

> **Technology Goal 2:** Develop and demonstrate low rolling resistance tires that can reduce vehicle rolling resistance and wheel weight for a Class 8 tractor-trailer. Demonstrate 35% reduction in rolling resistance from C_{rr}=8.2 kg/metric ton for drive wheels to a goal of C_{rr}=5.33 kg/metric ton. The

baseline for this goal is the EPA/NHTSA proposed baseline for a Class 8 tractor/trailer equipped with low rolling resistance dual tire drive wheel configurations having C_{rr}=8.2 kg/metric ton.

Technology Goal 3: Develop and demonstrate technologies that reduce essential auxiliary loads by 50% (from current 20 horsepower to 10 horsepower) for Class 8 tractor-trailers. The baseline for this goal is a Class 8 highway tractor/trailer with sleeper operating 5 day over-the-highway operations at 36,000 kg (80,000 pounds) CGVW.

Technology Goal 4: Develop and demonstrate lightweight material and manufacturing processes that lead to a 10% reduction in tare weight for a 15,500 kg (34,000 pounds) tractor/trailer combination. Establish a long-term stretch goal of reducing combined vehicle weight by 20%. The baseline for this goal is a Class 8 highway tractor/trailer with high roof sleeper and dry van trailer capable of 36,000 kg CGVW.

Technology Goal 5: Thermal Management & Friction and Wear. Increase heat-load rejected by thermal management systems by 20% without increasing radiator size. Develop and demonstrate parasitic friction reduction technologies that reduce driveline losses by 50%, thereby improving Class 8 fuel efficiencies by 3%. The baseline for this goal is a Class 8 highway tractor/trailer with sleeper operating at steady state 65 mph at 36,000 kg CGVW.

The technology goals are well stated, specific, and measureable. A timeline of 10 years is given in Section 3 of the *Roadmap and Technical White Papers*, which takes the goals to 2023 from the date of publication (21CTP, 2013). Comments on the goals are relegated to the sections in this chapter that address the different technical areas.

In many ways, the efforts of the four SuperTruck projects are addressing many of the opportunities to reduce vehicle power demands. Nevertheless, there are continuing efforts associated with individual projects, mostly funded by DOE, that address these areas as well and which are the focus of this chapter. As pointed out in Chapter 1, the committee is not charged with reviewing every project in the 21CTP portfolio, but it has instead addressed a subset of key projects that were presented to it at its meetings, as well as information gathered from the DOE Annual Merit Review and other publications. Table 5-1 lists the projects affiliated with the 21CTP that are addressed in this chapter, with associated estimates of DOE funding for 2012-2014.

AERODYNAMICS

Aerodynamic losses, expressed in such terms as horsepower, are directly proportional to the coefficient of drag (C_d), the frontal area, and the velocity of the vehicle cubed (NRC, 2010, 2012). At highway speeds, especially above 50-60 mph, such losses are pronounced. However, fleets report slower overall speeds. A study by the Federal Highway Administration, *Freight Performance Measurement: Travel Time in Freight-Significant Corridors* (FHWA, 2005), shows speeds less than 60 miles per hour. As one would expect, the speeds near urban centers are often less, as documented in *Freight Performance Measures Analysis of Freight-Significant Highway Locations–2013* (ATRI, 2013b). The distribution of speeds is significant for determining the potential for aerodynamic savings. The concept of a weighted aerodynamic-average speed (WAAS) (NRC, 2010) and the desire for real-world fuel savings (NRC, 2012) indicates the need for information on actual operation of vehicles.

The EPA SmartWay program turned 10 years old in 2014. In the spring of 2014, EPA announced a new program, SmartWay Elite Trailers (EPA, 2014). This moved the bar from a 5 percent reduction to a 9 percent reduction in fuel consumption from trailer aerodynamic devices and modified the test criteria to make them more repeatable and stringent. The EPA/NHTSA regulations for fuel consumption and GHG emissions could also address trailer aerodynamics and rolling resistance, as recommended in *Technologies and Approaches to Reducing the Fuel Consumption of Medium- and Heavy-Duty Vehicles* (NRC, 2010). New products are being introduced into the market, including ones with aerodynamic improvements.[2]

While dry-van trailers are the most popular trailers, the concept work done by DOE on tanker trailer aerodynamics is a good step toward understanding how other trailer types can be improved (Kambiz, 2014).[3] The project, Heavy Vehicle Aerodynamic Improvements (VSS006), at Lawrence Livermore National Laboratory (LLNL), had a budget of $900,000 in FY 2013 and $600,000 in FY 2014. Its objective was to provide guidance to industry to improve the fuel efficiency of Class 8 tractor-trailers and tankers through enhanced aerodynamics. Project VSS006 proposed an integrated tractor-trailer design, new from the ground up, that radically decreases aerodynamic drag and improves fuel efficiency. It designed a first-generation integrated tractor-trailer geometry and performed wind tunnel tests of selected aerodynamic devices to improve fuel efficiency. Plans are to conduct scaled experiments to design and validate the performance of aerodynamic treatments of an integrated tractor-trailer including improving the aerodynamics of tankers. Accomplishments to date include full-scale wind tunnel testing of two tractors, three trailers, and 23 devices. DOE has also completed testing on a track and collected on-the-road performance information. Funding for FY 2015 and beyond was not discussed. The proposed focus on improving the aerodynamics of trailer configurations other than dry vans is

[2] Vehicle OEMs have introduced new products with aerodynamic improvements in the last 2 years. Several announcements include DTNA's Freightliner Cascadia Evolution and its Western Star 5700 XE; the Kenworth T680; Peterbilt's 579 EPIQ; Volvo Trucks' 2016 VN; and the Navistar ES.

[3] Also, D. Anderson, "Vehicle and Systems Simulation and Testing," presentation to the committee, September 2014.

TABLE 5-1 DOE Funding for Selected 21CTP Projects Related to Vehicle Power Demand (dollars)

Public Review Project Title	Project ID	2012 Funding	2013 Funding	2014 Funding	Note
DOE/DOD Parasitic Energy Loss Collaboration	VSS005	250,000	200,000	170,000	
DOE's Effort to Reduce Truck Aerodynamic Drag through Joint Experiments and Computations	VSS006	650,000	900,000	600,000	
Development of High-Power-Density Driveline for Vehicles	VSS058	350,000	300,000	350,000	
CoolCab Test and Evaluation and CoolCalc HVAC Tool Development	VSS075	225,000	700,000	300,000	
Materials Approach to Fuel Efficient Tires	VSS084	186,000	675,000	167,000	Project to end in FY 2014
System for Automatically Maintaining Pressure in a Commercial Truck Tire	VSS085	571,189	713,810	161,535	Project will end in FY 2015
Aerodynamic Lightweight Cab Structure Components	LM060	365,000	280,000	65,000	
Improving Fatigue Performance of AHSS Welds	LM062	355,000	125,000	150,000	
Fleet DNA	VSS119	400,000	500,000	325,000	

NOTE: See Appendix D for complete list of projects associated with 21CTP. AHSS, advanced high-strength steel; HVAC, heating, ventilation, and air conditioning.

good and should include actions that can be taken to improve existing trailers, not just new trailers.

As for Technology Goal 1, which relates to aerodynamic drag, it could go further by taking into account the achievements of the SuperTruck program (a 40 percent improvement from a 2009 vehicle baseline) and current product offerings. While public information on the C_d for commercial vehicles is not readily available, with current regulations for GHGs and load-specific fuel consumption, it should be possible to estimate current values from the greenhouse gas emissions model (GEM) data, or to gather data on the C_d of commercial vehicles.

Response to Recommendations from NRC Phase 2 Review

R5-1 NRC Phase 2 Recommendation 5-1. Vehicle Aerodynamics: The Partnership should consider setting an aerodynamic drag stretch goal of 40 percent instead of 30 percent.

21CTP Response: The Partnership is aware of the NRC's recent work on heavy truck fuel consumption for EPA and NHTSA and the results of that work. The Partnership's work acknowledges the importance of trailers to the operational efficiency of the vehicle, and has made efforts to include trailer efficiency considerations in its SuperTruck research activities, from a vehicle systems perspective. The Partnership periodically reviews its goals and objectives to ensure they are in alignment with current technology progress and government agency research plans. SuperTruck research results will help inform future aerodynamic goal revisions. As information about the technology status of the aerodynamics work within SuperTruck becomes available, the Partnership will re-examine its goals for aerodynamics and adjust as necessary to provide the appropriate stretch targets.

Committee Comment on 5-1

The Partnership has not accepted a 40 percent stretch goal and has put the greatest part of its effort into the SuperTruck program. This program is focused on an idealized tractor-trailer combination rather than a real-world mix of tractors and trailers. It is indeed possible for improvements in the tractor aerodynamics to result in increased fuel consumption if the tractor is connected to a nonoptimized trailer. While not accepting the stretch goal, the Partnership's new goals are based on the EPA/NHTSA regulatory numbers, which set out the C_d and frontal area requirements in specific numbers.

Finding and Recommendation

Finding 5-2. The research on aerodynamics, in the SuperTruck program and at LLNL, has focused on idealized, integrated tractor-trailer configurations of both dry van and tanker configurations. It has provided useful data that are influencing current and future designs to achieve reduced aerodynamic drag and fuel consumption. Since trailers have long useful lives (15-20 years), research on modifications to existing trailers is needed to accelerate fuel consumption savings.

Recommendation 5-2. The technology goal and research aerodynamics should focus on achieving a C_d of 0.48 for a new high-roof sleeper tractor pulling a new or existing trailer that is certified to be SmartWay Elite.[4]

TIRE ROLLING RESISTANCE

Tires are critical to both safety and fuel consumption. Tire pressure, tread design, temperature effects, sidewall strength, durability, and materials are some of the factors that must be considered. Rolling resistance will continue to be a power demand for a Class 8 vehicle driving down the road at highway speeds. A reduction in fuel consumption with a 30 percent reduction in rolling resistance is possible based on research sponsored by NHTSA in support of the Phase 2 regulatory effort on fuel consumption and GHGs for MHDVs.[5]

A figure from Michelin included in the NRC Phase 2 report (2012) shows rolling resistance for tires in different axle positions on the vehicle. An update to this was presented to the NRC Committee on Technologies and Approaches for Reducing the Fuel Consumption of Medium- and Heavy-Duty Vehicles, Phase 2, in November 2013.[6] It shows the biggest improvement in drive axle tires, with a reduction in the coefficient of rolling resistance (C_{rr}) of 9.5 percent. However, it shows an increase in C_{rr} for the steer axle. This could be due to the regulations for reduced stopping distance, which put additional constraints on the steer axle weight. Figure 5-2 does suggest that wide-based single tires could achieve a C_{rr} of about 5.3, a 20 percent improvement over the C_{rr} of 6.6 listed for the drive axle. It appears that it will take considerable effort to achieve a C_{rr} of less than 5 kg/tonne. It should be emphasized that an important design trade-off is to understand the effect on stopping distance as the rolling resistance is reduced, along with consideration of other tire design factors.

Two projects related to tire rolling resistance that the Partnership notes fall under the 21CTP umbrella were reported at the DOE Annual Merit Review in 2014—namely, Project VSS084 (Martin et al., 2014) and Project VSS085 (Benedict, 2014).

The project VSS084, A Materials Approach to Fuel Efficient Tires, is exploring the use of both tire barrier coatings and tire filler. This project at PPG Industries is a way to improve the tire rolling resistance and overall fuel efficiency by at least 2 percent. Goodyear will be involved in the evaluation work. The goal of the coatings work is improved fuel efficiency by maintaining tire pressure. The goal of the filler work is improved tread wear without increasing rolling resistance. Currently candidate fillers are under evaluation by Goodyear. A barrier coating has been applied to tires. This project was started in October 2011 and ended in September 2014. The total project funding was $2,046,503; the DOE part is $1,485,851.

The project VSS085, System for Automatically Maintaining Pressure in a Commercial Truck Tire, aims to improve fuel use through maintenance of tire inflation. Other aims will be to extend tire life and improve safety. Technical accomplishments to date include component optimization, laboratory testing, and on-vehicle system testing. The project started in October 2011 with an end of May 2015 but it was extended through June 2016. The budget provided by DOE is $1,499,771; Goodyear provided $2,572,953 (Benedict, 2014).

Tire rolling resistance can easily vary ±5 percent when inflation pressures vary by ±20 percent (Figure 5-3). Kleffmann[7] also discussed the impact of rib pattern, winter tread pattern, tread depth, and footprint. Work is accordingly needed to address tire pressure maintenance and monitoring and inflation systems. The North American Council for Freight Efficiency (NACFE) has published a tire pressure systems confidence report (NACFE, 2013).

Response to Recommendations from the NRC Phase 2 Review

NRC Phase 2 Recommendation 5-2. Wide-Base Single Tires and Rolling Resistance Goal: The DOE should set the goal for reduced rolling resistance for the tires of the combination tractor-van trailer, rather than for the tractor drive wheels only, since improved-performance trailer tires are equally important to realizing the full benefit of reduced rolling resistance designs. This benefit can be achieved by combining the EPA base values

[4] A SmartWay Elite trailer is an EPA-designated 53-ft box dry-van or refrigerated trailer that achieves 10 percent or more fuel savings compared to a traditional trailer. Nine percent of the reduction comes from aerodynamic devices.

[5] T. Reinhart, SouthWest Research Institute, "Technologies for MD/HD GHG and Fuel Efficiency," presentation to NRC Committee on Assessment of Technologies and Approaches for Reducing the Fuel Consumption of Medium- and Heavy-Duty Vehicles, Phase 2, April 29, 2014.

[6] S. Lew, Michelin North America, Inc., "Test Methods for Truck Tire Rolling Resistance and Reducing Fuel Consumption of M-D and H-D Vehicles," Presentation to the Committee on Assessment of Technologies and Approaches for Reducing the Fuel Consumption of Medium- and Heavy-Duty Vehicles, Phase 2, on November 21, 2013.

[7] Jens Kleffmann, Continental, "Effect of Tire Inflation on Rolling Resistance," Presentation to NRC Committee on Assessment of Technologies and Approaches for Reducing the Fuel Consumption of Medium- and Heavy-Duty Vehicles, Phase 2, on November 21, 2013.

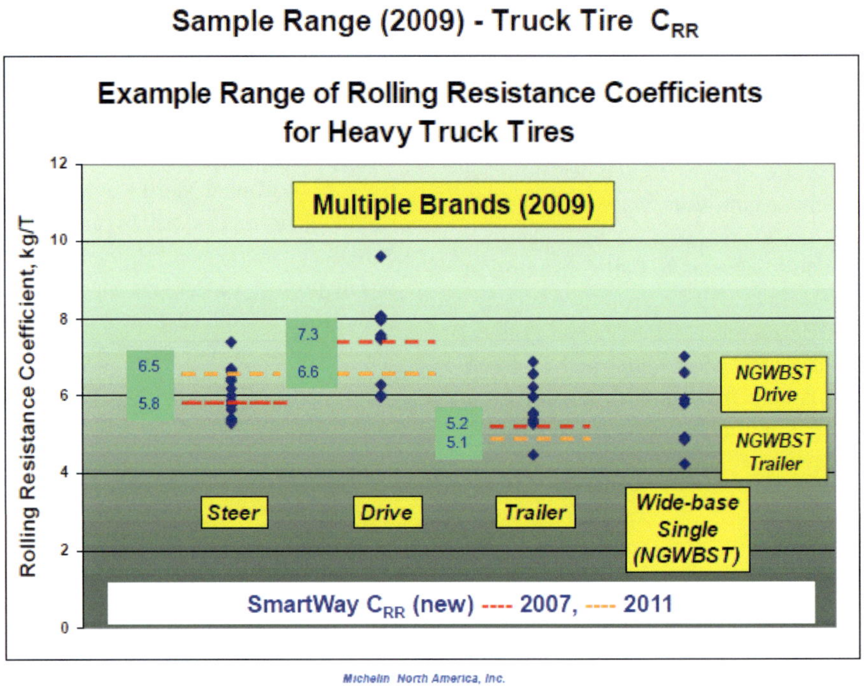

FIGURE 5-2 Some ranges of C_{rr}s for heavy truck tires. SOURCE: S. Lew, Michelin North America, Inc., "Test Methods for Truck Tire Rolling Resistance and Reducing Fuel Consumption of M-D and H-D Vehicles," Presentation to the Committee on Assessment of Technologies and Approaches for Reducing the Fuel Consumption of Medium- and Heavy-Duty Vehicles, Phase 2, on November 21, 2013. NOTE: The C_{rr} is dimensionless and can be expressed as kg/tonne or newton/kilonewton.

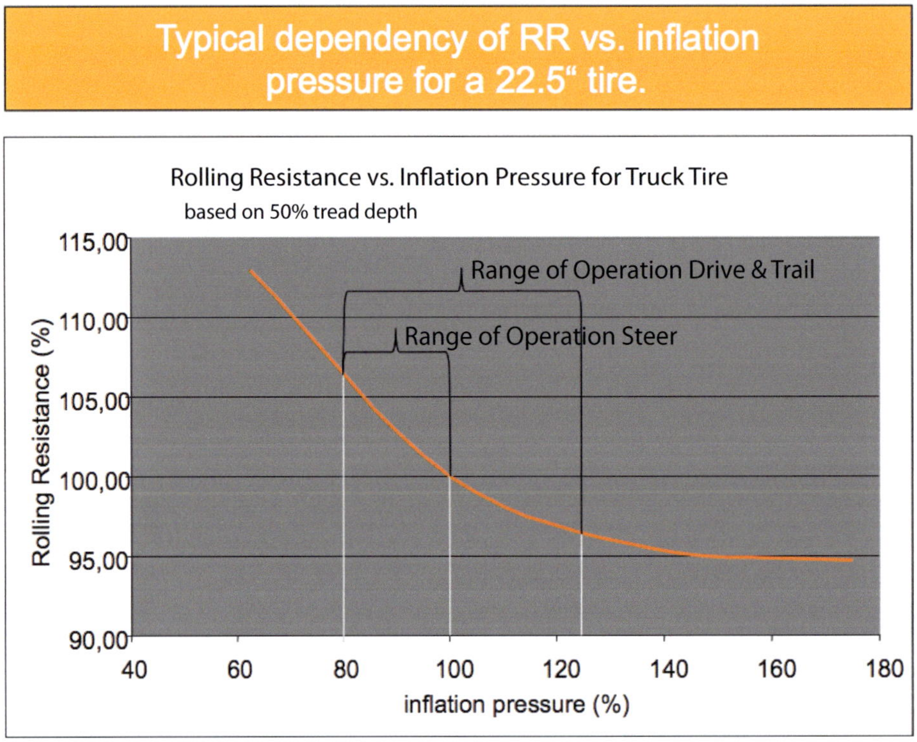

FIGURE 5-3 Typical dependency of rolling resistance on inflation pressure for a 22.5 in. tire. SOURCE: J. Kleffmann, Continental, "Effect of Tire Inflation on Rolling Resistance," Presentation to NRC Committee on Assessment of Technologies and Approaches for Reducing the Fuel Consumption of Medium- and Heavy-Duty Vehicles, Phase 2, November 21, 2013.

for steer and drive tires in the EPA/NHTSA GHG rule, with an assumed trailer tire C_{rr} value of about 0.0072.

21CTP Response: The Partnership concurs that a systems view of tire rolling resistance (including both tractor and trailer tires) is important to realizing the benefits of these tire technologies, and will take this into consideration when reviewing and revising Partnership goals. DOE, as a member of the Partnershi¡p, has initiated three tire technology projects in FY2012 (cross-cutting between light duty and heavy duty vehicles) that target 2% fuel consumption reduction for the full vehicle from rolling resistance improvements and automatic tire inflation.

Committee Comment on Response to 5-2

The Partnership has not specifically addressed trailer tires. However, the EPA SmartWay and SmartWay Elite programs do address trailer tires (Waltzer, 2014). At the 2015 SAE government/industry meeting, Anthony Erb of EPA presented test data confirming the relationship between reductions in rolling resistance and the amount of fuel decrease to be expected (Erb, 2015). Roughly a 10 percent reduction in rolling resistance provides a 2-3 percent improvement in fuel consumption on test road conditions according to SAE fuel consumption test procedures. Real-world savings will be less.

***

NRC Phase 2 Recommendation 5-3. Wide-Base Single Tire Retrofits: The 21CTP should consider producing a comprehensive summary that can be updated giving the prescriptions and precautions that carriers should consider when retrofitting NGWBSTs onto original equipment axles fitted with dual wheels and tires. This effect might best be managed in conjunction with the American Trucking Associations' (ATA's) Technology and Maintenance Council, which has drafted such a Recommended Practice and is a specialist in creating directives for ATA membership (ATA, 2007).

21CTP Response: The Partnership agrees that safety is extremely important when considering retrofits of NGWBS tires on existing trucks. The Partnership would encourage the use and promotion of Technology and Maintenance Council Recommended Practices to address this issue, and will consider addressing relevant safety concerns in the white papers and other 21CTP documentation addressing the use of next-generation wide-base single (NGWBS) tires.

Committee Comment on Response to 5-3

The Partnership expressed agreement with this recommendation, but specific examples of providing information are not known. The NRC (2010) report on medium- and heavy-duty vehicles addressed tires in general and tires for trailers in particular in Chapter 6. Next-generation wide-based single (NGWBS) tires are addressed, including barriers to further adoption of this technology. Based on presentations during that study, there has been improvement in the rolling resistance of dual tires as well.

***

NRC Phase 2 Recommendation 5-4. Wide-Base Single Tires Rolling Resistance Test Procedure: The 21CTP, strongly supported by DOT and EPA (the latter through its SmartWay program), should conduct an authoritative study of the several barriers (e.g., related to tread life, truck stability in blowouts, run-flat tires, and other topics) to the widespread carrier adoption of next generation wide base single (NGWBS) tires. The DOT should specifically support reduction of barriers to NGWBS tire acceptance by requiring the universal use by tire manufacturers of a rolling resistance test procedure like that in ISO (International Organization for Standardization) 28580, to ensure that comparative inter-laboratory data exist.

21CTP Response: The Partnership agrees that identifying and addressing barriers to NGWBS tire acceptance are critical in expanding the use of this technology to improve truck efficiency. The Partnership will consider the possibility of conducting a study of barriers, subject to available resources. Truck tire manufacturers at present do not correlate rolling resistance measurements among one another to any large extent: this may be due to the fact that rolling resistance has not been a specification provided to tire manufacturers by the vehicle OEMs. (In the case of light-duty tires, the vehicle OEM considers tire rolling resistance to be a very important performance requirement.) This may change as new truck fuel consumption regulations are imposed, and the need for lower rolling resistance tires increases. It should be noted that the ISO 28580 standard calls for a reference laboratory, but this has not yet been identified. The Partnership agrees that lack of consistent rolling resistance measurement could be a barrier to increased acceptance of NGWBS tires, along with the lack of education for fleets and owner-operators on the benefits of low rolling resistance tires. Absent any requirements to provide rolling resistance information at the point of sale, this information is not generally available to the tire purchaser.

Committee Comment on Response to 5-4

The Partnership acknowledged the issue of consistent determination of rolling resistance among manufacturers and using ISO 28580 as important, but it took no action. As a result of the NHTSA/EPA GHG 2014 regulations requiring rolling resistance as an input to the GHG emissions model (GEM) for MHDVs, vehicle OEMs responsible for these vehicles have worked with the tire manufacturers to resolve this issue (NRC, 2014).[8]

Finding and Recommendations

Finding 5-3. Tire technology and design are heavily invested in by the private sector and will continue to be worked on as NHTSA and EPA implement fuel consumption regulations for MHDVs.

[8] S. Lew, Michelin North America, Inc., "Test Methods for Truck Tire Rolling Resistance and Reducing Fuel Consumption of M-D and H-D Vehicles," Presentation to the Committee on Assessment of Technologies and Approaches for Reducing the Fuel Consumption of Medium- and Heavy-Duty Vehicles, Phase 2, on November 21, 2013.

Recommendation 5-3. Although fuel consumption can be reduced with reductions in tire rolling resistance, the limits to such reductions must be carefully weighed against tire traction and the ability of vehicles to attain safe stopping distances. Technology Goal 2 should be revised to include an analysis of the impact of reduced aerodynamic drag as well as a metric for the ability to safely stop the vehicle within current regulations. The Partnership should work to make tire rolling resistance data available to retail purchasers.

Recommendation 5-4. The Partnership should assess the current proprietary research being conducted by tire manufacturers to determine what gaps, if any, need to be filled by government-sponsored research. The focus should be on analytical tools that can be used to quantitatively assess results and identify directions for further improvements in low rolling resistance, traction for starting, stability control, stopping distance, tire inflation, life, and retreading.

AUXILIARY LOADS

Auxiliary and accessory loads will become more important in the future as regulations tighten and improvements are made in the engine and other parts of the vehicle. A clear definition of what constitutes an auxiliary load is needed to focus efforts and to avoid double counting improvements. Multiple definitions exist in the report *Technologies and Approaches to Reducing the Fuel Consumption of Medium- and Heavy-Duty Vehicles* (NRC, 2010). One definition related to Figure 4-1 in that report suggests that accessories are traditionally gear- or belt-driven by a vehicle's engine (examples include the water pump, air compressor, power-steering pump, cooling fans, and the air-conditioning system). Another reference noted in Figure 5-1 says that accessories are essential to engine operation, such as the fuel pump, water pump, and oil pump, while auxiliary loads are accessories used in a vehicle's operation, such as power steering, an air compressor, a cooling fan, and an air-conditioning compressor (NRC, 2010). This suggests that several items may be thought of as both an accessory and an auxiliary load.[9] Even though the engine is controlled by various electronics with sensors and actuators, the alternator is not part of this test, because the electronics are run from an off-board power source. In 2013, the Partnership introduced its roadmap and technical white papers, including the concept of "essential auxiliary loads" without a full description of the terminology (21CTP, 2013). The Truck and Engine Manufacturers Association (EMA) has been contacted to clarify the issue of what is included in the certification of an engine and what is an accessory (see regulatory reference quoted in the Annex at the end of this chapter). Of significance is the statement, "use good engineering judgment to simulate all engine work inputs and outputs as they typically would operate in use." This statement leaves open to estimates and engineering judgment several of the accessory loads on the engine and auxiliary loads on the vehicle.

Irrespective of whether something is an accessory or an auxiliary load, Project VSS133, "Cummins MD & HD Accessory Hybridization CRADA" (not included in Table 5-1) is addressing electrification of auxiliary loads (Deter, 2014; see Chapter 4). It was started in July 2013 and was scheduled to have ended in July 2015. It appears to be behind schedule as only 15 percent completion was reported in the 2014 Annual Merit Review (AMR), and a portion of spending is expected to be completed in the first half of 2015. Progress on this project is intertwined with the progress on Project VSS035, the Vehicle Systems Integration (VSI) project at Oak Ridge National Laboratory (see Chapter 4). The focus of that project is more to address hybrid certification testing options for hybrids for the next phase of regulations. It does not appear to address whether the models conform to real-world results.

Response to Recommendations from the NRC Phase 2 Review

R5-5 NRC Phase 2 Recommendation 5-5. Auxiliary Power Demands R&D: The Partnership should renew R&D efforts to further reduce fuel consumption related to auxiliary power demands.

21CTP Response: The Partnership should monitor auxiliary load improvements resulting from the SuperTruck projects.

Committee Comment on Response to 5-5

The Partnership has improved the description of its goals related to auxiliary loads by specifying that the 50 percent improvement is from a baseline of 20 horsepower to 10 horsepower and by updating the power use inventory. The SuperTruck program defines a 24-hour cycle of use that addresses auxiliary power demands related to hotel loads. Some SuperTruck teams have addressed other loads such as the air compressor and power steering system. More work is needed in this area. In particular, auxiliary power demands need to be better defined to avoid excessive claims of percentage improvements by suppliers.

Finding and Recommendation

Finding 5-4. The inconsistency in definitions of parasitic, accessory, essential auxiliary, and auxiliary loads creates confusion around assessments of where improvements can be made. The relative contribution of such loads will increase in the future.

[9] Dave Merrion, Chairman of Merrion Expert Consulting, LLC, suggests an accessory is a load needed to pass the engine certification test (personal communication, January 14, 2015).

Recommendation 5-5. The 21st Century Truck Partnership should work with EPA, NHTSA, and CARB to establish a clear definition of items included in the categories of accessory and auxiliary loads. Such clarification is needed to make an informed decision about which areas would be appropriate for government-sponsored research.

WEIGHT REDUCTION

Truck weight reduction affects fuel consumption by reducing tire rolling resistance and unrecovered energy used when accelerating or climbing a grade. The energy required to overcome resistance is approximately linearly dependent on the weight of the vehicle. A fully loaded tractor-trailer combination can weigh up to the standard federal highway limit of 80,000 lb (weights as high as 164,000 lb are allowed in some states). Reduction in overall vehicle weight could enable an increase in freight delivered on a ton-mile basis, enabling more freight to be delivered per truck and improving load specific fuel consumption (LSFC). New vehicle systems (such as hybrid power trains, fuel cells, and auxiliary power units) will present complex packaging and weight issues that will further increase the need for reductions in the weight of the body, chassis, and powertrain components in order to maintain vehicle functionality.

Opportunities for fuel efficiency impact vary considerably by truck type and class and duty cycle. Vehicle weight effects are more important for duty cycles with frequent starts and stops (NRC, 2010, Table 5-16). For urban delivery vehicles, a 10 percent reduction in weight can reduce fuel consumption by as much as 7 percent. Before FY 2007, numerous DOE projects had been aimed at vehicle weight reduction; several projects involving the national laboratories and industry led to useful examples of such reduction. Owing to budget reductions, no funding was provided for lightweight materials (other than propulsion materials) from FY 2007 through FY 2009 for meeting 21CTP goals. However, beginning in late 2010, funding was once again provided for weight reduction projects, primarily in the SuperTruck program (see Chapter 8).

Incentives for Vehicle Weight Reduction

Reducing the weight of Class 8 trucks is important as trucks have been adding weight, particularly with emissions-compliance devices; emissions control components have added as much as 400 lb to a typical tractor.[10] Aerodynamic devices, especially trailer devices, are growing in popularity, too, adding several hundred pounds in some cases. Weight reduction is also needed to offset other components such as auxiliary power units (~400 lb) added to reduce fuel consumption normally expended during idle. As selected truck-tractor technologies are expected to build on the EPA SmartWay configurations, some weight increase will be encountered—for example, with the use of tractor aerodynamic components and idle reduction components, as was just mentioned.

Weight reduction is beneficial if it can be reliably translated into more freight to be carried or reduced fuel consumption. Vehicles carrying freight in or on a trailer can either reach the maximum allowable weight for the vehicle or road before the available cargo space is filled (weigh out) or fill the available cargo space before reaching the weight limits (cube out). Vehicles that weigh out (before weight reduction) should see an improvement in LSFC, with more freight hauled for the same fuel consumed. Vehicles that cube out (before weight reduction) should see an improvement in load specific fuel consumption with the same freight hauled for less fuel consumed. Figure 5-4 indicates this bimodal split of gross vehicle weight in trucks on the road. Historically, bulk haulers and others that weigh out value the weight reduction more than those fleets that cube out.

In support of the overall goal to enable trucks and other heavy vehicles to be more energy efficient and to use alternative fuels while reducing emissions and remaining cost effective, the 21CTP seeks to reduce energy losses due to the weight of heavy vehicles without reducing vehicle functionality, durability, reliability, or safety. In addition, the 21CTP recognizes that improved materials may enable implementation of other technologies that can further improve the vehicle fuel efficiency (21CTP, 2013).

Weight Reduction Goals

Weight reduction goals vary according to the weight class of the vehicle; the targets for all classes range between a 10 percent and a 33 percent reduction in weight. Technology Goal 4 aims for a 10 percent reduction in tare weight for a 15,500 kg (34,000 lb) tractor-trailer combination, with a long-term stretch goal of reducing combined vehicle weight by 20 percent.

The weight targets for each vehicle class depend on performance requirements and duty cycle, with the targets reflecting the goal for total vehicle weight. The 21CTP recognizes that in some cases the weight reduction in the body and chassis will likely be higher. It is important to note that materials or technologies developed for a particular vehicle class are not necessarily limited to use in that class. For example, materials developed for lightweight frames for pickup trucks, vans, or sport utility vehicles (SUVs) will eventually be used in Classes 3-5 vehicles, and materials developed to meet the demanding performance requirements for Classes 7 and 8 trucks will find application in smaller vehicles (21CTP, 2013).

[10] Although MAP-21 (passed in 2012) allowed a federal exemption of up to 550 pounds for an APU, many roads are controlled by state regulations. A 400-lb exemption is the most common, as noted by the map of exemptions across the country found here: http://energy.gov/eere/vehicles/map-state-recognition-auxiliary-power-weight-exemption.

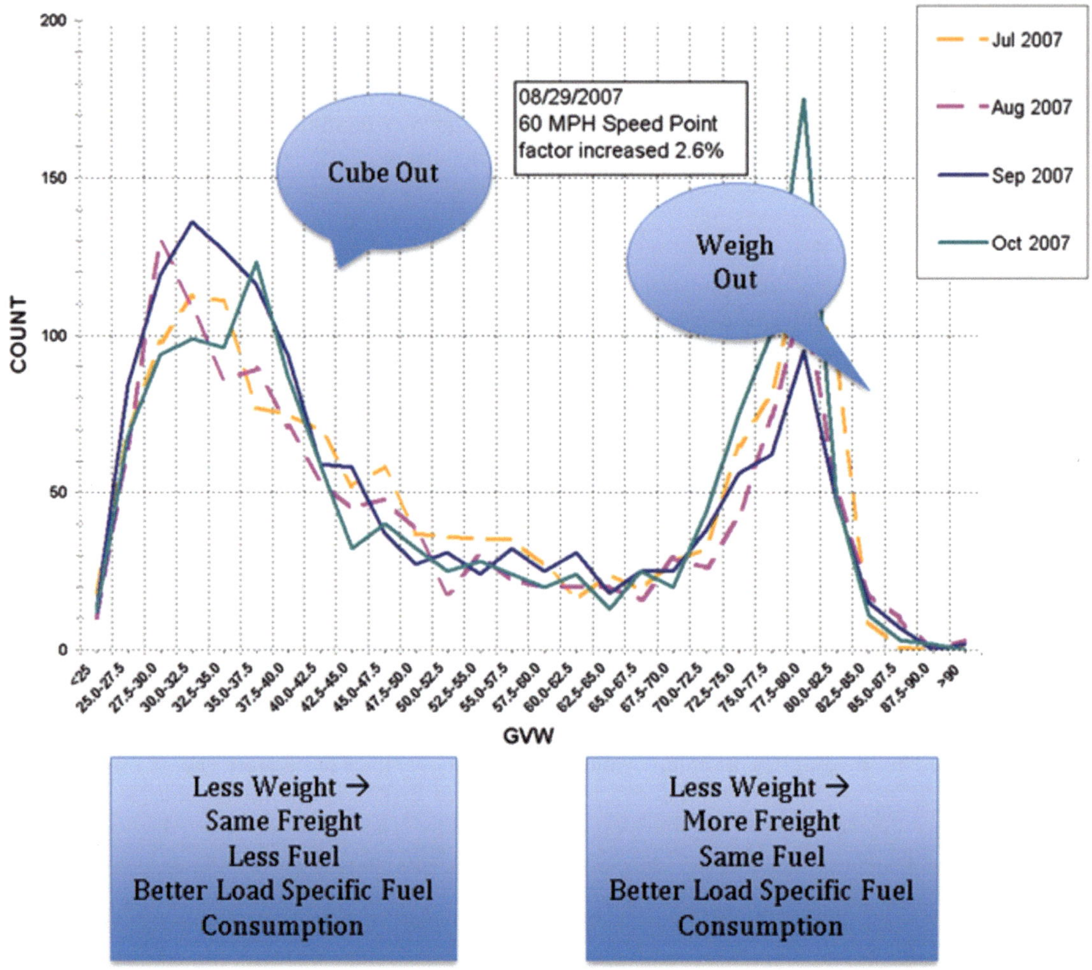

FIGURE 5-4 Bimodal distribution of truck gross vehicle weight (GVW) for five-axle vehicles from Weigh-In-Motion (WIM) data 421. SOURCE: Quinley (2010).

As noted by the Partnership, there has recently been "increased focus on manufacturing technologies that reduce the cost penalty associated with more expensive lightweight materials by conducting research in manufacturing technologies that are adaptable to the lower production volumes associated with heavy-duty commercial vehicles" (21CTP, 2013). In the committee's opinion, weight reduction must not sacrifice the durability, reliability, or performance of the vehicle. Achieving these goals by reducing inertial loading will lead to substantial benefits: increased fuel efficiency with accompanying reductions in emissions; increased available payload capacity for some vehicles; reduced rolling resistance; optimized cab and chassis mechanical structures for crash worthiness; and aerodynamic drag reduction systems.

Opportunities and Initiatives

Current materials in Class 8 trucks offer numerous opportunities for reducing vehicle weight by introducing aluminum alloys (25-55 percent weight reduction), magnesium alloys (40-70 percent weight reduction), carbon fiber composites (30-65 percent weight reduction), and high-strength steels (15-25 percent weight reduction), albeit often at a cost premium. The more obvious opportunities lie in the body structure (~19 percent of total tractor weight), the chassis/frame components (~12 percent), and wheels and tires (~10 percent). Truck manufacturers have been substituting lightweight materials for a number of components in the cab, chassis, and wheels. Examples include composite structure in the cab, aluminum panels, aluminum wheels, and aluminum fuel tanks. Weight-reduction opportunities will also be afforded by the extensive use of aluminum for both tractors and trailers (NRC, 2010, Figure 5-38).

For example, in one project, LM060, ("Aerodynamic Lightweight Cab Structure Components," from the 2013 VTO annual progress report (DOE, 2014), manufacturing methods for lightweight materials are being demonstrated that will increase the efficiency of Class 8 trucks by enabling more widespread use of weight-saving Al and enabling aerodynamic styling by a new approach to Al sheet forming. The project will ultimately develop forming technology that enables Al sheet to replace steel sheet, which, together with molded fiberglass-reinforced composite panels and components, will provide individual panel and component weight savings of approximately 40 percent (Smith, 2014). Pacific Car and Foundry Company (PACCAR), the Pacific Northwest National Laboratory (PNNL), and Magna are working on aerodynamic lightweight cab structure components. PACCAR Technical Center engineering staff completed the design review and material specifications for the production A-pillar component that will be used to demonstrate the advanced forming process for an aerodynamic cab structure. The A-pillar component consists of left- and right-hand parts, and the complexity of the part exceeds the conventional forming limits of Al sheet alloys. As a result, the current part is produced from sheet molding compound (SMC), which has approximately a 40 percent weight penalty compared to Al. PNNL and PACCAR completed the selection of a Tier 1 supplier to develop the prototype A-pillar forming process and placed a cost-shared subcontract with Magna's Stronach Centre for Innovation (SCFI). Magna will develop tooling and the forming process capable of producing the A-pillar component and deliver 25 each of left- and right-hand A-pillar parts to PNNL/PACCAR. Magna completed the forming analysis and simulation of a hybrid hot-forming/cold-stamping process capable of forming the A-pillar component in the X608 Al alloy. Based on the forming analysis, a prototype forming tooling has been designed and is being fabricated. In the coming years, fabrication of prototype aerodynamic formed components will be completed to deliver 25 left- and right-hand A-pillar parts for testing by PACCAR; and production feasibility for Al components in conjunction with Magna SCFI and PACCAR will be established.

In another example, DOE initiated a project that addresses advanced high-strength steel (AHSS) welds—ORNL Project LM062 (Improving Fatigue Performance of AHSS Welds). This project started in March 2011 and was scheduled to have ended in September 2014. The total project funding projection is $1,250,000 (DOE) and $650,000 (contractors). Downgaging of AHSS for weight reduction causes increased stress in the weld region. The objective of this project is weld residual stress mitigation to improve the fatigue performance of the weld joint of AHSS for high-volume vehicle production. Research has revealed the role of weld start-stop in controlling weld fatigue in short stitch welds. Microstructure-property modeling was used to increase understanding of the weld process. A special weld wire was developed, and weld fatigue life was improved by stress management. An in situ strain measurement technique was developed to directly measure the strain field during welding.

The SuperTruck projects, discussed in Chapter 8, are also developing and demonstrating several ways of reducing the weight of heavy-duty trucks by use of lightweight materials and advanced fabrication technologies. As discussed in the introduction to this chapter, current regulations for fuel consumption and greenhouse gas emissions are now based on load specific fuel consumption. Therefore, a key goal for weight reduction is translating it into usable freight that can be transported. This opens the possibility for increasing the amount of freight that can be moved within the confines of existing trailers. The industry has recognized this and has begun developing thinner walled trailers, wider doors, alternative flooring options, and techniques for stacking more freight into dry van trailers.

Findings and Recommendation

Finding 5-5. The SuperTruck teams achieved overall vehicle weight reductions, despite adding a number of fuel-saving features that increase weight (see Chapter 8). Unfortunately, many of the weight reduction technologies employed in SuperTruck are unlikely to prove cost effective. The Partnership goal of a 10 percent reduction in tare weight for a 15,500 kg tractor/trailer combination remains a good target, but cost effectiveness will be the challenge.

Finding 5-6. It is unknown what portion of full vehicles operate at the weight limit, what portion of vehicles are fully loaded but below the weight limit, and what percentage of vehicles are empty or only partly loaded. Understanding these factors is key to being able to determine the fuel savings available from weight reductions, and to understanding the cost-effectiveness of weight reductions or of increases in available cargo volume.

Finding 5-7. Weight reduction translates into more freight to be carried or fuel consumption reductions. Vehicles that weigh out (before weight reduction) see an improvement in load-specific fuel consumption (LSFC) with more freight hauled for the same fuel consumed. Vehicles that cube out (before weight reduction) see an improvement in load-specific fuel consumption with the same freight hauled for less fuel consumed.

Recommendation 5-6. The Partnership should initiate a study to determine what proportion of vehicle miles are run grossed-out, cubed-out, partly loaded, and empty. Understanding these proportions will help determine the value of potential weight reduction and cargo volume increasing features. Research should explore products and methods that allow more to be packed into a 53 ft dry van trailer, such as thinner walls, better flooring options, and double decking.

THERMAL MANAGEMENT

Management of temperature remains a challenge for vehicle and component OEMs. Listed below are areas where thermal management is needed:

- *Engine Compartment.* Managing the temperature in the engine compartment is important since increased heat has had side effects, some anticipated and some not expected. The impact of heat on major components of the engine itself and on the oil for lubrication is expected and addressed. However the impact on electrical insulation, seals on electronics, air compressors, and air hoses also need to be considered. The effect of waste heat recovery (WHR) or Rankine cycles is being addressed by the SuperTruck program.
- *Exhaust.* Exhaust temperatures can exceed 1,100°F (600°C), requiring special precautions (CVSA, 2010; Volvo Group North America, Inc., 2009). Insulation and protection must be provided to the pipes from the engine to the aftertreatment system and at the exhaust outlet. Some of this insulation and protection is to maintain the necessary temperature for the chemical reactions in the aftertreatment system, and some is to protect driver and service technician contact. Techniques for mitigating the temperature of the exhaust to protect personnel and materials external to the truck (such as hay under a truck in a field that is being harvested) were devised by several OEMs. Because vertical exhaust stacks create extra weight and aerodynamic drag, these high-temperature exhaust pipes are often designed to be closer to the ground.
- *Brakes.* Decreases in stopping distance over the last few years have required the brakes to do more work. This was accomplished in some cases by upsizing components of the braking system. Some vehicles have switched to disc brakes, but drum brakes remain the most common option, requiring cooling air flow to remain effective under heavy use, such as when descending mountains. The improvements in aerodynamics have driven reductions in the drag associated with the wheels and the wheel wells. The net effect has been to move the airstream away from the wheels, which can reduce air-cooling to the wheels. The improvements in aerodynamics also require the brakes to assume more of the work for stopping the vehicle.
- *Driver Personal Comfort.* The driver needs to be kept warm in the winter months and cool in the summer months. Blower motors and electrical resistance heaters (such as for mirrors) can be a measureable drain on the batteries and alternators. Some fleets have moved to using battery-powered air conditioners combined with diesel-fired heaters for comfort when not operating. However, the reliability and performance of the batteries have not always lived up to expectations. The DOE CoolCab and the Shorepower Truck Electrification Project (STEP, see Chapter 6) projects address this need.
- *Driver Food.* An often overlooked requirement in long-haul applications using sleepers is the need to maintain food in a refrigerator and to provide heating for the food.
- *Food Freight.* Recently, the Food and Drug Administration published regulations on food safety in trucking. While not as stringent as anticipated, they do highlight the need to control temperatures in refrigerated trailers. Trailer design and fleet operations have advanced to using trailers and box trucks with multiple refrigerated zones. This allows them to keep frozen food at lower temperatures than produce and to maximize the freight efficiency. The DOE STEP project, which is the same as the Interstate Grid Electrification project, ARRA VTO 70, addresses this issue through the transportation refrigeration unit (TRU) components.
- *Fuel.* The new emphasis on natural gas as a fuel for trucks has created the need for cryogenic temperature control on natural gas-powered trucks and tractors for maintaining liquefied natural gas.
- *Batteries.* Whether batteries are used for starting the vehicle, hotel loads for the driver, the air conditioning system, a hybrid power train, or a trailer refrigeration unit, thermal management of the battery is extremely important. Some of these issues, at least prototypes and analysis, are being addressed in the SuperTruck program. There is opportunity for more fundamental work to enable solutions to some of the thermal problems for batteries.
- *Solar.* Solar power has often been talked about for truck and passenger car applications. Generally, it has been perceived as too expensive for too little energy. However, it is now in production on a few passenger cars and is being tested in trucks. The SuperTruck program is addressing some of this. It is mentioned here only because the source of power is thermal in nature.

Projects for Thermal Management

The committee has identified four projects related to thermal management. Only the first project was identified as falling under the purview of the Partnership. Project VSS075, CoolCab Test and Evaluation and CoolCalc HVAC Tool Development, is discussed in Chapter 6 on engine idle reduction. It is mentioned here because the project is looking for ways to control the thermal energy for air-conditioning systems that provide driver comfort. This project at the National Renewable Energy Laboratory (NREL) addresses reducing the heating, ventilation, and air conditioning (HVAC) need as an approach to decreasing fuel use. The project has links with manufacturers and a good view of the end user. Anderson (2014) notes: "The goal is to demonstrate at least a 30% reduction in long-haul truck idle climate control loads with a 3-year or better payback period by 2015."[11] The program

[11] D. Anderson, 2014, "Vehicle and Systems Simulation and Testing," presentation to the committee, September 3, 2014.

plan is to develop and integrate technologies that address auxiliary load reduction and idle reduction to greatly improve commercial vehicle efficiency. To date, the CoolCalc modeling tool has been released to select industry partners, and testing is under way with instrumented cabs. Accomplishments include quantification of the impact of advanced paints, advanced glazings, sleeper microclimate evaluation, insulation, and auxiliary air-conditioning system. The project started in FY 2011 and is expected to be completed in September 2015. As might be expected, cab color makes a measurable difference in the heat load inside the cabin. The specific measurements will give guidance to manufacturers and fleets for the future selection of colors. Battery sizing will help to reduce the weight of the batteries and provide longer periods of comfort when the driver is off duty. The DOE budget (CoolCab/CoolCalc) is $1,060,000/$615,000, with the contractor's share $488,000.

Another project, VSS134, Vehicle Thermal System Modeling in Simulink, is focused on providing heat to the driver rather than cooling, but with the same emphasis on reducing fuel use when the driver is off-duty. This project was not included by the Partnership in its list of projects. The committee notes that VSS134 includes Daimler Trucks North America as one of its partners, which leveraged analysis capabilities being developed to assist on the SuperTruck project. The project also leveraged model results for the CoolCab project impact estimation.

Response to Recommendations from the NRC Phase 2 Review

R5-6 NRC Phase 2 Recommendation 5-6. Thermal Management: The Partnership should continue priority support of nano-fluid and high-efficiency under-hood cooling systems, as well as review other potential technical concepts, and validate them as an integrated system.

21CTP Response: DOE is planning to expand R&D on high efficiency HVAC systems. DOE agrees and is continuing support of nano-fluid and high-efficiency under-hood cooling systems. DOE will monitor other potential technology solutions to reach thermal management objectives.

Committee Comment on Response to 5-6

The Partnership indicates DOE is continuing to support R&D on nanofluids and underhood cooling systems. The Partnership also indicates DOE is planning to expand efforts for high-efficiency HVAC. No specifics were provided. The Daimler Trucks North America SuperTruck effort did include high-voltage HVAC developments.

Finding and Recommendation

Finding 5-8. The current research in thermal management is limited and more focused on idle reduction. The area of thermal management is in fact broader and encompasses a variety of areas applicable to medium- and heavy-duty vehicles.

Recommendation 5-7. The 21st Century Truck Partnership should establish a goal to reduce the energy required over a drive cycle for non-engine thermal loads by 50 percent and establish a research program focused on cooling for natural gas, trailer refrigeration, and batteries. The goal to increase heat load rejected by thermal management systems by 20 percent without increasing radiator size should continue to be pursued.

DRIVELINE POWER

The only focus on driveline power is to comment on the Partnership's response to an NRC Phase 2 recommendation.

R5-7 NRC Phase 2 Recommendation 5-7. Driveline Power Demand: The term "powertrain" should be removed from the 21 CTP Goal 5.b statement. In addition, the Partnership should update its study on the driveline power demand of 12 hp.

21CTP Response: The Partnership concurs: a subsequent revision to the Partnership's goal wording made after the completion of this review has removed the word "powertrain" from the subject goal, which will be published as part of the final white paper/roadmap document. The Partnership will review the current information on driveline power demand and consider updates to this study. The Partnership will review research results from the SuperTruck teams to gather current technology information for power demand, and revise assessments of power demand as appropriate.

Committee Comment on Response to 5-7

The term "powertrain" has been removed from the vehicle power demand goals in the 21CTP roadmap and technical white papers, Section 3 (2013), but not from the document's executive summary. The current goals for the Partnership are not specific in the driveline area.

FRICTION AND WEAR

As part of the effort for friction and wear, the project Development of a High Power Density Driveline, VSS058, was identified.[12] This project at Argonne National Laboratory (ANL) targets weight reduction through reduction in size and weight of the driveline systems. Driveline size reduction is to be achieved by developing coatings and lubricants for increasing the power density of the systems. The expectation is that these improvements would enable the design of smaller and lighter weight components without loss of performance. Good progress has been made in the development of a low viscosity lubricant formulation. Other work explored scuffing mechanisms and contact fatigue performance evaluation. None of the heavy-duty

[12] Ibid.

transmission manufacturers are involved in the work. The project started in October 2010 and is expected to end in FY 2015. The DOE funding is $870,000 and the contractor share is $120,000.

Most vehicle-level energy-balance studies indicate the driveline is about 98 percent efficient in transmitting energy from the engine to the wheels. Recent studies by Southwest Research Institute (SwRI) for line-haul applications indicate weight reduction may not be a good area for achieving improvements in fuel consumption even though it can result in load specific fuel reductions in fuel consumption as measured in gallons per ton-mile. Therefore, the proposed project's objective of a 5-7 percent improvement in fuel savings from lubrication and size/weight reduction of the driveline is not likely to be achieved. The connection between the lubricant and a 25 percent reduction in the size of driveline components is not clear.

The second project presented was Parasitic Engine Friction Collaboration, VSS005. This work deals with lubrication of the engine and should rightfully be classified with engine development for improved brake thermal efficiency (BTE). It should not be counted as vehicle power demand research. This project at ANL began in FY 2010 and is scheduled to be complete at the end of FY 2015, with a total budget of $1,887,000 for the 6 years. In this project, advanced engine friction models are applied to predict where parasitic friction losses occur and how advanced tribological concepts (lubricants, materials, additives, engineered surfaces) can reduce losses, component by component. By doing so, the project identifies potential pathways to reduce losses and thus improve both fuel consumption and reliability and durability.

Current government-sponsored research is focused on lubrication to achieve reductions in friction, primarily in the engine, with some work on the driveline. No work is perceived to be looking at friction and lubrication of power steering pumps, water pumps, air compressors, wheel ends, clutches, or gears. The industry is actively pursuing improvements in these areas and claiming as much as 2 percent improvement in fuel consumption.

OVERALL COMMENTS, FINDINGS, AND RECOMMENDATIONS

Before closing this chapter, there are few comments that apply to the general area of vehicle power demands and not necessarily to any of the individual areas that have been addressed in this chapter.

The Partnership also identified project VSS119, Fleet DNA, as under the umbrella of 21CTP. This project is led by NREL with many industry (NTEA/GTA, Cummins, PG&E, Oshkosh, Waste Management, Zonar, Parker), academic (Ohio State University, California State University, North Carolina State University, Calstart), and government/regulatory (ORNL, Clean Cities, South Coast Air Quality Management District [SCAQMD], CARB, EPA, ANL, City of Indianapolis) collaborations. The goal of this project is to define data capture and structure for a variety of drive cycles and conditions, and to then create a data storage warehouse to make this data available to a broad community of users. The project has also included an effort to incorporate models and data analysis tools. Highest priority has been given to Class 2 and Class 8 vehicle, and the current data have been gathered mostly from delivery vans and Class 8 trucks. Capturing such real-world data and drive cycles is extremely important to identifying areas of opportunity for reducing fuel consumption.

Response to Recommendations from the NRC Phase 2 Review

There was an NRC Phase 2 recommendation that applied broadly to the vehicle power demands area. The response to it by the Partnership and the committee's comment are noted as follows:

> **NRC Phase 2 Recommendation 5-8.** Although it is tempting to assume that the SuperTruck projects will address all of the technologies required to reduce tractor-trailer fuel consumption, in practice many technologies may be left behind, particularly those that are not yet very mature. The Partnership should carefully review the technologies that have been identified and determine whether any technologies to reduce vehicle power demand are not being adequately addressed by the SuperTruck program. The DOE should define projects and find funding to support the development of technologies beyond the scope of SuperTruck.
>
> **21CTP Response:** The SuperTruck projects are designed to develop combinations of advanced technologies into a Class 8 platform that can be commercialized in the near-term. In order to ensure commercial viability, the technologies are chosen by each industry team and not dictated by DOE. Technical approaches for reducing petroleum consumption that are not addressed by the SuperTruck projects may be appropriate for investigation through other pathways that address longer term technology development. Follow-on activities being considered post-SuperTruck may be used to address these technologies.

Committee Comment on Response to 5-8

The Partnership's response indicates limited acceptance of this recommendation but provides no specific suggestions for additional research. Now that the technology choices of the SuperTruck teams are available, the Partnership should review the situation and determine which technologies require new or additional R&D.

FINDING AND RECOMMENDATIONS

Finding 5-9. The industry has seen significant changes and improvements in fuel consumption since the Partnership was formed in 2002. As improvements in aerodynamic drag and tire rolling resistance have been made, the relative importance of accessory and auxiliary loads has increased.

The Partnership has repeatedly revised its goals to reflect the improvements and any changes in regulations.

Recommendation 5-8. The Partnership, in setting its goals, should use drive cycles, as it did for the SuperTruck progam. The cycles should take into account the many drive cycles that already exist, current operational regulations of the Federal Motor Carrier Safety Administration, the current and future regulations of EPA/NHTSA, and real world data that are being accumulated by such projects as the National Renewable Energy Laboratory's Fleet DNA project. Data on real-world weights of vehicles needs to be included.

Recommendation 5-9. Fleets either measure or monitor miles per gallon or gallons of fuel consumed. They rarely monitor load specific fuel consumption (LSFC) in gallons per 1,000 ton-mile, as in the regulations. Work is needed to change this practice. The Partnership, as part of DOE, should take the lead role, in combination with EPA, NHTSA, and CARB, in creating demand, perhaps through some sort of incentive, to produce LSFC information.

REFERENCES

21CTP (21st Century Truck Partnership). 2013. *Roadmap and Technical White Papers*. U.S. Department of Energy Office of Energy Efficiency and Renewable Energy. https://www1.eere.energy.gov/vehiclesandfuels/pdfs/program/21ctp_roadmap_white_papers_2013.pdf.

Ajayi, O.O., C. Lorenzo-Martin, A.C. Greco, and G.R. Fenske. 2014. Development of High Power Density Driveline for Vehicles (Developing Enabling Tribological Technologies). Argonne National Laboratory. DOE Annual Merit Review VSS058, Washington, D.C., June 19.

APS (American Physical Society). 2015. Energy Units. American Physical Society. Accessed March 10, 2015. http://www.aps.org/policy/reports/popa-reports/energy/units.cfm.

ATRI (American Trucking Research Institute). 2013a. An Analysis of the Operational Costs of Trucking: A 2013 Update. Arlington, Va.: American Trucking Research Institute.

ATRI. 2013b. Freight Performance Measures Analysis of Freight-Significant Highway Locations–2013. Arlington, Va.: American Trucking Research Institute.

Benedict, R. 2014. A System for Automatically Maintaining Pressure in a Commercial Truck Tire. The Goodyear Tire & Rubber Company. DOE Annual Merit Review VSS085, Washington, D.C., June 19.

Birky, A.K. 2014. Modeling for Market Analysis: HTEB, TRUCK, and LVChoice. TA Engineering, Inc. DOE Annual Merit Review VAN012, Washington, D.C., June 18.

Cox, J. 2014. Plug-in Hybrid Electric Commercial Fleet Demonstration and Evaluation. Principal Investigator: Matt Miyasato, South Coast Air Quality Management District. DOE Annual Merit Review VSS083, Washington, D.C., June 19.

CVSA (Commercial Vehicle Safety Alliance). 2010. Inspection of Vehicles Equipped with 2007 and 2010 EPA Certified Engines. North American Standard Inspection Program. Accessed March 11, 2015. http://goo.gl/OhYubo.

Delgado, O., and N. Lutsey. 2014. The U.S. SuperTruck Program: Expediting Development of Advanced HDV Efficiency Technologies. Washington, D.C.: International Council on Clean Transportation (ICCT). http://www.theicct.org/us-supertruck-program-expediting-development-advanced-hdv-efficiency-technologies. Accessed April 23, 2015.

Deter, D. 2014. Cummins MD & HD Accessory Hybridization CRADA. Center for Transportation Analysis and ORNL. DOE Annual Merit Review VSS133, Washington, D.C., June 16-20.

DOE (U.S. Department of Energy). 2014. Lightweight Materials R&D Program, FY 2013 Annual Progress Report. Vehicle Technologies Office. DOE/EE-1039. http://energy.gov/sites/prod/files/2014/04/f15/2013_lightweight_materials_apr.pdf. Accessed August 19, 2015.

EPA (U.S. Environmental Protection Agency). 2014. EPA-designated "SmartWay Elite" Trailers. EPA SmartWay Technology Program. http://epa.gov/smartway/forpartners/documents/aerodynamic/420f15008.pdf. Accessed March 10, 2015.

Erb, A. 2015. The Place of Low Rolling Resistance Tires in EPA's Greenhouse Gas Reduction Strategy. U.S. Environmental Protection Agency. 2015 SAE Government/Industry Meeting, Washington, D.C., January 23. http://www.sae.org/events/pdf/gim/2015_gim_guide.pdf.

Fadler, G. 2013. Navistar Fuel Economy and Emissions. Navistar, Inc. Presentation to the NRC Committee on Assessment of Technologies and Approaches for Reducing the Fuel Consumption of Medium- and Heavy-Duty Vehicles, Phase 2, Washington, D.C.

Fenske, G., A. Matthews, and N. Demas. 2014. DOE/DOD Parasitic Energy Loss Collaboration. Argonne National Laboratory. DOE Annual Merit Review VSS005, Washington, D.C., June 16.

FHWA (Federal Highway Administration). 2005. Freight Performance Measurement: Travel Time in Freight-Significant Corridors. U.S. Department of Transportation. http://www.ops.fhwa.dot.gov/freight/freight_analysis/perform_meas/fpmtraveltime/traveltimebrochure.pdf.

Gonder, J. 2014. ICEER: Internal Combustion Engine Energy Retention. National Renewable Energy Laboratory Team: E. Wood and S. Lopp. DOE Annual Merit Review VSS126, Washington, D.C., June 18.

Heavy Duty Manufacturers Association. 2014. Transformative Technology for Tractor Driver Behavior. Heavy Duty Manufacturers Association Commercial Vehicle Outlook Conference, Dallas, Tex., August 20.

Kambiz, S. 2014. DOE's Effort to Improve Heavy Vehicle Fuel Efficiency through Improved Aerodynamics. Lawrence Livermore National Laboratory. DOE Annual Merit Review VSS006, Livermore, Calif., June 19.

Karbowski, D., A. Rousseau, S. Pagerit, N. Kim, and D. Lee. 2014. Trip Prediction and Route-Based Vehicle Energy Management. Argonne National Laboratory. DOE Annual Merit Review VSS125, Washington, D.C., June 18.

Kim, N., and A. Rousseau. 2014. Vehicle Level Model and Control Development and Validation Under Various Thermal Conditions. Argonne National Laboratory. DOE Annual Merit Review VSS127, Washington, D.C., March.

Lustbader, J. 2014. CoolCab Test and Evaluation & CoolCalc HVAC Tool Development. National Renewable Energy Laboratory Team: C. Kreutzer, M. Jeffers, J. Tomerlin, R. Langewisch and K. Kincade. DOE Annual Merit Review VSS075, Washington, D.C., June 19.

Lustbader, J., and T. Kiss. 2014. Vehicle Thermal System Modeling in Simulink. National Renewable Energy Laboratory. DOE Annual Merit Review VSS134, Washington, D.C., June 16.

Martin, J., M. Bowman, and P. Votruba-Drzal. 2014. A Materials Approach to Fuel-Efficient Tires. PPG Industries. U.S. Department of Energy Annual Merit Review VSS084, Washington, D.C., June 19.

Meyer, J. 2014. Advanced Climate Systems for EV Extended Range (ACS forEVER). Halla Visteon Climate Control Corp., Presenter: J. Schneider. DOE Annual Merit Review VSS135, Washington, D.C., June 19.

NACFE (North American Council for Freight Efficiency). 2013. Tire Pressure Systems—Confidence Report. http://nacfe.org/wp-content/uploads/2014/01/TPS-Detailed-Confidence-Report1.pdf. Accessed April 24, 2015.

NRC (National Research Council). 2010. *Technologies and Approaches to Reducing the Fuel Consumption of Medium- and Heavy-Duty Vehicles*. Washington, D.C.: The National Academies Press.

NRC. 2012. *Review of the 21st Century Truck Partnership, Second Report*. Washington, D.C.: The National Academies Press.

NRC. 2014. *Reducing the Fuel Consumption and Greenhouse Gas Emissions of Medium- and Heavy-Duty Vehicles, Phase Two: First Report.* Washington, D.C.: The National Academies Press.

Pearl, I. 2014. Natural Gas Vehicles in Hazard Classification Areas. Natural Gas Vehicles Task Force Meeting. http://www1.eere.energy.gov/cleancities/pdfs/ngvtf14_pearl.pdf. Accessed April 4, 2015.

Quinley, R. 2010. WIM Data Analyst's Manual, FHWA Report IF-10-018, June 2018. http://www.fhwa.dot.gov/pavement/wim/pubs/if10018/if10018.pdf.

Rask, E. 2014. Advanced Technology Vehicle Lab Benchmarking-Level 2 (in-depth). Argonne National Laboratory. DOE Annual Merit Review VSS031, Washington, D.C., June 17.

Rugh, J.P. 2014. Electric Drive Vehicle Climate Control Load Reduction. National Renewable Energy Laboratory. DOE Annual Merit Review VSS097, Washington, D.C., June 17.

Smith, M. 2014. Aerodynamic Lightweight Cab Structure Components. Pacific Northwest National Laboratory. DOE Annual Merit Review LM060, Washington, D.C., June.

Stephens, T. 2014. Impact Analysis: VTO Baseline and Scenario (BaSce) Activities. Argonne National Laboratory. DOE Annual Merit Review VAN001, Washington, D.C., June 18.

Technology and Maintenance Council. 2014. Fuel Efficiency and Vehicle Emissions Regulations: Now and for the Foreseeable Future. S11 Tech Session, March 11. http://www.tech-i-m.com/insights/. Accessed March 27, 2015.

Volvo Group North America, Inc. 2009. Volvo's SCR No Regen Engine. http://www.volvotrucks.com/SiteCollectionDocuments/VTNA_Tree/ILF/Products/2010/09-VTM075_NoRegen_SS_041609.pdf. Accessed March 11, 2015.

Walkowicz, K. 2014a. Medium Duty ARRA Data Reporting and Analysis. National Renewable Energy Laboratory Team: A. Duran and A. Ragatz. DOE Annual Merit Review VSS124, Washington, D.C., June 17.

Walkowicz, K. 2014b. Fleet DNA. National Renewable Energy Laboratory Presenter: A. Duran. DOE Annual Merit Review VSS119, Washington, D.C., June 16.

Waltzer, S. 2014. U.S. EPA SmartWay Technology Program. Presentation to the Technology and Maintenance Council, Portland, OR, March 11.

Woodrooffe, J., and T. Vachon. 2000. Revision 3 Typical Large Truck-Tractor Trailer: Distribution of Energy Requirements (kWh) 65 mph, Two load conditions, Level Road for one hour.

ANNEX

Regulations Related to Certification of Engines for Discussion of Accessory versus Auxiliary Loads

§1065.110 Work inputs and outputs, accessory work, and operator demand.[13]

(a) *Work.* Use good engineering judgment to simulate all engine work inputs and outputs as they typically would operate in use. Account for work inputs and outputs during an emission test by measuring them; or, if they are small, you may show by engineering analysis that disregarding them does not affect your ability to determine the net work output by more than ±0.5% of the net expected work output over the test interval. Use equipment to simulate the specific types of work, as follows:

(1) *Shaft work.* Use an engine dynamometer that is able to meet the cycle-validation criteria in §1065.514 over each applicable duty cycle.

(i) You may use eddy-current and water-brake dynamometers for any testing that does not involve engine motoring, which is identified by negative torque commands in a reference duty cycle. See the standard setting part for reference duty cycles that are applicable to your engine.

(ii) You may use alternating-current or direct-current motoring dynamometers for any type of testing.

(iii) You may use one or more dynamometers.

(iv) You may use any device that is already installed on a vehicle, equipment, or vessel to absorb work from the engine›s output shaft(s). Examples of these types of devices include a vessel›s propeller and a locomotive›s generator.

(2) *Electrical work.* Use one or more of the following to simulate electrical work:

(i) Use storage batteries or capacitors that are of the type and capacity installed in use.

(ii) Use motors, generators, and alternators that are of the type and capacity installed in use.

(iii) Use a resistor load bank to simulate electrical loads.

(3) *Pump, compressor, and turbine work.* Use pumps, compressors, and turbines that are of the type and capacity installed in use. Use working fluids that are of the same type and thermodynamic state as normal in-use operation.

(b) *Laboratory work inputs.* You may supply any laboratory inputs of work to the engine. For example, you may supply electrical work to the engine to operate a fuel system, and as another example you may supply compressor work to the engine to actuate pneumatic valves. We may ask you to show by engineering analysis your accounting of laboratory work inputs to meet the criterion in paragraph (a) of this section.

(c) *Engine accessories.* You must either install or account for the work of engine accessories required to fuel, lubricate, or heat the engine, circulate coolant to the engine, or to operate aftertreatment devices. Operate the engine with these accessories installed or accounted for during all

[13] See 70 FR 40516, July 13, 2005, "Engine-Testing Procedures", as amended at 73 FR 37292, June 30, 2008. e-CFR data current as of April 22, 2015. Available at http://www.ecfr.gov/cgi-bin/text-idx?SID=28840d4abb0e4aa4074d081c70105ee5&node=se40.33.1065_1110&rgn=div8.

testing operations, including mapping. If these accessories are not powered by the engine during a test, account for the work required to perform these functions from the total work used in brake-specific emission calculations. For air-cooled engines only, subtract externally powered fan work from total work. We may ask you to show by engineering analysis your accounting of engine accessories to meet the criterion in paragraph (a) of this section.

(d) *Engine starter.* You may install a production-type starter.

(e) *Operator demand for shaft work.* Operator demand is defined in §1065.1001. Command the operator demand and the dynamometer(s) to follow a prescribed duty cycle with set points for engine speed and torque as specified in §1065.512. Refer to the standard-setting part to determine the specifications for your duty cycle(s). Use a mechanical or electronic input to control operator demand such that the engine is able to meet the validation criteria in §1065.514 over each applicable duty cycle. Record feedback values for engine speed and torque as specified in §1065.512. Using good engineering judgment, you may improve control of operator demand by altering on-engine speed and torque controls. However, if these changes result in unrepresentative testing, you must notify us and recommend other test procedures under §1065.10(c)(1).

(f) *Other engine inputs.* If your electronic control module requires specific input signals that are not available during dynamometer testing, such as vehicle speed or transmission signals, you may simulate the signals using good engineering judgment. Keep records that describe what signals you simulate and explain why these signals are necessary.

6

Engine Idle Reduction

INTRODUCTION

Engine idle reduction was discussed in the National Research Council (NRC) Phase 2 report (NRC, 2012); this report provides a brief update. Engine idling, in sleeper tractors alone, uses 2 billion gallons of diesel fuel according to a recent study (NACFE, 2014). The engine is idled (1) while waiting in queues at weigh stations, toll booths, ports and depots, (2) to maintain temperature in the cab for the comfort of the driver both day and night, (3) to power electrical appliances such as refrigerators and microwave ovens, (4) to maintain the charge level of batteries, and (5) to maintain the temperature of the engine oil and fuel during cold weather. Unnecessary idling leads to increased fuel consumption as well as emissions of carbon dioxide (CO_2), criteria pollutants, and noise. Reducing idling and associated fuel consumption can be avoided with idle reduction technologies, attention to policies affecting freight efficiency, and changes in driver behavior, e.g., turning off the engine when not needed. Over a decade ago, it was common for trucking companies to report idle time in excess of 50 percent. Fleets with good operations would often report 35 percent idle time, and a benchmark number was below 20 percent. Today, it is more common for these same fleets to report worst case numbers around 35 percent, and averages near 20 percent, with benchmark numbers suggesting as low as 5 percent (NACFE, 2013, 2014). Unfortunately, there is no specific and agreed-on method for determining idle time. The electronic control modules (ECMs) on the engines are the source of this information for fleets. How an individual engine manufacturer decides to calculate idle time varies. Some may exclude time waiting in traffic, while others do not. Some may exclude initial idle at start before the vehicle moves. Time limits are often used in the ECMs to differentiate between idle and nonidle operation of the engine.

California remains a leader in regulations related to idling. As noted by the California Air Resources Board (CARB), "pursuant to state regulation, operators of diesel-fueled trucks with a gross vehicle weight rating greater than 10,000 lb are not to idle for more than 5 minutes when stopped within California's borders. As of January 1, 2008, this restriction also applies to sleeper berth trucks. Consequently, many operators are now required to use some form of idle reduction technology to provide cab comfort services during periods of sleep and rest" (CARB, 2014). Beginning in 2008, idling for more than 5 minutes on most commercial vehicles was allowed, but required a Certified Clean Idle sticker. According to CARB, a Certified Clean Idle label is for vehicles that use an engine that has been certified to an optional NO_x idling emission standard of 30 g/hr.[1] Additional information about California's commercial vehicle idling regulations and idle reduction technologies are given by the California Environmental Protection Agency's CARB (2013, 2014).

The clean idle engine reduces nitrogen oxide emissions compared to older engines, but still burns fuel. While these new, clean idle engines burn less fuel while idling than older engines, estimates for the amount of fuel burned while idling vary from about 0.13 gallons per hour for a diesel-powered auxiliary unit, to as much as 1 gallon per hour for the main engine at certain speeds (Curran et al., 2013; Detroit Diesel Corporation, n.d.). Engine manufacturers do not regularly include a specification for idling fuel consumption in their product literature.

Anti-idling regulations around the country have been driven by the need to reduce fuel use, emissions, and noise. A patchwork of regulations at the state and municipality level has been created. The National Idling Reduction Network News website run by the U.S. Department of Energy (DOE) is updated monthly with information on idling; Figure 6-1 is taken from the August 2014 issue. It shows the changes in coverage of idling regulations throughout the 50 states. Note that there are more areas with "jurisdictional" regulations than with statewide regulations.

[1] See http://www.arb.ca.gov/enf/advs/adv376.pdf.

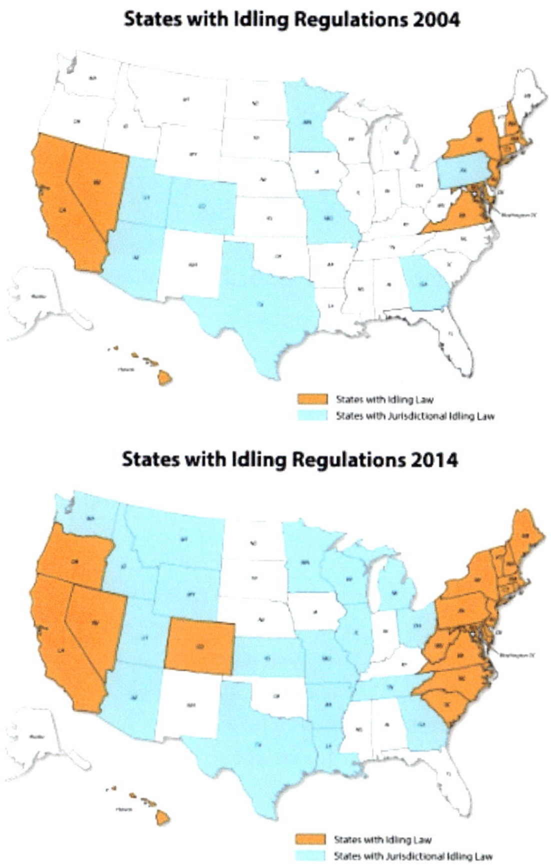

FIGURE 6-1 Extent of truck idling regulations in the United States, 2004 and 2014. In 2014 there were idling regulations in 20 states and the District of Columbia, compared to 10 states and the District in 2004. There are now jurisdictional laws in 18 states compared to 8 in 2004. SOURCE: DOE (2014).

The North American Council on Freight Efficiency (NACFE) in conjunction with the Carbon War Room published a report on idle reduction technologies in 2014 (NACFE, 2014); this document provides an update on currently available products and technologies.

21CTP IDLE REDUCTION GOALS

The current stated goals of the Partnership related to idle reduction were provided in a document in November 2014 (DOE, 2014). Compared to some of the other goals of the Partnership, these goals are general in nature rather than specific and time limited. The goals and some activities and accomplishments have been identified and are discussed below:

(1) *Promote the incorporation of idle reduction (IR) equipment on new trucks as fuel-saving devices, just as they are so identified by the DOE SuperTruck Initiative.* The four SuperTruck teams have all decided to use battery-powered auxiliary power units (APUs) for the energy needs during the rest portions of the 24-hr cycle of the program. The Partnership suggests it would be worthwhile to conduct a comparison of performance and lifetime costs of battery-powered APUs to those of diesel-fueled APUs and to publish the results in the IdleBox[2] web tool run by Argonne National Laboratory.

(2) *Establish a nationwide multimode IR education program.* Argonne National Laboratory created IdleBox, an electronic resource of idling reduction materials. It includes a calculator that fleets can use to assess idle reduction needs. IdleBase is a component of IdleBox that provides a compendium of every state's idling reduction laws. ANL also publishes the online monthly *National Idling Reduction Network News* for DOE, which contains important information on funding sources, changes in regulations, and updates to weight exemptions for APUs.

(3) *Work with OEM truck manufacturers to obtain data on the number of new trucks being ordered with IR options.* Because many vehicle original equipment manfacturers (OEMs) consider information on sales data to be confidential and proprietary, the Partnership was unable to get complete data for this goal. Figure 6-2 shows the data reported by some manufacturers. The Auto Engine Start/Stop item is a feature available that automatically starts and runs the engine for a limited amount of time if the cab temperature drops, the engine oil temperature drops, or the battery charge, as measured by voltage, gets low.

(4) *Conduct a fleet survey to gather data on the amount of in-use idling hours that are accumulated by each type of heavy-duty vehicle.* This goal has not been met for a combination of reasons such as lack of funding and need for Office of Management and Budget (OMB) approval for surveys involving more than 10 people. Alternatives based on information from the National Renewable Energy Laboratory's (NREL's) Fleet DNA project and a revival of the Vehicle Inventory and Use Survey (VIUS) have been discussed.

(5) *Analyze data from the EPA SmartWay Transport Partnership to measure fuel savings and emissions reductions associated with the various types of IR equipment available.* No information was made available with which to evaluate this goal.

(6) *Develop improved IR systems to minimize fuel required, cost, and weight to meet hotel functions in sleeper cabs.* The SuperTruck program; the CoolCab Thermal Load project; the CoolCalc heating, ventilation, and air conditioning (HVAC) model; and STEP

[2] See IdleBox Toolkit for Idle-Reduction Projects at http://www1.eere.energy.gov/cleancities/toolbox/idlebox.html. Accessed February 18, 2015.

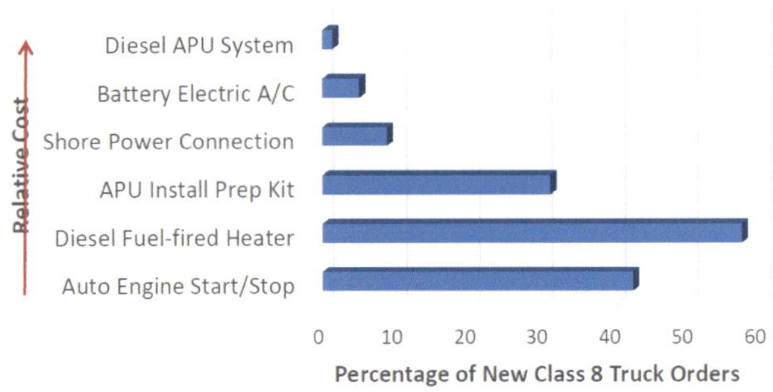

FIGURE 6-2 Typical order rate for idle reduction devices in new trucks. SOURCE: DOE (2014).

are activities that support this goal.[3] Each of these activities is briefly discussed below.

PROJECTS AND ACTIVITIES

The 21st Century Truck Partnership (21CTP) has difficulty identifying specific projects and funding devoted to all idle reduction activities and projects. Idle reduction activities are found in the SuperTruck program, the Clean Cities program, and elsewhere. In a presentation to the committee on September 3, 2014, "Update on Idling Reduction Activities," G. Keller identified four specific projects (CoolCab, CoolCalc, the ShorePower Truck Electrification Project [STEP], and SuperTruck) as being R&D projects in this area. Funding information was reported for the first two projects, CoolCab and CoolCalc, as $575,000 in 2012, $1,200,000 in 2013, and $500,000 in 2014. The January 2015 issue of the *National Idling Reduction Network News* reports up to $211 million having been spent on idle reduction activities in the "last decade" by the Partnership organizations, the private sector, and the states.

All four teams for the SuperTruck program (Daimler, Volvo, Cummins-Peterbilt, and Navistar) have chosen battery-powered APUs (see Chapter 8). For the Cummins-Peterbilt team, the NRC Phase 2 report recommended abandoning the solid oxide fuel cell (SOFC) approach for the APU (NRC, 2012). The team and the Partnership continued for a period of time with the SOFC approach but eventually stopped that effort because the SOFC approach was found to be heavier, more costly, and to take longer to warm up than specified in the goal. Currently it is using a battery-powered APU with a 13.2 kWh capacity lithium-ion pack and electrically controlled HVAC. The pack can be recharged in 6 hours of driving. Over the 24-hr duty cycle of the program, a 7 percent savings in fuel has been reported compared to the baseline chosen for the SuperTruck program.[4] The Daimler team is using a hybrid drivetrain configuration that includes a lithium-ion battery pack and electrically controlled HVAC. To reduce the need to cool the cab, it is using solar reflective paint, similar to the CoolCab project. They are reporting a 3 percent fuel savings compared to the baseline chosen for the SuperTruck program.[5] Volvo Trucks is also using a battery-powered APU and electrically controlled HVAC. It is in the process of installing solar panels to assist in charging.[6] However, during a visit to Volvo, the committee learned that the solar panels would only be able to support the electrical load of a fan.

The CoolCab and CoolCalc projects at NREL are normally classified as thermal management projects by DOE. They have investigated advanced insulation technologies to reduce the load, the ability of different paints to better reflect sunlight, and techniques to cool the occupant rather than the whole cab. Reducing these thermal loads increases the potential for using battery-powered APUs rather than idling the engine. The energy balance for recharging the batteries while the vehicle is operating has not been reported.

An analytical HVAC system model and test methods called CoolCalc will be useful to further these investigations. The projects were expected to have been completed in 2015.

STEP, also referred to as Interstate Grid Electrification project ARRA VTO 70, is a separate project supported by American Recovery and Reinvestment Act (ARRA) funds. Fifty truck stops have been outfitted with electrical power outlets only, sufficient to support 1,252 vehicles. The build-out of the system took longer than anticipated owing to varying municipal regulations at the different sites as well as prolonged negotiations with some truck stop owners. The Cascade Sierra Solutions project was replaced by STEP in 2014. Data collection is under way and is reported by NREL. Utilization of the pedestals and electrical connections was reported as less than 20 percent in 2014 status reports. STEP provides only electrical power and does not provide for as many amenities for the driver as might be desired. For instance, IdleAir is a supplier providing HVAC externally, internet connectivity, satellite television, and 120-V electricity. Recently, Con-Way Truckload announced a dedicated IdleAir facility in its Laredo, Texas, facility (Owens and Bachman, 2014). A final report is due in 2015.

RESPONSE TO RECOMMENDATIONS FROM THE NRC PHASE 2 REPORT

NRC Phase 2 Recommendation 6-1. DOE, EPA, and DOT should develop a consolidated list of the funding provided for the idle reduction projects, review the effectiveness of these projects, and formulate a coordinated and consistent plan to encourage the adoption of idle reduction technologies to meet the goal of reducing fuel use and emissions produced by idling engines by at least two-thirds by 2017. The EPA and DOT

[3] G. Keller, ANL. "Update on Idling Reduction Activities," presentation to the committee, September 4, 2014, and D. Anderson, "Vehicle Systems Simulation and Testing," presentation to the committee on September 3, 2014.

[4] Ibid.

[5] Ibid.

[6] Ibid.

should work to find incentives for states to promulgate uniform anti-idling regulations.

21CTP Response: Presently, the National Idling Reduction Network News publication reports the various sources for funding idling reduction programs culled from press releases. We agree that a consolidated listing of these projects would be useful. The recommendation calls for a more coordinated effort between the DOE, EPA, and DOT to maintain the momentum begun in the application of idling reduction (IR) technologies, and to ensure minimal overlap of these programs across agencies. Establishing a more structured approach to the introduction of IR devices would be conducive to introducing objective measures to monitor the effectiveness of these various programs. Further, we agree that achieving nationwide uniformity of anti-idling regulations needs to be accomplished soon, and that the EPA and DOT could be instrumental in developing incentives to states to pass such rules.

Committee Comment on Response to 6-1

The Partnership response, while supportive of the Phase 2 Recommendation 6-1, does not suggest action will be taken in this area. The Partnership was asked for and did provide some input on the idle reduction report of NACFE. In consideration of their goals in this area, more work should have been done apart from the SuperTruck program to accomplish idle reduction for current vehicles now in the field. Going forward, with the completion of the SuperTruck program, the Partnership will need to readdress its activities in this area.

***

NRC Phase 2 Recommendation 6-2. The DOE should conduct a study that includes wide ranges of truck models, ages, and fleets to determine payback periods for the range of commercially available add-on idle reduction systems. The DOE should continue to encourage the deployment of add-on idle reduction systems through communications to manufacturers and end users.

21CTP Response: The 21CTP agrees with NAS that it would be valuable for DOE and EPA to fund a comprehensive study to verify the performance and payback claims of add-on idle reduction systems across a variety of popular trucks and climate regions. Such a study would be extremely valuable to the trucking community in helping to identify the most cost-effective add-on systems to invest in for their particular applications. The DOE could share these study results with quarterly updates as an addendum to the National Idling Reduction Network News publication.

Committee Comment on Response to 6-2

While the Partnership agreed with Phase 2 Recommendation 6-2, there is no evidence that anything has been done that addresses this recommendation. The number of options available for anti-idle have increased, but their effectiveness is still a question. The report from the North American Council for Freight Efficiency provided a payback calculator, but more work among the Partnership, DOE, EPA and NACFE is needed to validate or refine this model.

***

NRC Phase 2 Recommendation 6-3. The DOE should reassess the viability of the Solid Oxide Fuel Cell (SOFC) APU, particularly for application to the SuperTruck program.

21CTP Response: The information presented during the 21CTP NAS review was based upon one of our initial A-Level prototype units. Since then, we have made significant progress and are now assembling our B-Level prototype units. These units should be capable of demonstrating the targeted goal of 35% efficiency and output of 3kW. On the SOFC stack, Delphi has completed more than 10,000 hours of durability testing. Additionally, we have accumulated thousands of hours of on-truck, real-work application data. We are scheduled to deliver a B-Level unit during Q1 '12 to a national fleet for use on one of their regular in-service long haul trucks. Currently, our start-up time is ~2 hours. The 5-hour example reported on represented a given demonstration. Our goal is to be at operating temperature in under 1 hr. Current costs reflect laboratory built prototype units. Delphi is investing in production intent tooling to drive down overall unit cost. Funding to date has allowed Delphi, as well as other fuel cell developers, to move their products from concept design to real-world demonstrations. Congress has recently reinstated funding for SECA and other fuel cell programs. Delphi will use the re-funded SECA program to further improve the power output and durability of its SOFC stack.

Committee Comment on Response to 6-3

The Partnership rejected the Phase 2 Recommendation 6-3. Subsequently, the SOFC was eliminated from the plans of the SuperTruck teams (both Daimler Trucks North America and Cummins-Peterbilt explored this). At the International Automobile Association (IAA) show in September 2014, a diesel-fueled SOFC APU was on display; Eberspaecher is planning commercial introduction of SOFC APUs in Class 8 trucks in 2016-2017 (Eberspaecher, 2014). Activities for fuel cells continue at AVL and Volvo (Rechberger et al., 2013). Based on developments in Europe, there is a need to reevaluate the future of fuel cells for idle reduction and other auxiliary loads on commercial vehicles.

***

NRC Phase 2 Recommendation 6-4. The 21CTP should review and potentially revise its idle reduction plans and goals in view of the fact that the proposed 2017 fuel efficiency standards provide an incentive for the adoption of idle reduction technologies as a means for achieving these standards for Class 8 long-haul trucks with sleeper cabs.

21CTP Response: The 21CTP agrees that the EPA's rulemaking to establish fuel efficiency standards for heavy-duty truck fleets provides an incentive to look beyond Class 8 long-haul trucks with sleeper cabs to other types of trucks for additional opportunities to apply idle reduction technologies. We feel that a substantial improvement to the idle reduction goal would include support to establish a program to address the fuel wasted in work day idling of all types of vocational trucks.

Committee Comment on Response to 6-4

The Partnership response indicates agreement with the Phase 2 Recommendation 6-4, but no action has been taken in this area. As the Partnership looks beyond the SuperTruck program, the value of start/stop, battery-operated or fuel-cell operated booms on electric utility trucks, and options for heating and cooling without running the primary engine needs to be investigated.

***

NRC Phase 2 Recommendation 6-5. The 21CTP should revise its new idle reduction goals to include metrics, funding, and timing for the overall goal of reducing fuel use and emissions produced by idling engines. The associated "action items" should be supportive of these goals.

21CTP Response: The 21CTP agrees with the NAS recommendation that the inclusion of a progressive and measurable program for idle reduction goals development is needed along with the year-to-year funding necessary to develop data to enable such an approach.

Committee Comment on Response to 6-5

The response to the Phase 2 Recommendation 6-5 indicates agreement in this area but based on the reference to needed funding it would seem to lack signs of commitment.

FINDINGS AND RECOMMENDATIONS

Finding 6-1. Improvements have been made in reducing idle time for long haul trucks over the last 10 years. The growth of fleet management systems and efforts by the Partnership, DOE, EPA, and the California Air Resources Board (CARB) have helped fleets understand the impact and the methods available to reduce idle time. No consistent way to measure and track idle time has been found, since algorithms are independently determined by the developers of the software for engine controls and fleet management systems.

Finding 6-2. The Partnership has focused much of its effort on the SuperTruck program and made progress in highlighting the impact of idling by including a 24-hr cycle in its fuel use evaluation. The stated goals for data acquisition and data analysis have not been met.

Recommendation 6-1. The Partnership, in collaboration with EPA, the National Highway Traffic Safety Administration (NHTSA), and CARB should review the North American Council for Freight Efficiency (NACFE) payback calculator and establish a consistent way to measure and track idle time for both over-the-road and vocational vehicles. As a follow-on, the Partnership should run a program for field data acquisition and analysis, leveraging the resources of NACFE and its fleets.

Recommendation 6-2. As the Partnership looks beyond the SuperTruck program, the value of start/stop systems, battery-operated or fuel-cell operated auxiliary loads, and options for heating and cooling without running the primary engine should be investigated.

Recommendation 6-3. The Partnership should establish goals, specific plans, and funding to reduce the nation's consumption of fuel for idle by over 50 percent by 2025. The baseline should be from the estimate the Partnership generates for total 2016 idle fuel usage based on information to be acquired from the field using DOE's Fleet DNA project and working with fleets, industry associations, and vehicle OEMs.

REFERENCES

CARB (California Environmental Protection Agency Air Resources Board). 2004. Airborne Toxic Control Measure to Limit Diesel-Fueled Commercial Motor Vehicle Idling. http://www.arb.ca.gov/msprog/truck-idling/2485.pdf.

CARB. 2013. Facts about California's Commercial Vehicle Idling Regulations. http://www.arb.ca.gov/msprog/truck-idling/factsheet.pdf.

CARB. 2014. Idle Reduction Technologies for Sleeper Berth Trucks. http://www.arb.ca.gov/msprog/cabcomfort/cabcomfort.htm.

Curran, S., K. Settles, and K. Keel-Blackmon. 2013. ORNL Operations Best Practices Guide: Idle Reduction. Oak Ridge National Laboratory Sustainable Campus Initiative. https://sustainability-ornl.org/documents/ORNL%20Idle%20Reduction%20Guide.pdf.

Detroit Diesel Corporation. n.d. Fuel Economy: Creating Optimal Performance. http://www.demanddetroit.com/pdf/performance/5927-3_DD_FuelEcon_FINAL.pdf.

DOE (U.S. Department of Energy). n.d. Vehicle Technologies Office: National Idling Reduction Network News. http://energy.gov/eere/vehicles/vehicle-technologies-office-national-idling-reduction-network-news.

DOE. 2014. 21CTP Idle Reduction Goal Status. Submitted to the NRC Committee for Review of the 21st Century Truck Partnership, Phase 3, November 17.

Eberspaecher. 2014. Eberspaecher at the IAA Commercial Vehicles 014: Diesel fuel-cell APU revolutionized energy management in the truck. Press release, July 17. https://www.eberspaecher.com/index.php?eID=download&file=fileadmin/data/corporatesite/presse/pressemitteilungen/2014_presse/pressemitteilungen/PR_IAA_CV_2014_preliminary_report.pdf.

NACFE (North American Council for Freight Efficiency). 2013. Anti Idling Technology Overview. http://nacfe.org/wp-content/uploads/2013/08/Trucking-Efficiency-Anti-Idling-Tech-Overview-072413.pdf.

NACFE. 2014. Confidence Report on Idle-Reduction Solutions. http://www.carbonwarroom.com/sites/default/files/reports/Idle-Reduction_Confidence_Report.pdf.

National Research Council (NRC). 2012. *Review of the 21st Century Truck Partnership, Second Report.* Washington, D.C.: The National Academies Press.

Owens, K., and K. Bachman. 2014. Con-way Truckload Opens Idle Air Facility at Laredo, Texas, Terminal. Con-way Press Release. http://www.con-way.com/en/about_con_way/newsroom/press_releases/Dec_2014/2014_dec_09.

Rechberger, J., A. Kaupert, C.G. Greisen, J. Hagerskans, and L. Blum. 2013. Fuel-cell APUs show promise for heavy-duty truck anti-idling. *SAE Off-Highway Engineering Online.* http://articles.sae.org/12659/.

7

Safety

INTRODUCTION

> Safety is a central element in the 21CTP vision—and truck manufacturers have stated on numerous occasions that safety is their number one priority. The public has also placed a high premium on safety with concern about driver distraction, driver fatigue, truck aggressivity, and risks associated with exposure to heavy trucks. While truck safety statistics show steady improvement, crashes involving heavy trucks still account for about one out of ten motor vehicle fatalities in the United States. (21CTP, 2013)

Although the 21st Century Truck Partnership (21CTP) focuses on the development of technologies to reduce fuel consumption, an important consideration in the development of any vehicle technology is to maintain or improve safety for the driver and other motorists. 21CTP has recognized the relationship between vehicle safety and the introduction of a new vehicle technology. The Partnership also states that it supports the development and early adoption of safety technology with the objective of

> [Promoting] the development and early adoption of technologies and processes to improve truck safety, resulting in the reduction of fatalities and injuries in truck-involved crashes, thus enabling benefits related to congestion mitigation, emission reduction, reduced fuel consumption, and improved productivity" (21CTP, 2013).

Truck and bus manufacturers, industry suppliers, and federal agencies that participate in 21CTP are working to ensure that as fuel consumption improvements are pursued through advances in technology, safety remains uncompromised.

A priority of the Department of Transportation (DOT) is to pursue solutions that help prevent crashes altogether, through collision warning systems, automatic vehicle control intervention technologies, and/or enhanced vehicle inspection and enforcement systems that help to identify and correct mechanical or operational conditions that could compromise safety.

> Implementation of such technologies and systems is expected to help substantially in reducing fatalities and injuries, and will also have secondary benefits of reducing congestion and idling—thereby reducing fuel consumption and improving overall productivity of the trucking industry (21CTP, 2013).

This safety chapter focuses on a review of the safety-related responses of the Partnership to the NRC Phase 2 report, an assessment of the Partnership's progress toward the safety goals, and a discussion of truck safety activities undertaken by the DOT. These activities include the following:

- Summary of federal government (primarily DOT) activities related to truck safety
- Safety technology, including brakes, roll and electronic stability control, and forward collision avoidance technology and cab crashworthiness.

The 21CTP goals related to safety are as follows:

(1) Ensure that advancements in truck design and technology to improve fuel efficiency do not have any negative impacts on safety and
(2) Ensure that efforts to improve safety do not reduce efficiency—and, where possible, actually contribute to improvements in overall motor carrier industry system efficiency.

Goal 1 is addressed in Chapter 8 in the discussion of technologies implemented in the SuperTruck projects. A general summary of truck safety was presented in the previous NRC Phase 2 report (NRC, 2012). To progress with the assessment, this review will focus more specifically on safety matters as it reviews and evaluates federal safety activities and safety technologies considered by 21CTP.

SUMMARY OF FEDERAL GOVERNMENT ACTIVITIES RELATED TO TRUCK SAFETY

The DOT regulates heavy-duty vehicle safety in the United States under three separate administrations. The National Highway Transportation Safety Administration (NHTSA) has responsibility for new vehicle safety requirements focused at the level of the original equipment manufacturers (OEM). The Federal Highway Administration (FHWA) governs vehicle size and weight, including gross vehicle weight (GVW), axle weight, and vehicle length, width, and height, which are key vehicle design parameters. The Federal Motor Carrier Safety Administration (FMCSA) has responsibility for vehicle and fleet operating regulations, including vehicle operator matters such as hours of service.

The following is a brief synopsis of federal DOT activities related to safety and their distribution across these three agencies.

NHTSA has examined the effectiveness of systems such as the Electronic Stability Control system and the Roll Stability Control system, Forward Collision Avoidance and Mitigation (F-CAM), and Lane Departure Warning (LDW).[1] In a presentation to the committee, NHTSA outlined its efforts to improve truck crashworthiness, rear-underride guard improvements, and truck cab crashworthiness in particular. Pilot studies and research related to vehicle-to-vehicle (V2V) communications are also ongoing.

The FHWA is conducting the Map-21 Truck Size and Weight Study, which has a substantial safety component and also has very significant implications for specific fuel consumption and emissions by virtue of the increased vehicle cargo capacity (Hughes Raymand, 2014). In her presentation to the committee, Ms. Hughes Raymand[2] outlined the Smart Roadside Initiative (SRI), focused on improving the efficiency and safety of U.S. roadways by providing for the exchange of important safety and operational information among the users and operators of the system, including commercial vehicles and roadside and central office systems (RITA, 2014).

As part of the program, DOT is overseeing the development of several prototypes that were scheduled to have been deployed in early 2015 at multiple weigh stations and other strategic points along commercial vehicle routes across the country. These prototypes will demonstrate the integration of multiple technologies that together will facilitate the following:

- Exchange of driver, carrier, vehicle identification, and status information between commercial vehicles and commercial vehicle management systems at highway speeds.
- Integration of roadside applications with external information systems to seamlessly share information on commercial vehicle safety history, inspection status, and credential status.
- Determination of truck weight by weigh-in-motion technology, which uses dynamic weigh scales imbedded in the traffic lane that measure vehicle axle weights as it drives at highway speeds.
- Roadside access for law enforcement to information that supports the identification of the driver, vehicle, and motor carrier.

FHWA has funded recent efforts that use technology to improve truck parking in the United States. The lack of suitable parking for trucks has safety implications—for example, when truck drivers have exhausted their hours of service and must still park their vehicles, if there is no available space at parking facilities they sometimes park on the side of a road, which presents a significant safety risk to the truck and motoring public. The technology being developed uses electronic systems to inform the driver of parking space availability at designated sites so the driver can confidently navigate to a parking facility that still has space.

The FMCSA has supported the development of a web-based course to train commercial vehicle inspectors on how to detect leaks from natural gas and propane trucks and buses and another web-based course to familiarize commercial vehicle inspectors with the safety aspects of electric-drive commercial vehicles. Updates have been made to the Federal Motor Carrier Safety Regulations to address electric-drive commercial vehicles.

The above discussion of activities summarizes the current government-sponsored work being performed having to do with truck safety. Table 7-1 provides a summary of safety-related project expenditures funded by DOT.

Override and Underride Issues

A clinical review of the Large Truck Crash Causation (LTCC) database was undertaken as an exploratory evaluation of front override and side underride in serious truck crashes for NHTSA (Blower and Woodrooffe, 2012). The goals were to determine the incidence of front override and side underride (i.e., whether there is a significant safety problem) and to develop an understanding of the data elements needed to determine the best way to address the problem.

Overall, in front and side impact crashes, some underride was identified in 53.9 percent of the crashes, and passenger compartment intrusion (PCI) was coded in 44.2 percent. The rate of override/underride in side impacts is lower than the rate when the front of the truck is involved. There was some override or underride in 72.0 percent of front impacts, compared with 53.9 percent when the truck side is struck. Rates

[1] A. Svenson, NHTSA, "Heavy Vehicle Safety Research," Presentation to the committee on May 14, 2014.

[2] C. Hughes Rayman, "Supporting Safe and Efficient Goods Movement on the Nation's Highways: An Overview of Research and Data Programs," Presentation to the committee on November 18, 2014.

TABLE 7-1 Summary of DOT Expenditures on Safety-Related Projects by Fiscal Year (dollars)

Project	Sponsor	2012 Funding	2013 Funding	2014 Funding	Total	Recipient
Safety Systems	NHTSA	500,000	600,000	500,000	600,000	Various
Crash Avoidance	NHTSA	700,000	1,100,000	1,000,000	2,800,000	Various
System for Automatically Maintaining Truck Pressure in a Commercial Truck Tire	DOE	571,189	713,810	161,535	1,446,534	Goodyear and The Rubber Company
Update FMCSA regulations to address electric-drive commercial vehicles	FMCSA			150,000	150,000	Various
Web-based course to train commercial vehicle inspectors on how to detect leaks from natural gas and propane trucks and buses	FMCSA				150,000	Various
Web-based course to familiarize commercial vehicle inspectors with safety aspects of electric drive commercial vehicles	FMCSA				150,000	Various
Intelligent Transportation Systems	NHTSA	2,500,000	2,000,000	3,700,000	8,200,000	Various
Heavy Vehicles	NHTSA	2,100,000	2,400,000	2,100,000	6,600,000	Various
Wireless Roadside Inspection	FMCSA		3,000,000		3,000,000	Various

of light vehicle PCI are also lower in side impact crashes, with PCI identified in 65.4 percent of front impacts but only 48.5 percent of side impacts. Underride and PCI could not be determined in 7.9 percent and 7.3 percent of front and side impacts, respectively.

Impacts to truck fronts and to the sides of trailers tended to result in override or underride at higher rates than impacts to the sides of truck cabs or to straight truck cargo bodies. When the truck front was involved, there was identifiable override in 72.0 percent of the impacts. Similarly, impacts on trailer sides resulted in underride in 68.9 percent of the crashes. Side impacts to truck or tractor cabs resulted in underride in 43.5 percent of cases, and side impacts to the cargo body area of straight trucks resulted in underride in about 52.6 percent of such crashes.

In frontal impacts, truck bumper height appears to have a linear relationship with the probability of override. Override occurred in 87.3 percent of frontal impacts where the bottom of the front bumper was above the axle, 72.4 percent when the bumper was at the axle, and only 57.7 percent when the bottom of the bumper was below the axle.

Front axle setback did not appear to affect the incidence of override, but there did appear to be some effect on PCI, such that there was somewhat more PCI identified for setback front axles than for axles set forward. In side impacts, the important elements were cargo bed height and whether the striking vehicle hit the axles. Only low cargo beds were associated with lower probabilities of underride (about 30.0 percent). Standard height (about dock height, or 48-50 in.) and high cargo beds had statistically indistinguishable rates of underride.

Light vehicles hit the truck's axles in 73.9 percent of side impacts, and overall light vehicles that hit the truck's axles actually underrode the truck at higher rates than light vehicles that did not. However, it was found that the geometry of the crash had a significant effect on whether striking the truck's axles would prevent underride. In crashes in which the light vehicle was going in the same direction as the truck and sideswiped it, and in crashes where the light vehicle struck the truck at about a 90 degree angle, hitting the truck's axles prevented underride in about 35 percent of cases. But when the light vehicle was going in the opposite direction as the truck and moved into it at a shallow angle, hitting the axles prevented underride in only about 20.7 percent of crashes.

The review of LTCC cases produced evidence that front override and side underride are significant problems in serious crashes between heavy trucks and light vehicles. Front override and side underride were found in most of the crashes examined. Preliminary estimates from this review are that override occurs in almost three-quarters of crashes involving the front of the truck and in over half of the crashes when the sides of the trucks were struck (Blower and Woodrooffe, 2012).

SAFETY TECHNOLOGIES BEING CONSIDERED BY THE PARTNERSHIP

Several potential countermeasures to reduce deaths and injuries related to truck crashes have been identified by the 21CTP. These include various crash avoidance technologies as well as crashworthiness initiatives that improve occupant protection in the event of an incident. The 21CTP has identified several areas of accident avoidance and these are excerpted below.

> Crash avoidance initiatives fall into six primary categories: (1) improved braking performance including roll and stability control systems; (2) collision mitigation technologies that directly intervene to warn drivers and/or take control of the vehicle in collision imminent situations; (3) diagnostic technologies that improve the ability to maintain safety-critical systems; (4) human factors research to improve the driver-vehicle interface, identify sources of distraction and enhance driver performance through a variety of technology and operational strategies; (5) SmartRoadside; a program to improve how state, local and federal officials interact with commercial vehicle operators and drivers at the "roadside" to reduce down-time associated with vehicle inspections, port operations, border crossings and other venues. Components of this program include wireless roadside inspections, size and weight compliance, and other state-based programs; and (6) cross-cutting research related to dedicated short-range communications (DSRC)-based systems—a set of technologies and applications focused on establishing standardized wireless communications between vehicles to support safety, mobility and efficiency within the motor carrier industry. (21CTP, 2013)

Of these primary categories, items 1 through 4 have safety implications at the vehicle level and will be discussed in this section of the report.

Crash Avoidance Technologies

Improved Braking Performance

Brake performance remains a long-standing challenge for heavy vehicles. New crash avoidance systems rely on well-adjusted brakes to function properly, and this requirement strongly favors disc brake technology. The Partnership notes that NHTSA Final Rule FMVSS No. 121 requires a 30 percent reduction in stopping distance for new commercial tractors. The Partnership's Roadmap identifies disc brakes and more powerful drum brakes as the most likely strategy to meet new standard. The superior performance of disc brakes and their ability to remain in adjustment suggest that they would be a better choice than drum brakes. Disc brakes have better heat rejection characteristics than drum brakes, which is important given that aerodynamic drag is greatly reduced in fuel-efficient vehicles, thus requiring brakes to extract more energy. The current FMVSS No. 121 does not address "brake out of adjustment," which is arguably the most critical brake issue facing the trucking industry.

Therefore, the 21CTP has stated that "increased research and analysis on the use of disc brake systems for tractor trailers is supported by the 21CTP." According to the Roadmap, disc brake systems offer increased reliability, shorter stopping distance, and opportunities for mass reduction since they are lighter and less expensive to maintain than drum brake systems.

Air disc brakes offer a proven alternative to drum brake designs. When compared to drum brakes, air disc brakes have a number of advantages, including these:

- *No exaggeration of friction coefficient differences.* This results in improved side-to-side consistency between left and right brakes.
- *Reduced fade.* Consistent contact between the friction surfaces remains, even with brake disc warm-up and radial expansion.
- *High thermal load capacity.* Heat dissipation is efficient for internally vented brake discs. As such, it is possible to maintain high braking performance, even under demanding conditions.
- *Minimal and consistent hysteresis.* This is due to the high efficiency of the actuating mechanism.
- *Servicing ease when changing brake pads.* Compared to drum brakes, disc brakes require only a fraction of the service time.

Unfortunately, the air disc brake system market penetration rate in the United States remains low, so that mechanisms to encourage industry acceptance of this foundation brake technology may be required.

21CTP considers electronically controlled braking systems (EBS) to be important technologies. These systems replace the pneumatic brake activation signal with an electronic activation signal. The main benefit of this system is reduced lag time between operator execution and brake response time, which reduces stopping distance. EBS offers more precise brake control and will provide the platform for the advanced safety systems of the future. Furthermore, the elimination of signal lag ensures that every wheel-end brakes at the same time, which improves vehicle control.

21CTP considers improved brake systems as an enabler for other safety technologies. Having well-adjusted brakes with reliable braking performance is essential for the operation of many of the advanced collision avoidance technologies. In a presentation to this committee, NHTSA noted that out-of-service brake problems are detected in 20-30 percent of trucks inspected.[3] Most of these problems are associated with out-of-adjustment brakes and of these, virtually all are related to antiquated drum brake design, which is inherently

[3] L. Loy, "FMCSA Research and Technology," Presentation to the committee on November 18, 2014.

susceptible to adjustment problems but is nonetheless used on most tractor-trailer combinations in North America. As mentioned previously, out-of-adjustment is an important characteristic, since electronic stability control (ESC), roll stability control (RSC), and forward collision avoidance and mitigation (F-CAM) systems all rely on properly functioning brake systems to maximize safety performance. When the brake systems are out of adjustment or compromised, crash avoidance system performance suffers. It should also be noted that in Europe, air disc brakes have experienced high market penetration and out-of-adjustment problems are less common (Marmy, 2015).

Roll Stability Control and Electronic Stability Control

RSC systems are designed to reduce the probability of vehicle rollover in a curve by sensing lateral acceleration and reducing speed when threshold limits are exceeded. ESC provides rollover prevention similar to that provided by RSC, with the added ability to address vehicle loss of control (LOC) due to understeer or oversteer through selective braking at the tractor. The overlapping characteristic of these technologies centers on the ability of ESC to manage LOC scenarios as well as to replicate the functionality of an RSC system for curve-related roll stability cases.

RSC and ESC technologies are able to assess vehicle mass by monitoring engine torque and vehicle acceleration performance on a continuous basis. An onboard algorithm uses this data to set the lateral acceleration threshold and establish mass-related braking strategies for vehicle deceleration during challenging curve maneuvers. The technology has the ability to override driver power commands to the engine and can activate the vehicle retarder/engine brake as well as the foundation brakes. The degree of intervention depends on the amount of lateral acceleration (over-speed in a curve) that the vehicle experiences. RSC and ESC technologies perform almost identically when controlling for excessive speed in a curve with the exception that ESC can apply the foundation brakes (all tractor axle and trailer axle brakes), including the tractor steer axle, while RSC can apply the foundation brakes but not the tractor steer axle.

RSC and ESC are both mature commercially available technologies that cannot be retrofitted. DOT has initiated rulemaking to require ESC to be fitted on new Class 8 trucks. The University of Michigan Transportation Research Institute (UMTRI) conducted a study for NHTSA to quantify performance and estimate the benefits of this technology (Woodrooffe et al., 2009).

Collision Mitigation Technology

The 21CTP identifies advances in collision warning and avoidance systems as an area of research that has a potentially high payback when it comes to improving safety. The Roadmap identifies the following crash avoidance systems for future research: lane departure warning (LDW), forward collision warning (FCW), side object detection (blind spot monitoring, or BSM), lane change/merge (LCM), and rear object detection and collision warning FCW and Mitigation F-CAM systems are defined as forward-looking radar-based systems that combine FCW with automatic collision mitigation braking (CMB) capability. The FCW feature generates visual, audible, and/or haptic warnings for the driver when/if a lead vehicle comes within a predefined distance and closing rate with the subject vehicle (i.e., the F-CAM equipped vehicle). If the driver does not respond to the warning with a braking input, and if the threat continues to worsen, then the F-CAM system applies foundation brakes at a point when the collision is determined to be "imminent" (i.e., not avoidable through an evasive steering or lane change maneuver). Driver warnings and automatic braking actions of current production systems could be effective at helping to mitigate crash severity or to avoid the crash altogether. It should be noted that F-CAM technology is not meant to convey or imply an adaptive cruise control (ACC) feature, even though all commercial vehicles offering F-CAM systems do in fact include ACC capability. F-CAM systems address truck striking rear-end collisions, which are the most common crash type on the divided highway network.

The estimated reduction in fatalities and injuries related to collisions with a truck striking the rear end of another vehicle was computed in Table 7-2 (Woodrooffe et al., 2013).

Woodrooffe et al. (2013) show that the current generation, commercially available technology will reduce fatalities by 24 percent, injuries by 25 percent, and property-damage-only crashes by 9 percent (see Table 7-2). The data also suggest that the second- and third-generation versions of the system will bring substantially greater benefits. The second -generation system is able to detect stationary threat objects in the roadway, typically through the fusion of radar and vision systems. The third generation has more aggressive automated braking deceleration, achieving 0.6 *g*. This is highly relevant

TABLE 7-2 Reduction in Injury Severity by Collision Mitigation Capability for Tractor Semitrailers (F-CAM components) (percent)

Capability	Fatal	Injury	No Injury
F-CAM subsystem contribution			
FCW only	31	27	11
CMB only, 2nd generation	26	32	10
CMB only, 3rd generation	44	42	19
Complete F-CAM system contribution			
Current generation	24	25	9
Second generation	44	47	20
Third generation	57	54	29

NOTE: The benefits assume that all tractor semitrailers operating in the United States were fitted with the technology. Also note: "No injury" means property damage only. SOURCE: Woodrooffe et al. (2013).

to 21CTP as it represents an area where the needed research and development have indicated the potential for substantive safety improvement.

Safety System Diagnostic Technologies

A vehicle with a safety system that is not functioning properly is a safety risk to both its occupant and society generally. DOT is supporting the development of systems that inspect, monitor, and diagnose the vehicle components and technologies that influence vehicle safety. The 21CTP Roadmap identifies tire pressure monitoring systems (TPMS) and brake system out-of-adjustment diagnostics as two distinct areas that will benefit from onboard diagnostics (OBDs).

Tire Pressure Monitoring Systems

The 21CTP's vision for the future truck includes an efficient, accurate, and cost effective tire pressure monitoring system. Tire pressure monitoring systems can improve safety while also reducing operating costs for the vehicle owner. These systems continually measure the air pressure for all tires and relay that information to the operator via a dashboard-mounted OBD interface or transmit the data to a fleet's manager, or both. Having properly inflated tires will reduce blowouts, vehicle sliding in inclement weather, and improve vehicle handling. Properly inflated tires will also reduce fuel consumption directly since a properly inflated tire will provide lower fuel consumption than an improperly inflated tire because of the additional power that is required to move a vehicle with one, or many, underinflated tires (see Chapter 5, section on "Tire Rolling Resistance").

Brake System Sensors and Diagnostics

As previously mentioned, many of the new crash avoidance and mitigation technologies rely on well-adjusted brake systems. DOT is sponsoring research on more reliable and accurate brake system diagnostic systems. It identifies the leading brake diagnostic system as the on-board stroke monitoring system, which is a strong indicator of the state of drum brake adjustment. These systems will enable the driver or the fleet to receive real-time information on the condition of the vehicle's braking system. These systems also increase safety by notifying the operator, or fleet manager, of vehicles with brake systems that are out of adjustment.

Cab Crashworthiness

A "fatal truck crash" is defined as a crash in which someone (an occupant of the truck, an occupant of another vehicle, or a pedestrian) is killed. In the early 2000s, 700 to 800 hundred truck drivers were killed in truck crashes each year. In recent years, the number of truck driver fatalities has decreased, in large part owing to a general reduction in truck fatal crashes. However the proportion of drivers killed in relation to the number of fatal truck crashes has remained between 14 and 16 percent over the years. In 2003 and 2004, 700 truck drivers were fatally injured in crashes, and the number increased substantially in each of the next 3 years. The trend in the number of truck drivers killed began to decline after 2007, possibly due to a reduction in truck travel brought on by the recession. In 2007, a total of 796 truck drivers were killed in 5,049 fatal truck crashes, a 15.8 percent occurrence (Jarossi et al., 2012). In 2008, there were 639 truck drivers killed in 4,352 fatal truck crashes (14.7 percent); in 2009, there were 487 drivers killed in 3,450 fatal truck crashes (14.1 percent); and in 2010, 540 truck drivers were killed in 3,699 fatal crashes (14.6 percent). While the number of truck drivers killed in traffic crashes has fluctuated over the period, the ratio of drivers killed in relation to fatal truck crashes shows little change, indicating that crash safety for drivers is not improving.

It is estimated that 757 truck occupant fatalities occur per year; about 3,000 A-injuries[4] and about 7,700 B-injuries. Most of the fatalities occurred in truck-tractors, with an average of 425 per year. Single unit trucks (SUTs) had an average of 324 fatalities annually.

A recent UMTRI study (Woodrooffe and Blower, 2013) found that rollover and frontal impact were identified as the collision types associated with the most serious driver injuries. Rollover and frontal impact in collisions accounted for 72.7 percent of all tractor-trailer driver fatalities and A-injuries in crashes. Rollover is the dominant crash mode, accounting for 44.5 percent of fatalities, and everyday A-injuries frontal collision events account for 28.2 percent. No other crash event comes close to the share of these two crash types.

In events where the truck rolls over, one in eight truck drivers dies or receives incapacitating injuries. In contrast, in crashes where the truck does not rollover one in 167 drivers die or receive incapacitating injuries.

Rollover events with belted drivers account for 37 percent of all injured truck drivers while unbelted drivers account for 50 percent. Focusing on the risk associated with rollover, one in nine belted drivers die or receive incapacitating injuries while one in three unbelted drivers die or receive incapacitating injuries. Seat belts were shown to be particularly effective at reducing fatalities and incapacitating injuries in rollover events by a factor of three.

Ejection is highly associated with the most severe injuries. Among SUT drivers, almost 39.9 percent of ejected drivers suffered fatal injuries, and almost 24.6 percent were coded with A-injuries. Among tractor-trailer drivers, 25.4 percent

[4] A-injuries: incapacitating, which prevent the injured person from walking, driving or normally continuing the activities he was capable of performing before the injury occurred. B-injuries: nonincapacitating injury other than a fatal injury or an incapacitating injury, which is evident to observers at the scene of the accident in which the injury occurred.

of ejected drivers suffered fatal injuries and an additional 19.0 percent suffered A-injuries. Ejection accounted for 35.0 percent of SUT driver fatalities and 22.6 percent of tractor-trailer driver fatal injuries.

Seat belt use was shown to virtually eliminate complete ejection for both SUT and tractor-trailer drivers (though a small percentage of belted drivers are partially ejected in some crashes). Furthermore, rollover accounts for almost 65 percent of ejected tractor-trailer drivers in fatal crashes.

There are challenges to the acceptance of safety technology in the heavy commercial vehicle industry. While vehicle manufacturers offer safety technology beyond that required by the Federal Motor Vehicle Safety Standards (FMVSS), the commercial uptake for these technologies are for the most part very low. Given that most commercial drivers have no influence on the vehicle purchasing process, including specifying vehicle safety content, this may tend to slow the adoption of safety protection available to heavy vehicle occupants.

Several potential countermeasures have been identified:

- Measures to increase seat belt usage may include the installation of enhanced seat belt warning systems that activate a visual and audible warning when truck drivers and other vehicle occupants fail to use their seat belt.
- Increasing the integrity and robustness of cab structures and the protection of cabs particularly with respect to rollover.
- The installation of side curtain air bags to prevent occupant ejection through the side windows and head trauma.
- Increasing occupant head space during rollover events through installation of automatic pull-down seats.

The regulation of safety content of commercial vehicles has not progressed to the same degree as it has in light-duty vehicles. For example, air bag systems are not mandatory in heavy trucks and to date, only one vehicle manufacturer offers front air bags as standard. No manufacturers offer side curtain air bags, which counteract partial and full driver ejection during rollover events, a major cause of driver injury and death (Woodrooffe and Blower, 2013).

Given the 21CTP goals for improved safety and the interagency cooperation that defines the program, it is not unreasonable to expect that DOT should have a safety related program along the lines of SuperTruck, with its own focus on improving safety for truck drivers involved in accidents. Areas of potential emphasis include improved cab structural integrity and prevention of driver ejection during rollover events.

RESPONSE TO RECOMMENDATIONS FROM THE NRC PHASE 2 REPORT

The following section discusses this committee's evaluation of the Partnership's responses to the recommendations of the NRC Phase 2 report.

NRC Phase 2 Recommendation 7-1. The Partnership should review the wording of its safety goals and consider rewording them so as to unambiguously state that safety will not be compromised in reducing fuel consumption.

21CTP Response: The Partnership will review wording of safety goals to ensure appropriate emphasis is placed on safety—and that safety is not compromised in achieving fuel efficiency goals.

Committee Comment on Response to 7-1

The Partnership agreed with the Phase 2 committee's recommendation that wording of the safety goals should be clarified to emphasize that safety will not be compromised to achieve reductions in fuel consumption. However, this correction of the wording has not occurred to date. Additionally, it appears that the roadmap section on safety was revised in the most recent 2013 publication, but this specific goal was never revised. The Partnership has indicated that it is open to ideas from the NRC committee on how to reword this goal to clarify the intended meaning (21CTP, 2013).

***

NRC Phase 2 Recommendation 7-2. The committee supports the emphasis that the DOT and the 21CTP are giving to crash-avoidance technologies and recommends that crash-avoidance technologies continue to be given high priority and technical support.

21CTP Response: The Partnership agrees with the committee's observations and recommendations.

Committee Comment on Response to 7-2

The Partnership responded favorably to the recommendation for giving crash avoidance technologies a high priority and has made modest gains in this area. Partnership assessments were conducted on truck-related crash-avoidance technologies such as ESC, RSC and F-CAM systems (described in earlier sections). Of the technologies assessed by the Partnership, it is this committee's opinion that the F-CAM systems show the best potential for further development. Disc brake technology can also be viewed as a crash-avoidance technology since it brings about shorter stopping distances with improved thermal capacity compared to conventional braking systems. Implementation of disc brake systems will also reduce the chronic problems of conventional brakes requiring constant readjustment. The Partnership also supports continued research in the following crash-avoidance technologies:

- Lane departure warning (LDW)
- Forward collision warning (FCW)
- Side object detection (blind spot monitoring, BSM) and lane change merge (LCM)
- Rear object detection and collision warning

***

NRC Phase 2 Recommendation 7-3. The DOT should evaluate the conclusions and recommendations of the TRB study Achieving Traffic Safety Goals in the United States: Lessons from Other Nations of highway safety in other nations, and consider the possibility of establishing more aggressive initiatives and goals for highway safety in general. The DOT should also consider establishing more aggressive goals for heavy-duty truck safety.

21CTP Response: DOT will review the TRB study (Achieving Traffic Safety Goals in the United States: Lessons from Other Nations). DOT regularly re-evaluates its safety goals each year, and will take into consideration information from this study, as well as the special circumstances impacting traffic safety in United States.

Committee Comment on Response to 7-3

There is no evidence that this particular recommendation has been addressed at this time.

***

NRC Phase 2 Finding 7-4. Some of the potential safety improvements considered by the committee may have negligible impact on fuel consumption and, in some cases, appear to have positive implications. However, further study of the potential highway safety impact of high productivity vehicles is warranted.

Partnership Response: USDOT will launch a major study of this issue based on direction given in MAP-21; specifically, Section 32801 requires completing a "Comprehensive Truck Size and Weight Limits Study." The scope of this study can be found in the authorizing legislation.

Committee Comment on Response to 7-4

The Map-21 Truck Size and Weight Study is currently under way and attempting to address this question, albeit in a limited manner.

FINDINGS AND RECOMMENDATIONS

Finding 7-1. Many safety technologies could be effectively evaluated and demonstrated in a safety-focused program—for example, a Safety SuperTruck similar to the DOE fuel consumption reduction SuperTruck program.

Recommendation 7-1. DOT should consider implementing a Safety SuperTruck program to develop, integrate, and evaluate safety technologies such as cab structural integrity, side curtain airbags, advanced forward warning and collision mitigating systems to help industry attain a more integrated and complete safety package with a view to generating greater purchaser acceptance of safety technology not mandated by law.

Finding 7-2. Properly performing and well-adjusted braking systems form an essential platform for the crash avoidance technologies being assessed by the 21CTP. In terms of stopping distance, braking control, brake adjustability, and thermal capacity, disc brake systems are superior to drum brakes. Disc brakes provide a better foundation than drum brakes for future technologies dependent on reliable brake performance.

Recommendation 7-2. 21CTP should assess ways to encourage industry to adopt disc brakes and measures should be taken to encourage broad adoption of these superior brake systems.

Finding 7-3. The current generation of commercially available Forward Collision Avoidance and Mitigation systems should reduce fatalities in truck striking rear-end collisions by 24 percent, injuries by 25 percent, and property damage only crashes by 9 percent. Second- and third-generation versions of the system will bring substantially greater benefits. Second-generation systems will be able to detect stationary threat objects in the roadway through the fusion of radar and vision systems, while third-generation systems have more aggressive automated braking deceleration, achieving 0.6 *g*.

Recommendation 7-3. 21CTP should assess future generation Forward Collision Avoidance and Mitigation system development to identify barriers to development and establish incentives to foster commercialization.

REFERENCES

21CTP (21st Century Truck Partnership). 2013. Roadmap and Technical White Papers. U.S. Department of Energy Office of Energy Efficiency and Renewable Energy. https://www1.eere.energy.gov/vehiclesandfuels/pdfs/program/21ctp_roadmap_white_papers_2013.pdf.

Blower, D., and J. Woodrooffe. 2012. Heavy Vehicle Crash Data Collection and Analysis to Characterize Rear and Side Underride and Front Override in Fatal Truck Crashes. UMTRI Report UMTRI-2012-31.

FHWA (Federal Highway Administration). 2014. MAP-21 Comprehensive Truck Size and Weight Limits Study. http://www.ops.fhwa.dot.gov/freight/SW/map21tswstudy/index.htm.

Jarossi, L., D. Hershberger, and J. Woodrooffe. 2012. Trucks Involved in Fatal Accidents: Codebook 2010. Center for National Truck and Bus Statistics, University of Michigan Transportation Research Institute. UMTRI-2012-30, November.

Marmy, J. 2015. International Road Transport Union. Geneva, Switzerland.

NRC (National Research Council). 2012. *Review of the 21st Century Track Partnership: Second Report*. Washington, D.C.: The National Academies Press.

RITA (Research and Innovative Technology Administration). 2014. Smart Roadside. http://www.its.dot.gov/factsheets/pdf/Smart%20Roadside.pdf.

Woodrooffe, J., C. Flannagan, D. Blower, S. Bogard, P.E. Green, and S. Bao. 2013. Automated Control and Brake Strategies for Future Crash Avoidance Systems–Potential Benefits. SAE 13CVA-0024.

Woodrooffe, J., and D. Blower. 2013. Heavy Truck Crashworthiness: Injury Mechanisms and Countermeasures to Improve Occupant Safety. UMTRI Report 2013-41.

Woodrooffe, J., D. Blower, T. Gordon, P.E. Green, B. Liu, and P.F. Sweatman. 2009. Safety Benefits of Stability Control Systems for Tractor-Semitrailers. NHTSA DOT HS 811.

8

SuperTruck Program

INTRODUCTION

The SuperTruck program was designed to provide a full Class 8 tractor-trailer vehicle demonstration of a wide range of technologies, many of which were developed at the component or subsystem level under previous 21st Century Truck Partnership (21CTP) projects.[1] The SuperTruck program aligns with findings and recommendations set out in the NRC Phase 1 review of 2008 (NRC, 2008). For example, NRC Phase 1 Recommendation 1-2 stated that project "goals should be clearly stated in measurable engineering terms." Recommendation 3-1 called for demonstrating 50 percent brake thermal efficiency (BTE) at an operating point representative of a 65 mile per hour (mph) cruise operation. Recommendation 3-8 called for completing the 50 percent engine BTE demonstration before embarking on a 55 percent effort. The SuperTruck program had already started at the time of the NRC Phase 2 review in 2011, and the project plans were reviewed in Chapter 8 of the NRC Phase 2 report (NRC, 2012).

Four project teams were awarded projects under the SuperTruck program. All of the teams were given the same basic fuel-saving targets, along with a requirement to maintain "comparable vehicle performance":

- Achieve 50 percent BTE from the engine at a highway cruise operation speed/load point;
- Demonstrate a path to 55 percent BTE from the engine (see Chapter 3); and
- Demonstrate a 50 percent increase in freight efficiency, measured in freight ton-miles per gallon, on a long-haul drive cycle.

In addition to these targets, the Cummins team added a target to measure the effectiveness of its auxiliary power unit (APU) systems, which handle hotel loads when the vehicle is parked. The Daimler team also evaluated freight efficiency over a somewhat different 24-hour cycle. The Cummins target is to demonstrate a 68 percent increase in freight efficiency on a 24-hr duty cycle (drive cycle plus overnight hotel load).

The teams were given some flexibility regarding the targets. For example, each team defined its own long-haul drive cycle(s), and each team derived the cruise operation point for the 50 percent BTE demonstration from its selected drive cycle. The teams also had broad flexibility regarding the selection of specific technologies used to achieve the targets. All four projects were set up with a time frame of approximately 4 years. Table 8-1 compares the overall technical approaches of the four SuperTruck teams.

PROJECT BUDGETS AND RELEVANCE TO 21CTP

The Cummins-Peterbilt project was funded by the Department of Energy (DOE) using money from the American Recovery and Reinvestment Act of 2009 (ARRA).[2] DOE's funding is $38.8 million, with the industrial partners contributing $38.8 million, for a project total of $77.7 million. The Daimler SuperTruck project was also funded by DOE with ARRA money. DOE committed $39.6 million to the Daimler project, and the industrial partners also put in $39.6 million, for a total project funding of $79.1 million. As shown in Table 3-1, the project budget was split approximately 60/40 between vehicle and engine work.

The Volvo SuperTruck program was funded by DOE at $18.9 million, from regular DOE appropriations. Volvo also contributed $19.1 million to its program, for a total project funding of $38 million. The Volvo project is smaller in scale, because Volvo is also participating in a European government-funded program along the lines of SuperTruck.

[1] D. Anderson, 21st Century Truck Partnership, "Vehicle Technologies Office Vehicle and Systems Simulation and Testing," presentation to the committee, September 3, 2014.

[2] Budget information provided by Ken Howden, DOE, to the committee in March 2015. Also see Table 3-1.

TABLE 8-1 SuperTruck Team Technical Approaches

Technology	Cummins-Peterbilt	Daimler-Detroit Diesel	Volvo	Navistar
Aerodynamics	Many features—tractor and trailer are integrated. Retractable trailer skirts.	Many features—tractor and trailer are independent. Fixed trailer skirts.	Many features—tractor and trailer are independent. Fixed trailer skirts.	Many features. Final configuration TBD. Considering active ride height and pitch control.
Transmission	Eaton 10-speed AMT with narrow step between the top two gears.	12-speed Daimler DT-12 AMT with e-Coast.	Volvo 12-speed DCT.	Eaton 10-speed AMT with Precise Lube with direct top gear.
Hybrid powertrain	No.	120 kW motor, 2.4 kWh, 360 V battery. Serves as starter, handles hotel loads.	No.	Started with a 360 kW series hybrid, deleted for poor results. Now 48 V microhybrid.
Rolling resistance	Low RR, wide-base singles,	Low RR, wide- base singles.	Low RR wide-base singles.	Michelin low RR wide-base singles.
Axles	6 × 2 smart axles.	6 × 2 smart axles with active oil level control.	6 × 2 smart axles with active oil level control.	6 × 2 smart axle with tall 1.91:1 ratio to enable direct top gear.
Idle management	Li-ion battery APU, 13.2 kWh, 240-amp engine-driven alternator, 6 hr recharge, 400 lb.	Hybrid system used for idle management.	Energy-dense batteries, improved cab insulation.	TBD.
Route management	GPS-based cruise, route management system, driver display.	GPS-based cruise, integrated with hybrid system, driver display.	GPS-based cruise, route management system, driver display.	GPS-based cruise with driver coaching features.
Weight reduction	Aluminum fifth wheel and driveshaft, lighter axles and wheels, silicon carbide infused aluminum brake drums, magnesium cross members, lightweight air suspension, no lead-acid batteries.	Aluminum frame and cross members, lightweight air suspension.	Aluminum frame, cross members, and driveshaft, lightweight axles, suspension, and wheels.	MMC brakes, lighter axles, driveshaft, and wheels, aluminum engine and transmission Mounts, variable gage/sandwich frame rails, Al cross members, lighter suspension, smaller and lighter fuel tanks, composite cab, polycarbonate glass, aluminum/composite trailer.
Solar panels	No.	On trailer, to charge hybrid battery.	On cab roof to power fan to extract hot air from cab.	No.
Base engine	15 L Inline 6, no downsizing.	10.7 L Inline 6, downsized from 15 L baseline.	11 L Inline 6, downsized from 13 L baseline.	12.6 L Inline 6 baseline and SuperTruck.
rpm at 65 mph	~1,180.	~1,300.	Data not provided.	~1,050 or 1,125.
Engine efficiency features	High-efficiency turbo, low friction seals, lower power oil pump, low- viscosity oil, cylinder kit friction reduction, higher PCP, cal. optimization, overall 30% FMEP reduction.	turbo match, optimized liner cooling, variable speed water pump, low viscosity oil, piston friction reduction, 15% higher PCP, cal. optimization.	High-efficiency turbo, variable coolant and oil pumps, reduced friction pistons, rings, and liners, low-viscosity oil, improved thermal management.	High-efficiency turbo, elevated coolant temp, low-friction power cylinder, thermal insulation, reduced air flow restrictions, variable displacement oil pump.
Fuel system	HPCR with reduced parasitic fuel pump.	Amplified HPCR.	HPCR (converted from unit injector baseline).	Amplified HPCR.
Combustion refinement	Very high CR, piston bowl, injector match, 4.3 g/hp-hr engine-out NO_x, conventional diffusion burn.	High CR, piston bowl, low EGR, injector match, conventional diffusion burn, higher engine-out NO_x, model-based controls.	Increased CR, advanced piston bowl design, conventional diffusion burn, same engine-out NO_x as US 2010.	Looking at 6 g engine-out NO_x, higher injection press, revised piston bowl and high CR, evaluating diesel and dual fuel options, low swirl.

continued

TABLE 8-1 Continued

Technology	Cummins-Peterbilt	Daimler-Detroit Diesel	Volvo	Navistar
Electric drive components		Electric HVAC.	Electric dual-zone HVAC.	Electric HVAC, 48 V.
Waste heat recovery	Rankine cycle, R245 working fluid, mechanical drive, uses EGR and exhaust heat, turbine expander.	Rankine cycle, ethanol working fluid, electric drive, uses EGR and exhaust heat, scroll expander.	Turbocompound plus Rankine cycle with ethanol working fluid, mechanical drive, uses EGR & exhaust heat.	Turbocompound, Rankine cycle, and e-turbo are being evaluated.
Aftertreatment	High conversion efficiency, low back pressure.	High conversion efficiency, low back pressure.	High conversion efficiency, low back pressure.	High conversion efficiency, low back pressure.
Turbo technology	High-efficiency VG.	Asymmetric.	High efficiency.	Possible e-turbo.
Exhaust gas recirculation loop	Reduced flow rate and reduced flow restriction HPL.	HPL.	Reduced flow HPL.	Reduced flow rate and reduced flow restriction HPL.
Variable valve actuation	No.	No.	No.	Being evaluated.
Cooling system	Conventional cooling package, engine-driven fan, optimized to minimize fan-on time.	Angled cooling package, hydraulic motor fan drive, active grill shutters.	Variable speed engine-driven fan, variable speed cooling pump.	Three-speed engine-driven fan, ECM-controlled thermostat, high coolant temp., variable-speed cooling pump, variable coolant pressure.
Auxiliary power demand		Clutched air compressor with active controls; clutched power steering pump with reservoir; cab insulation and solar reflective paint for reduced A/C power demand.	Clutched air compressor with active controls; low-energy power steering; look-ahead smart alternator; LED lighting; cab insulation for reduced A/C power demand.	Clutched air compressor with intelligent dryer control, accessories run on deceleration / coast.

NOTE: TBD, to be determined; AMT, automated manual transmission; DCT, dual clutch transmission; RR, rolling resistance; MMC, metal matrix composite; PCP, peak cylinder pressure; HPCR, high-pressure common rail; CR, compression ratio; HVAC, heating, ventilation, and air conditioning; EGR, exhaust gas recirculation; VG, variable geometry (turbocharger); HPL, high pressure loop; GPS, global positioning system; rpm, revolutions per minute; LED, light-emitting diode; ECM, electronic control module; FMEP, friction mean effective pressure.

The Volvo team started about 1 year after the Cummins and Daimler teams, and the project is expected to be complete about a year after Cummins and Daimler. Finally, the Navistar program was assigned $35.8 million in regularly appropriated DOE funding, with $40.4 million in funding committed by the contractor, for a total Navistar project funding of $76.2 million. Navistar put its SuperTruck project on hold for approximately 2 years to deal with the conversion of its engine products from an EGR-only approach to the use of selective catalytic reduction (SCR) for compliance with 2010 emissions standards. Navistar resumed participation in SuperTruck in November 2014, and its completion is expected to be approximately 1.5 years behind that of the Cummins and Daimler teams.

The SuperTruck programs are highly relevant to the goals of the Partnership. As noted in the introduction to this chapter, these programs address recommendations from the NRC Phase 1 report, and the Phase 2 review in 2011 provided a generally positive review of the SuperTruck project plans (NRC, 2008, 2012, see Chapter 8 of the latter report).

CUMMINS-PETERBILT PROJECT

Members of and suppliers to the Cummins-Peterbilt team are listed in Table 8-2. Table 8-2 also lists some of the technology content used in the demonstration truck. One noteworthy aspect of the team structure was that Peterbilt actually did the design and fabrication work for the aerodynamic upgrades of the trailer used for its SuperTruck rather than depending on the trailer manufacturer for this support. The trailer manufacturer was responsible for weight reduction efforts on the trailer. This approach appears to be common to all the SuperTruck projects and is driven by the modest technical capabilities of trailer manufacturers.

TABLE 8-2 Cummins-Peterbilt Project Team Members and Suppliers

Team Member	Area of Responsibility
Cummins	Engine, waste heat recovery system, overall program management
Peterbilt	Tractor
Eaton	Transmission
Delphi	Solid oxide fuel cell APU (dropped from program)
Corvus	Lithium ion APU
Utility Trailer Manufacturing	Lightweight trailer
US Xpress (truck fleet)	End user feedback
Supplier	
Cooper Bussmann	Power distribution for the battery APU
Continental	Route display
SAF Holland	Aluminum fifth wheel
Dana	Aluminum driveshaft and 6 × 2 axle
Bendix	Traction control for 6 × 2 axle
Hendrickson	Lightweight steer axle and trailer tandem axles
Alcoa	Advanced lightweight wheels
Firestone	Integrated air suspension bags
Metalsa	Variable gage frame rails
Meridian	Magnesium cross members
Century, Inc.	Ceramic brake drums
Modene	Cooling package and heat exchangers
Oak Ridge National Laboratory	Advanced sensor development
Purdue University	Advanced combustion analysis; premixed charge compression ignition (PCCI) combustion modeling

Cummins-Peterbilt Technical Approach

Cummins selected the 15 L ISX engine for this program. This is the largest displacement engine to be used in the program. Cummins opted to avoid downsizing the engine in order to take advantage of efficiency technologies that can only be applied to a relatively low brake mean effective pressure (BMEP) engine, such as a very high compression ratio and a combustion strategy that leads to high cylinder pressures. Cummins did aggressively downspeed the engine, which causes the engine to operate at a higher BMEP for a given road load under cruise conditions. Using a large displacement engine increases the challenge of packaging the engine, the waste heat recovery (WHR) system, the exhaust aftertreatment system, and the cooling system into a highly aerodynamic vehicle layout. A large displacement engine also means that efforts to reduce engine friction are particularly important. Finally, a large displacement engine imposes a weight penalty. These disadvantages must be considered in light of the engine efficiency achieved.

According to the Cummins 2014 annual merit review (AMR) presentation (Koeberlein, 2014), conventional diesel combustion was retained, but with revisions to the combustion system (compression ratio, piston bowl, injector specification, and calibration). Turbocharger efficiency was improved, and the EGR circuit was optimized to minimize pumping work (and thus the pressure differential required to drive EGR flow). Extensive efforts were made to reduce friction, and a 30 percent reduction in friction mean effective pressure (FMEP) is claimed. The aftertreatment system design was modified to achieve higher conversion efficiency, which allows higher engine-out oxides of nitrogen (NO_x), and it was increased in size to provide less back pressure on the engine, which reduces the engine's pumping work. The engine was calibrated to produce 4.3 g/hp-hr NO_x on the Supplemental Emission Test (SET) cycle. A WHR boiler bypass is integrated into the exhaust system. The bypass allows reduced vehicle heat rejection in situations where the cooling fan would otherwise have to be used. Since cooling fan power is greater than the contribution of the WHR system, minimizing fan-on time is important to achieving low on-highway fuel consumption.

Peterbilt elected to use a production Model 579 tractor as the basis for the final SuperTruck demonstration tractor. The Cummins annual merit review presentation shows that the SuperTruck demonstrator has a 46 percent reduction in the coefficient of drag (C_d) from the baseline tractor-trailer configuration, based on computational fluid dynamic (CFD) calculations (Koeberlein, 2014). If the mirrors could be eliminated and replaced with cameras, the benefit would increase to 49 percent. In contrast, a completely clean sheet tractor design achieved a 49.6 percent C_d reduction without mirrors. Given the results of this evaluation, Cummins-Peterbilt decided that there is no need to consider an all-new tractor design. However, the Cummins-Peterbilt tractor and trailer are designed as a matched set. This means that the Peterbilt SuperTruck tractor cannot pull a conventional trailer, and the Cummins-Peterbilt SuperTruck trailer is not compatible with a conventional tractor.

Because WHR introduces an additional heat rejection demand to the vehicle cooling system, the team put effort into improving airflow through the cooling system and under the hood during fan-off operation. Slide 17 of the 2014 Cummins merit review shows some results from a CFD evaluation of cooling package airflow (Koeberlein, 2014).

A diagram of the Cummins WHR system used in the Demonstrator 1 vehicle is shown in Figure 8-1. According to Cummins, the final SuperTruck demonstrator vehicle used a WHR system that includes energy recovered from the engine coolant and lubricant circuits (Koeberlein, 2014). No details were provided on whether these circuits were run at an elevated temperature in order to provide higher quality waste heat. Also, the final demonstrator system fed power back into the front accessory drive belt rather than into the crankshaft through a gear train.

The original project plan called for the use of a solid oxide fuel cell APU to handle hotel loads during the stationary portion of the 24-hr duty cycle. According to Cummins, several problems were encountered with the fuel cell, including a long warm-up time, low thermal efficiency, high weight, and

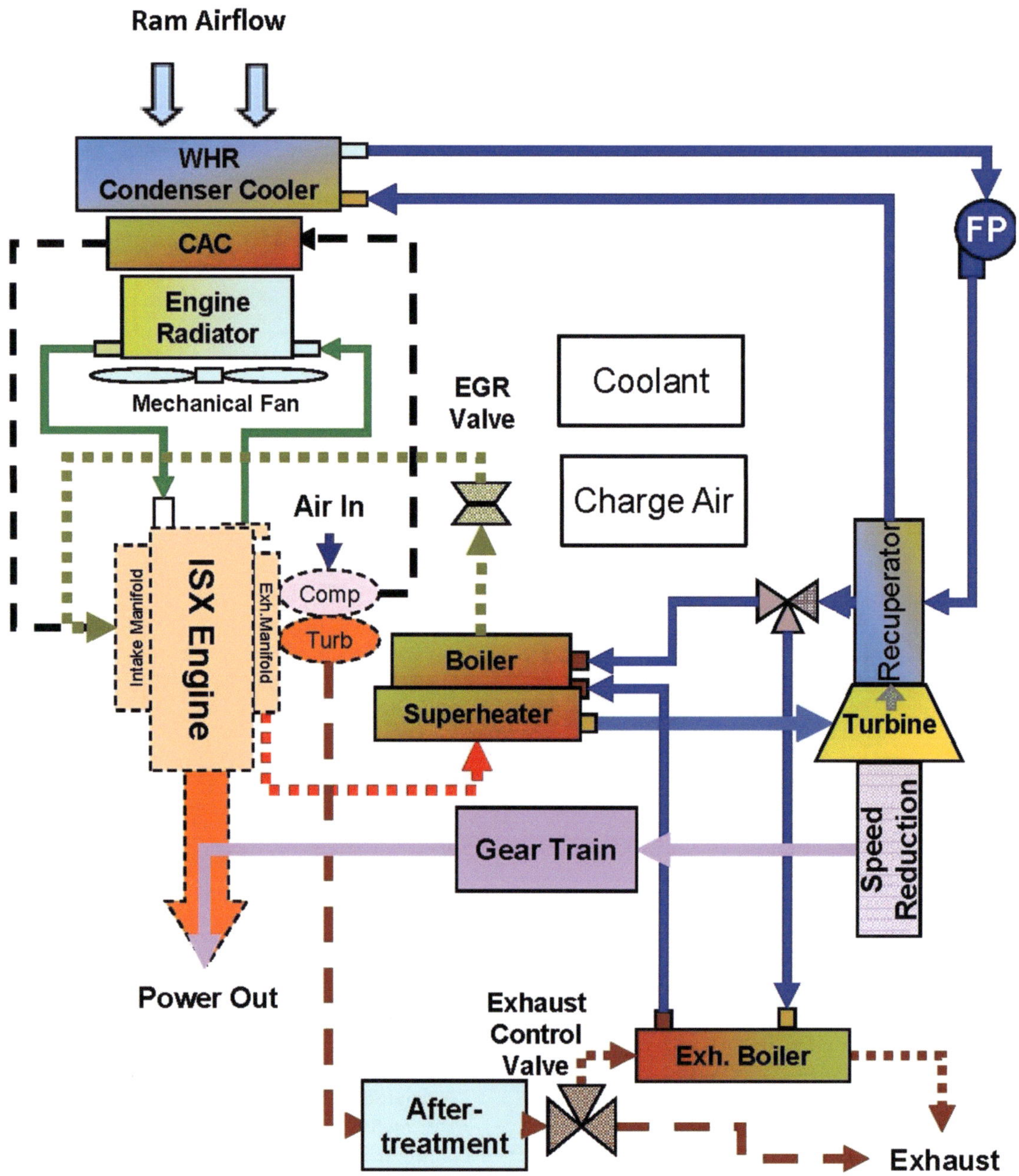

FIGURE 8-1 Cummins Demonstrator 1 WHR system layout. SOURCE: Koeberlein (2014).

a low peak power capacity (Koeberlein, 2014). After efforts to resolve these issues proved unsuccessful, the fuel cell was replaced with a custom-designed lithium ion battery-based APU. This system has a 13.2 kWh storage capacity, of which 12.2 kWh is used during the hotel load portion of the 24-hr duty cycle. The battery is recharged by a 240 amp engine-driven alternator, and a full system recharge takes 6 hr of driving time. The heating, ventilation, and cooling (HVAC) system was converted to electric power to work with the APU. The Cummins SuperTruck battery APU system weight is about 400 lb, which is comparable to the weight of diesel-engine driven APU systems that are widely used in the field today.

The original project plan called for a dual clutch transmission. In the end, a single clutch automated manual transmission was used. Optimization work was performed on the transmission ratios and to improve the mechanical efficiency of the transmission. Another feature planned early in the program was a turbocharger with its own continuously variable transmission (CVT) (Fleet Owner Magazine, 2010). The idea was to improve transient response by accelerating the turbocharger using crankshaft power when needed, and to generate power to the crankshaft when more than adequate turbine power was available. The VanDyne turbocharger proved to be not adequately developed for use in a vehicle application, so a more efficient conventional (variable geometry) turbocharger was employed instead. The Cummins approach to the 55 percent BTE requirement is described in Chapter 3.

Cummins-Peterbilt selected a test route running from Denton, Texas, to Vernon, Texas, and back.[3] The route is primarily rural, with a couple of traffic lights. Most of the route follows US 380 and US 287, which combine sections of divided highway with stretches of two- and four-lane undivided road. The route length is 311 miles, with 550 feet in elevation change. A figure on Slide 15 of Cummins' presentation to the committee shows that the distribution of grades on this route is very close to the national average grade distribution of interstate highways in the lower 48 states.[4]

Cummins-Peterbilt Project Results

Table 8-3 shows information on the Cummins-Peterbilt 2009 model baseline tractor-trailer. These data are based on results provided by Cummins-Peterbilt in the AMRs, along with information provided to the committee by Cummins. Note that the 24-hr cycle includes 8 hr of idling or APU use to support cab hotel loads. Table 8-4 summarizes the SuperTruck project results to date.

Based on the results in Table 8-4, some calculations can be made to determine the relative contribution of engine efficiency and vehicle power demand. Assuming that the 2009 baseline Cummins engine had a BTE of 42 percent (Koeberlein, 2014), the reduction in vehicle fuel consumption due to improved engine efficiency is about 17.6 percent. Vehicle fuel consumption on the long-haul cycle was reduced by 40 percent, so about 22.4 percent fuel consumption reduction is due to vehicle power demand reduction. These same data can be expressed in another way: Of the total vehicle fuel savings on the long-haul drive cycle, about 44 percent is due to engine efficiency improvements, and 56 percent is due to vehicle power demand reduction. If the increase in cargo allowed by empty weight reduction is taken into account, the engine efficiency/vehicle efficiency split is 41/59 percent.

Additional results provided by the Cummins-Peterbilt team include:

- Reduction in C_d: 46 percent
- Weight increases:
  —WHR + aftertreatment upgrades: 950 lb
  —Aerodynamic devices: 2,500 lb
  —Idle reduction system: 400 lb
- Weight reductions:
  —Reduced fuel load, 400 lb
  —Tractor weight reduction, 2,400 lb
  —Trailer weight reduction, 2,355 lb
- Net vehicle weight reduction/payload increase: 1,305 lb
- Cummins-Peterbilt SuperTruck freight efficiency on driving cycle (no idle): 178 ton-mi/gal
- Cummins-Peterbilt SuperTruck load specific fuel consumption (LSFC) on driving cycle: 5.64 gal/1.000 ton-mile
- Share of total fuel savings attributed to:
  —Engine improvements and WHR, 42 percent
  —Tractor aerodynamic improvements, 14 percent
  —Trailer aerodynamic improvements, 28 percent
  —Driveline and tire improvements, 15 percent

TABLE 8-3 Cummins-Peterbilt 2009 Baseline Data

2009 Baseline Vehicle Configuration	Peterbilt 386, ISX engine with 450 hp, 13-speed Eaton Ultrashift AMT, aerodynamic hood and fairings, 63-in. sleeper, 6 × 4 with dual steel wheels and tires, 42-in. trailer gap, 20-in. gap from sleeper extender to trailer
Tractor-trailer empty weight	33,729 lb
Test weight	65,000 lb
Test payload	31,271 lb
mpg and gal/100 mi on drive cycle	6.45 mpg, 15.5 gal/100 mi
ton-mpg and gal/1,000 ton-mi	101 ton-mpg, 9.90 gal/1,000 ton-mi
mpg and gal/100 mi on 24-hr cycle	5.4 mpg, 18.5 gal/100 mi
ton-mpg and gal/1,000 ton-mi (24 hr)	84.4 ton-mpg, 11.8 gal/1,000 ton-mi

SOURCE: Koeberlein (2014).

The data presented above show one significant issue introduced by the application of fuel-saving technologies: weight increase. In the Cummins-Peterbilt SuperTruck, the weight increase associated with the fuel saving technologies totals 3,850 lb. To achieve the reported increase in payload of 1,305 pounds, the empty weight of the tractor and trailer had to be reduced by 5,155 pounds, or 15.8 percent. Achieving weight reductions of this size in a production truck is likely to be extremely expensive and require extensive analysis and development. Once requirements for cost-effectiveness are applied, it is likely that the application of efficiency technologies will push up empty weight. This has a slight negative impact on fuel consumption but a larger negative impact on freight efficiency, since the maximum legal payload will be reduced.

The vehicle-level demonstration tests were conducted in December 2013. Temperatures were low and winds were high (Damon, 2014). For the 24-hr duty cycle, the average

[3] E. Koeberlein, Cummins-Peterbilt, "Cummins SuperTruck Program: Technology and System Level Demonstration of Highly Efficient and Clean, Diesel Powered Class 8 Trucks," presentation to the committee, May 14, 2014.

[4] Ibid.

TABLE 8-4 Cummins-Peterbilt Project Results to Date[a]

Target	Status	Complete?
50% engine BTE at cruise	51% engine + WHR BTE demonstrated	Yes
50% freight efficiency increase on a long haul drive cycle (33% reduction in load specific fuel consumption)	76% demonstrated, plus a 66% increase in mpg (43% reduction in gallons per ton-mile, 40% reduction in gallons per mile)	Yes
68% freight efficiency increase on a 24-hr duty cycle (40.5% reduction in load specific fuel consumption)	86% demonstrated, and a 75% increase in mpg (46% reduction in gallons per ton-mile, 43% reduction in gallons per mile)	Yes
55% engine BTE	49.4% engine-only BTE demonstrated, more work planned	Q2 2015

NOTE: Results are based on a comparison to the reference truck, run on the same route, with the two trucks about 1 min apart.

[a]The results presented in this table are based on data collected by the committee prior to May 2015. At the DOE Annual Merit Review in June 2015, David Koeberlein of Cummins presented the following results: "Developed framework and analysis for 55% thermal efficiency, completed analytical roadmaps for both diesel and dual fuel approaches, and completed targeted engine tests to validate roadmaps" (Koeberlein, 2015). However, the committee was not in a position to review these results from the June 2015 AMR.

wind during the driving portion was 14 mph, with gusts to 28 mph. For the basic drive cycle comparison, the average wind was 13 mph, with gusts to 33 mph. These high winds had the effect of driving up fuel consumption for both the baseline reference truck and the SuperTruck test vehicle. The difference in performance between the baseline and SuperTruck vehicles is likely to be at least slightly related to wind conditions, so a test in milder conditions might give slightly different results. The way the two vehicles' C_d values vary with yaw angle will account for any difference in wind sensitivity.

The headline trip result of 10.7 mpg was achieved during a later test, where the average wind was 6 mph, with no gusts. This compares to a result of 9.9 mpg on the windy day. Additional results obtained with the Demo 2 (final version) SuperTruck tractor are summarized in Table 8-5. These tests are outside the scope of the program requirements, but they provide valuable insight.

The results shown in Table 8-5 reveal the very modest sensitivity of fuel consumption to weight increases above 65,000 lb but a more significant sensitivity to lower payload. Giving up the aerodynamic SuperTruck trailer and replacing it with a standard trailer with a skirt causes a fuel consumption increase of about 11 percent, which is very significant. Note that these results are affected by variations in wind conditions. Also note that the Table 8-5 results do not look at differences in performance on a given day, where the test truck is compared to a reference truck run at the same time, which is the most accurate way to determine differences.

The Cummins-Peterbilt team had its fleet partner, US Xpress, drive the Demo 1 level vehicle with a commercial load on a 950-mi route in Texas, using US Xpress drivers. US Xpress evaluated vehicle drivability and loading/unloading issues. As a result of this test, the access panel covering the trailer tandem was modified so that it could open 180 degrees for tire inspection. Another change implemented as a result of the fleet test was to make the trailer skirt pneumatically retractable for access to loading docks and for crossing crowned areas such as railroad tracks.

Cummins-Peterbilt Project Management

A committee subgroup visited with Cummins-Peterbilt and Eaton in Columbus, Indiana, on August 28, 2014. With one exception, the subgroup found the project to be well organized and well run. The only exception was that the project team was unaware of the flexibility they had under the contract to change plans. This caused a delay in moving away from use of a fuel cell APU after it became clear that the fuel cell's issues could not be resolved within the project

TABLE 8-5 Additional SuperTruck Tractor Fuel Economy Results

Vehicle Configuration	Wind	gal/1,000 ton-miles	gal/100 mi	mpg
SuperTruck trailer at 65,000 lb	6 mph, steady	5.72	9.3	10.7
SuperTruck trailer at 80,000 lb	6.5 mph, steady	4.04	9.6	10.4
SuperTruck trailer at 32,500 lb (empty)	23 mph, gusts to 40	Infinity (no payload)	7.9	12.7
SuperTruck Tractor pulling a standard 53 ft trailer with skirts, 65,000 lb	9 mph, gusts to 18	6.52	10.6	9.4

scope. The subgroup was impressed with the way the project team took advantage of the program to improve their modeling and simulation capability. There were several situations where simulation and test results did not match, and the team reviewed these in detail to understand and resolve the differences. The improved simulation capabilities that resulted had an impact on the program, and they will also increase the ability of the team to deliver improved products in the future.

The subgroup asked the project team about its experience with DOE research programs. Cummins-Peterbilt emphasized the value of long-term (3- to 4-year) system integration projects. These projects are long enough to enable the contractors to recover and change plans when issues are encountered, and still successfully meet the project goals. In a short-term project, companies must take less risk in order to succeed. The project team felt that long-term system integration projects also provide extensive learning opportunities, develop useful relationships among partners, speed the progress of technology development, and build momentum toward production.

The committee asked Cummins about the relative value of one large, longer duration project, compared to several smaller projects. Cummins told the committee that they see more value in large, long-duration projects, even where limited funding means a lot of competition to win these projects.

DAIMLER-DETROIT DIESEL PROJECT

Table 8-6 lists the Daimler team members and their responsibilities.

Daimler used the 14.6 L DD15 engine for the 2009 baseline vehicle, the 12.8 L DD13 engine for its A-Sample truck, and a 10.7 L engine for the B-Sample (final demonstration) truck.[5] Note that the 10.7 L engine is not currently offered in the North American market. It is sold in Mercedes trucks in Europe, under the name OM470. The 10.7 L OM470 represents a significant downsizing from the baseline DD15 engine. The smaller displacement brings weight, heat rejection, and packaging advantages over the baseline engine. On the other hand, power and torque are substantially reduced, and durability in long-haul applications is likely to be compromised. Power is down from 455 hp and 1,550 lb-ft in the baseline truck to 390 hp and 1,400 lb-ft with the 10.7 L engine, according to the Detroit Diesel 2014 AMR presentation (Singh, 2014) and subsequent communication from Daimler. Daimler did a study of vehicle performance on the selected SuperTruck routes and ensured that the SuperTruck achieved comparable performance on the following metrics: overall trip time, time spent at less than 30 mph, and time spent in top gear.

TABLE 8-6 Daimler Project Team Members

Company	Area of Responsibility
Daimler	Engine, transmission, and vehicle, overall program management
Massachusetts Institute of Technology (MIT)	Engine lubrication
Atkinson LLC	Engine controls development
Delphi	Solid oxide fuel cell APU (dropped from program)
Bowman	Electric turbocompounding
Air Squared	WHR expander development
Oak Ridge National Lab	Dual fuel combustion development for 55% BTE
NREL	Energy management
Oregon State University	Energy management, weight reduction
Telogis	Energy management
A123	Hybrid system batteries
Eaton	Hybrid system
US Hybrid	Hybrid system
Miasole	Solar system
Itk Engineering	Hybrid system
Auto Research Center	Scale model wind tunnel testing
CD-adapco	Computational fluid dynamics
TitanX	Aerodynamics/cooling system
Modine	Cooling package and heat exchangers
Mekra-Lang	Aerodynamics/cooling system
Freight Wing	Aerodynamic devices
Corning	Aftertreatment
Eberspaecher	Aftertreatment
Johnson Matthey	Aftertreatment
Strick	Trailer and weight reduction
Inmagusa	Weight reduction
Toray	Weight reduction
Michelin	Tires
ConMet	Powertrain/parasitics
Bendix	Powertrain/parasitics
Accuride	Suspension
Ashland	Powertrain/parasitics
Parker	Powertrain/parasitics
Schneider National	Fleet
Walmart	Fleet

Based on these criteria, Daimler believes that the SuperTruck provides comparable performance to the baseline truck. Reduced vehicle power demand and assistance from the hybrid system are able to compensate for the lower engine power. However, certain aspects of performance, such as vehicle speed on a long grade, are still likely to suffer.

Conventional diffusion burn diesel combustion was retained, but with revisions to the combustion system

[5] D. Kayes, D. Rotz, and S. Singh, Daimler Truck North America LLC, "Super Truck Team Presentation: Daimler," presentation to the committee, May 15, 2014.

(compression ratio, piston bowl, injector specification, and calibration) (Singh, 2014). Turbocharger efficiency is not discussed, but the EGR circuit was modified to reduce EGR flow, and the turbocharger was reoptimized to match the lower EGR rates. Changes were made to reduce cylinder kit friction. A variable-speed water pump is used to reduce parasitic power (as in one version of the current production DD15), and low viscosity oil is employed at a higher than normal operating temperature in an effort to reduce friction. The aftertreatment system was upgraded to allow for higher engine-out NO_x and lower backpressure. The engine was calibrated to produce higher engine-out NO_x on the SET cycle than the baseline engine, although the engine-out NO_x level is not specified. A waste heat recovery boiler bypass is integrated into the exhaust system. The bypass allows reduced vehicle heat rejection in situations where the cooling fan would otherwise have to be used. Since cooling fan power can be greater than the contribution of the WHR system, minimizing fan-on time is important to achieving low on-highway fuel consumption.

Singh (2014) mentions that engine downsizing and some of the engine efficiency measures result in challenges for cylinder block and head design, and for long-term high-load durability. These issues are driven by increased peak cylinder pressures (PCPs) and reduced oil film thicknesses. Noise, vibration, and harshness (NVH) and emissions are also mentioned as potential issues with the SuperTruck combustion system. Detroit Diesel Corporation worked with MIT on modifications to reduce heat rejection through the cylinder liners. Considerable effort has been made to implement model-based controls for the engine and the hybrid system.

Daimler made extensive modifications to the production truck cab to create the final SuperTruck demonstration tractor. The cab was widened to match the sleeper width, and the windshield has been raked back. The shape of the hood has been extensively modified to optimize aerodynamic efficiency with the smaller engine and angled cooling system. Mirrors were retained, but at the legal minimum size (about 1/3 of the baseline size) and with extensive aerodynamic optimization. The mirrors are supplemented by cameras. Daimler told the committee that the overall wind-averaged C_d (a weighted calculation of drag based on time spent at each yaw angle and wind speed)[6] of the final demonstrator truck, including both the tractor and trailer, is about 50 percent lower than the baseline. Most of the improvement in C_d came from the trailer, despite extensive development work on the tractor.

If waste heat recovery works on an energy stream (such as EGR flow) that must be cooled anyway, there is no additional heat rejection demand on the vehicle. However, when exhaust energy downstream of the aftertreatment is captured by a WHR system, this represents heat that otherwise would simply flow out of the exhaust pipe. A portion of this heat will be converted into work, but the remainder must be rejected by the tractor's cooling system. Because WHR and the cooling system for the hybrid battery place an additional heat rejection demand on the tractor's cooling system, the team put significant effort into improving airflow through the cooling system and under the hood during fan-off operation. The cooling package is mounted at an angle, with the top of the package being farther back in the tractor. This makes an engine-driven fan impractical, so a hydraulic motor-driven fan is used. In addition, Daimler added active shutters to limit cooling air flow under the hood when it is not required. According to Slide 10 of the Daimler 2014 AMR (Rotz, 2014), the shutters improve C_d by 6 percent at zero yaw when they are closed, while reducing underhood airflow by 60 percent. It should be noted that the 6 percent represents C_d improvement measured on SuperTruck only, with its unique hood, cooling package, and underhood configuration. The shutter system is not expected to produce the same results on a production vehicle.

A diagram of the Detroit Diesel WHR system used in the final demonstrator vehicle is shown on Slide 11 of the Daimler 2014 AMR (Rotz, 2014). The basic layout is similar to the Cummins system shown in Figure 8-1, but there are a number of detail differences. Detroit Diesel uses ethanol rather than a refrigerant as the working fluid, and the Detroit Diesel system does not use a recuperator. The Detroit Diesel system uses a scroll expander rather than the turbine used by Cummins. The condenser on the Detroit Diesel system is an ethanol to water heat exchanger. The water then goes to the front of the truck to reject heat. In the Cummins system, the working fluid condenser is part of the cooling package at the front of the truck. The Detroit Diesel system description does not mention using energy from the coolant and lube circuits. Finally, the Detroit Diesel WHR system power output is used to generate electricity rather than being fed back to the engine crankshaft. The electricity can be used to power the vehicle HVAC system and other hotel loads, to power the hybrid motor, or to recharge the hybrid system batteries. It is worth noting that to the extent the SuperTruck engines reduce EGR flow (as they increase engine-out NO_x), the supply of high-quality heat for the WHR system is reduced. This has the effect of reducing the potential power that the WHR system can generate.

The Daimler hybrid system is a parallel system with a 120 kW motor and a relatively small 2.4 kWh battery. The motor power rating falls approximately midway between ratings for so-called "mild" and "full" hybrid drive systems for a Class 8 truck (NRC, 2010), while the battery capacity falls toward the mild end of the hybrid spectrum. As a result, the hybrid system represents an engineering compromise solution that prevents the battery from capturing most of the truck's available braking energy, but it is significantly lighter, smaller, and less expensive than a comparable full hybrid drive system for this vehicle.

[6] See http://www.uwal.org/download/temp/Heavy%20Duty%20Truck%20Aerodynamics.pdf.

Daimler made a significant effort to get as much value as possible from the hybrid system. This was accomplished by a high level of integration of the hybrid drive into the vehicle's overall system design, making it possible to use the same equipment for several different purposes. The hybrid system in the Daimler SuperTruck plays a role in engine starting, waste heat recovery, idle reduction, enabling electric air conditioning, and providing torque fill during transmission shifting. Torque fill is a feature that uses the electric motor to drive the axle during a shift event, while power from the diesel engine is interrupted, thus providing smoother vehicle acceleration.

Daimler Trucks uses the hybrid system battery to handle hotel loads during sleeper use. The 2.4 kWh battery cannot cover hotel loads for the entire night, so the engine will restart automatically when required to recharge the battery. The hybrid system battery also provides engine start power, eliminating all but one of the 12 V lead acid batteries. The hybrid battery has four sources of power: regenerative braking, the waste heat recovery system, solar panels on the trailer roof, and, finally, engine power fed to the motor/generator. Engine power is not used to charge the battery on the highway because of the power transformation losses involved. Despite the extensive system integration effort, Daimler told the committee that the hybrid system was not cost-effective, given its high cost, its modest fuel savings on a long-haul cycle, and competition from the low-cost eCoast functionality. The eCoast feature provides a significant portion of the fuel savings that can be achieved by a hybrid in a long-haul application, but eCoast is only a software control feature, with no additional hardware required.

The Daimler SuperTruck uses a production 12-speed automated manual transmission (AMT) made in-house by Daimler. The production design was modified to allow incorporation of the hybrid system motor. The calibration of the AMT was revised to accommodate the hybrid system and to keep the engine closer to its optimum operating point. A feature called eCoast has been added (Rotz, 2014, Slide 6). This feature disconnects the engine from the driveline when the vehicle power demand is zero and allows the engine to drop to idle. eCoast comes into play on gentle downhill grades and any time the driver wants to gradually slow the truck. The use of e-Coast preserves vehicle inertia, since the energy is not used to spin the engine. That inertia then reduces the fuel required the next time the vehicle demands power. In other words, the vehicle is its own energy storage device, in which energy is stored in the form of kinetic energy. Volvo Trucks currently offers a similar feature in some of its production trucks under the trade name Eco-Roll (Volvo Trucks, n.d.), and Cummins has announced a similar system for 2015 production availability.

The Detroit Diesel approach to the 55 percent BTE requirement is described in Chapter 3. Daimler Trucks and Detroit Diesel selected two routes on which to evaluate their SuperTruck vehicles (Rotz, 2014). A third route was added later (conference call with Rotz, 2015). The first route runs from Dallas to San Antonio in Texas, using I-35. This route has frequent minor grade fluctuations of ±0.5 percent to 1 percent) due to numerous underpasses, along with a few more significant hills of up to ±5 percent grade. The I-35 route is run at 65 mph, which tends to favor aerodynamic improvements. This route was given a 64 percent weighting on a time basis by Daimler, meaning that 64 percent of the driving portion of the 24-hour drive cycle consisted of the I-35 route. The second route is from Portland to Canyonville in Oregon, using I-5. This route has significant segments that are flat or nearly flat, along with some hills with grades up to about 6 percent. The Oregon route is run at 58 mph, just above the state speed limit of 55 mph. This route was given a 28 percent weighting, since there are more states with a truck speed limit of 65 mph or greater than states with a 55 mph limit. Daimler Trucks did not compare the grade distribution on these routes to that of the national road network, but they appear to the committee to represent reasonably typical routes. The third route, which was given an 8 percent weighting, used local urban highways in the Portland area. This route was meant to represent getting from the loading dock out to the long-haul route.

Because the Daimler SuperTruck includes a hybrid system, the results reported below are based on having the battery in the same state of charge at the beginning and end of each test. The headline test result of 12.2 mpg on the San Antonio to Dallas I-35 route is an average of 5 round-trip runs of approximately 400 miles each, all made at a cruise speed of 65 mph. This result was publicly announced at the Mid-America Truck Show in March 2015. The 12.2 mpg number translates to 206 ton-mi/gal, or 4.85 gal/1,000 ton-mi. These are the best results reported so far in the SuperTruck program, but remember that test routes and test conditions vary, so results are not directly comparable.

Daimler-Detroit Diesel Project Results

Based on Table 8-7, some calculations can be made to determine the relative contribution of engine efficiency and vehicle power demand. Assuming that the 2009 baseline Detroit engine had a BTE of 44 percent, the reduction in vehicle fuel consumption due to improved engine efficiency is about 12.4 percent. Average vehicle fuel consumption on the two long-haul cycles was reduced by 47.4 percent, so about 35 percent fuel consumption reduction is due to vehicle power demand reduction, including the effect of the hybrid system. These same data can be expressed in another way: Of the total vehicle fuel savings on the long-haul drive cycle, about 26 percent is due to engine efficiency improvements and 74 percent to vehicle power demand reduction. Additional results include these:

- Reduction in C_d: 54 percent (final demonstrator, scale model wind tunnel result) (Rotz, 2014, Slide 14).

TABLE 8-7 Daimler Trucks and Detroit Diesel Project Results to Date[a]

Target	Status	Complete?
50% engine BTE at cruise	50.2% engine + WHR BTE demonstrated	Yes
50% freight efficiency increase on a long-haul drive cycle (33% reduction in LSFC)	Freight efficiency/LSFC results: 96.3% increase in freight efficiency (49% reduction in LSFC) demonstrated on Oregon route, 119.5% increase in freight efficiency (54.5% reduction in LSFC) demonstrated on Texas route. Fuel economy/fuel consumption results at 65,000 lb: 80.1% increase in mpg (44.5% decrease in FC) on the Oregon route and 101.3% increase in mpg (50.3% reduction in FC) on the Texas route	Yes
68% freight efficiency increase on a 24-hr duty cycle (40.5% reduction in LSFC). Note that this goal is not a contract requirement	115% (53.5% LSFC reduction) Demonstrated on weighted combination route plus idle cycle. 97.4% mpg improvement (49.3% reduction in fuel consumption)	Yes
55% engine BTE	Work under way; no plan for full engine test	No

NOTE: FC, fuel consumption.

[a]The results presented in this table are based on data collected by the committee prior to May 2015. At the DOE Annual Merit Review in June 2015, Sandeep Singh of Detroit Diesel Corporation presented the following results: "Achieving 55% BTE is expected to require advanced combustion strategies such as DF-LTC (dual fuel and low temperature combustion), plus additional improvements in parasitic reductions, component efficiencies, WHR, etc. beyond those achieved during SuperTruck. Daimler and ORNL look to continue DF-LTC efforts beyond SuperTruck to address these issues" (Singh, 2015). However, the committee was not in a position to review these results from the June 2015 AMR.

- Net vehicle weight reduction/payload increase: 1,550 lb (A-Sample truck) (Rotz, 2014, Slide 12) and 2,800 lb on the final demonstrator (conference call between Tom Reinhart, committee member, and Derek Rotz, Daimler, April 19, 2015)
  —Details of weight reduction/increases were not provided.
- Distribution of aerodynamic drag reduction:
  —Trailer aerodynamic improvements: 72 percent of the total vehicle improvement, with 1/3 of the aerodynamic engineering effort.
  —Tractor aerodynamic improvements: 28 percent of the total vehicle improvement, with 2/3 of the aerodynamic engineering effort.

It is somewhat surprising that Daimler's wind tunnel test results show that tractor and trailer aerodynamic benefits are independent of each other (Rotz, 2014, Slide 14). In a simple test that involved swapping between the baseline and final demonstration levels of tractor and trailer, no synergy was found for the highly aerodynamic final demonstration tractor and trailer. Other researchers have found that tractor and trailer aerodynamics do depend on each other, so this result may represent an unusual case. It is also worth noting that the Daimler Trucks trailer design is compatible with standard tractors, unlike the Cummins-Peterbilt trailer design, where the tractor and trailer form a matched set.

In many cases, when individual technologies are combined into a package, the fuel saving is less than the sum of the individual contributions. This is always the case where multiple technologies target the same source of energy loss, but it can also occur when unrelated technologies are combined. For example, if average vehicle power demand is reduced by lowering aerodynamic drag and tire rolling resistance, the engine may spend more time at light load, where it is less efficient. On the other hand, Daimler reported at least one instance where the whole benefit from a package of features is greater than the sum of the parts. Reducing vehicle power demand has the effect of increasing the performance of the eCoast feature, since as aerodynamic drag and tire rolling resistance are reduced, the vehicle power demand drops below zero more often. In other words, it takes less of a negative grade to produce a zero power demand condition as C_d and C_{rr} go down, so the eCoast feature has more kinetic energy to work with.

Daimler-Detroit Diesel Project Management

A committee subgroup visited Daimler Trucks North America in Portland, Oregon, on October 17, 2014. Another subgroup visited Detroit Diesel Corporation on November 24, 2014. The subgroup that visited Daimler Trucks was impressed with the way the project team was taking advantage of the program to improve its modeling and simulation capability for vehicle aerodynamics, and its fuel economy test capability. The improved simulation and test capabilities had an impact on the program, but they will also increase the ability of the team to deliver improved products in the future.

Detroit Diesel feels that the single operating point 55 percent BTE target is not as attractive as improving efficiency over a realistic on-road duty cycle. It also pointed out the advantages of a system integration project like SuperTruck compared to component-level development projects. When integrating the complete system, Daimler made some discoveries about the real-world performance of certain technologies that were quite different than projections from simulation and test cell operation. SuperTruck drove the integration of many engine and vehicle technologies that had been previously considered only on a stand-alone basis.

Some Daimler suggestions for future research opportunities are these:

(1) Development of WHR systems to achieve lower cost, weight, and complexity, along with higher performance;
(2) Development of advanced combustion technologies such as dual fuels; and
(3) Development of friction reduction technologies for a high-cylinder-pressure engine.

Overall, Daimler was very enthusiastic about the SuperTruck program. One manager called it "one of the best government projects ever." By doing a complete vehicle, Daimler learned about the potential of many technologies, as well as issues and limitations that stand in the way of introducing some of these technologies.

VOLVO PROJECT

Because the Volvo SuperTruck project did not begin until June 2011, it was not included in the NRC Phase 2 report on 21CTP. The project is expected to be completed by June 2016. Volvo elected to divide the project into two phases. Phase 1 delivered a concept evaluation vehicle (VEV-1, or mule). This vehicle was used to validate candidate technologies during 2013. Phase 2 will deliver a final SuperTruck demonstrator that will include the technologies validated in Phase 1, as well as additional technologies and refinements.

The total cost of the project is projected to be $38 million, with 50 percent cost sharing. This total is about half of the budget for the other SuperTruck projects. As of September 2014 the total spent was $24 million; as of June 2015, $26 million was spent (Amar, 2015). A list of the key team members and their contributions is provided in Table 8-8.

TABLE 8-8 Volvo SuperTruck Collaborators and Partners

Organization	Key Contribution
Volvo Technology of America	Project lead and concept simulations
Volvo Group Truck Technology	Power train, vehicle integration, and testing
Ridge/Freight Wing	Trailer aerodynamics
Grote	Advanced LED lighting systems
Ricardo	Rankine cycle WHR Generation 1 development
Hendrickson	Lightweight trailer axle and suspension components
Alcoa Wheels	Lightweight wheels
Michelin	Advanced low-friction tires
Metalsa	Ultralight aluminum frame assembly
ExxonMobil	Advanced fuels and lubricants
Chalmers University of Technology	55% BTE testing
Penn State University	55% BTE simulation and testing
University of Michigan	55% BTE simulation and testing
Drexel University	WHR topology simulation

Phase 1 Technologies and Results

The tractor selected for the Phase 1 mule (VEV-1) was a model VNL 670 (Figure 8-2). The wheel fairings were modified for reduced drag. LED lighting was used for both internal and external lights to reduce electrical power demand and weight. The trailer was also equipped with LED lighting. Volvo calculates that equipping both tractor and trailer with efficient LED lighting in place of the standard incandescent system can save up to 120 gal/yr of fuel (DOE VTO, 2013). The use of LED lighting allowed the use of light gauge wiring for weight reduction.

According to Volvo's 2013 FY progress report for vehicles systems simulation and testing, VEV-1 was upgraded with a new and lighter 6 × 2 axle configuration, a prototype proprietary lighter weight suspension, lightweight aluminum wheels, and wide-based low rolling resistance tires (DOE VTO, 2013). In addition, the standard two-piece driveshaft was replaced with an aluminum one-piece unit that provided a 25 percent reduction in weight. The total weight savings from these components was approximately 900 lb.

FIGURE 8-2 Volvo VEV-1 preliminary demonstration vehicle. SOURCE: Amar (2013).

The Volvo team used complete vehicle CFD simulations to design and optimize aerodynamic parts or add-on devices for the tractor and the trailer. Freight Wing's latest designs for trailer aerodynamic devices were used as a starting point for the vehicle aerodynamic simulations. In parallel with the aerodynamic optimization activities, Freight Wing explored opportunities to make the trailer add-on devices more practical from an operational perspective. In particular, new methods for enabling the tail fairing geometry to fold and provide convenient access to cargo were investigated. Different materials including reinforced composite panels were also evaluated for opportunities to improve product durability. The intent was to make the aerodynamic geometry that has proven to be effective in prior work as practical as possible for real-world utilization and production.

Prototype parts corresponding to the designs simulated through CFD were fabricated and installed on VEV-1. Primarily these included tractor wheel skirts, trailer skirts, and a trailer tail, which were tested under real-world conditions during the fuel economy tests. Tests results demonstrated a 13 to 15 percent fuel consumption reduction for the complete vehicle, compared with the MY 2009 baseline (DOE VTO, 2013). This correlated very well with the simulated aerodynamic drag reduction of 30 percent for the corresponding geometry, and confirmed the accuracy of the CFD methods.

For Phase 1, Volvo chose its 13 L engine (DOE VTO, 2013). This choice was made because the 13 L is Volvo's highest volume production engine, and the 2009 baseline vehicle was equipped with a 13 L engine. The basic engine improvements included high-pressure common rail fuel injection (in place of the production engine's unit injectors), revised piston bowl geometry, reduced-friction power cylinder components, advanced lube and coolant pumps that reduce parasitic losses, and a mechanical turbocompound system.

A Generation 1, Rankine-cycle WHR system was also included in Phase 1. The WHR system exceeded previous performance in steady-state operation, despite the addition of a more efficient combustion chamber and turbocompounding, both of which reduced the heat available to the system. Energy recovery was possible during nearly all positive power operation, with interruptions during coasting or engine brake operation. The advanced power train system installed in the VEV-1 chassis successfully completed multiple on-road tests with varying route profiles and vehicle loads (DOE VTO, 2013).

The lower exhaust temperatures resulting from the turbocompound and improved combustion efficiency created some additional challenges for efficient function of the exhaust aftertreatment system (EATS). A chemical model of the EATS was developed for the unit delivered as part of VEV-1, allowing for transient evaluation of the EATS in the SuperTruck environment.

TABLE 8-9 Volvo 2009 Baseline Test Results

2009 baseline vehicle configuration	NVL 670, D-13 engine with 485 HP and 1,650 lb-ft, full chassis aero treatment, 44" trailer gap, cab and roof aero fairings, 2.65:1 overall ratio in top gear
Tractor-trailer empty weight	33,300 lb
Test weight	65,000 lb
Test payload	31,700 lb
mpg and gal/100 mi at 65 mph	7.2 mpg, and 13.9 gal/100 mi
ton-mpg and gal/1,000 ton-mi	114 ton-mpg and 8.77 gal/ 1,000 ton-mi

The transmission choice for Phase 1 was a 12-speed dual clutch unit (DOE VTO, 2013). A dual clutch transmission eliminates power interruptions during shifts. The elimination of power interruptions enables performance improvements for both WHR and turbocompounding, since they no longer need to deal with the rapid changes in engine gas flow and temperature that are characteristic of normal shifting with a manual transmission or an AMT. The dual clutch transmission also allows for further engine downspeeding, since shifts are both more comfortable and without power interruption. As a result, a driver will tolerate higher shift frequency with a dual clutch than with an AMT.

Advanced lower-friction lubricants were used in all power train components as well as axle bearings. These lubricants include synthetic low viscosity oil for the transmission and axle. Volvo provided the committee with baseline test results for its 2009 model truck. These results were measured at a steady speed of 65 mph on level ground. Results are averaged for two directions to minimize the impact of wind and minor grade fluctuations. These results, shown in Table 8-9, cannot be directly compared to results from the SuperTruck test program, which are run over a defined on-highway drive cycle. However, the baseline test results do give a good view of the starting point for the Volvo SuperTruck work. SuperTruck results against the program goals for Phase 1 are shown in Table 8-10. Phase 2 (final) results of the Volvo program are expected in 2016.

Volvo Phase 2 Plans and Progress

The Phase 2 vehicle is referred to as VEV-2 or as the demonstrator. A very aggressive approach was taken for weight reduction for the demonstrator. A prototype ultralightweight frame assembly was designed in 2013 and delivered on schedule in the first quarter of 2014 (see Figure 8-3). The weight savings achieved was 800 lb, exceeding the internal target of 40 percent compared with the equivalent steel frame ladder. The subsequent assembly of axles and chassis-mounted components uncovered no issues with the design. Work is under way to build a second vehicle for track evalu-

TABLE 8-10 Volvo SuperTruck Phase 1 Results[a]

Target	Status	Complete?
50% engine BTE at cruise	48% engine + WHR BTE demonstrated	Planned for 2015
50% freight efficiency increase on a long-haul drive cycle (33% reduction in load specific fuel consumption)	43% demonstrated over 6,000 mi of road tests	Planned for 2015
55% engine BTE	Work under way; no plan for full engine test	Planned for 2015

SOURCE: Amar (2014).

[a]The results presented in this table are based on data collected by the committee prior to May 2015. At the DOE Annual Merit Review in June 2015, John Gibble of Volvo presented the following results: "50% BTE engine component development is complete. System integration and test is ongoing" (Gibble, 2015). Volvo also reported simulation results suggesting that 56.2% BTE could be possible, and 48% BTE without waste heat recovery. The final demonstrator vehicle build is under way. However, the committee was not in a position to review these results from the June 2015 AMR.

ation and data collection and to perform further analysis on the chassis assembly.

For the Phase 2 (final) demonstrator truck, Volvo selected its 11 L engine, which is 400 lb lighter than the 13 L used in VEV-1 (Gibble, 2014). The 11 L is calibrated to produce 425 hp, compared to 485 hp for the 2009 baseline 13 L engine. Volvo expects the 11 L to at least match the 1,650 lb-ft torque of the 13 L. It projects that the lower vehicle power demand will result in comparable vehicle performance, although there may be slight misses in low-speed acceleration rate and in speed on steep grades. Similar fuel efficiency improvement technologies that were incorporated into the 13 L engine in Phase 1 are included in the Phase 2 11 L engine: high-pressure common rail fuel injection, revised piston bowl geometry, reduced friction power cylinder components, advanced lube and coolant pumps, and turbocompounding. The final demonstrator vehicle will also include an aftertreatment system with new developments to improve conversion efficiency and reduce package size.

For the Phase 2 WHR system, Volvo is focusing its efforts on weight and cost reduction and improved reliability. Working toward the goal of 50 percent BTE, Volvo has operated three engines in test cells, as well as six component-level test stands. These activities are maturing the various technologies in parallel.

In order to maximize overall HVAC system efficiency, cab insulation was increased, and the efficiency of the heating and cooling systems was improved. The energy management system is designed to always select the most efficient energy source/storage system to power typical hotel mode loads (DOE VTO, 2013).

To investigate the potential for various idle reduction concepts, it was necessary to understand the detailed energy usage and balance over a 24-hour period. Several shorter road cycles were combined with a number of stops and engine-off events to form 24-hour cycles. The proportion of the different types of roads (flat, hilly, etc.) was verified by Volvo to be representative of typical North American long-haul operation. A 24-hr electrical load consumption profile was developed using representative electrical configurations, historical weather conditions, and other factors, in order to help size and optimize the hotel load systems. With such a

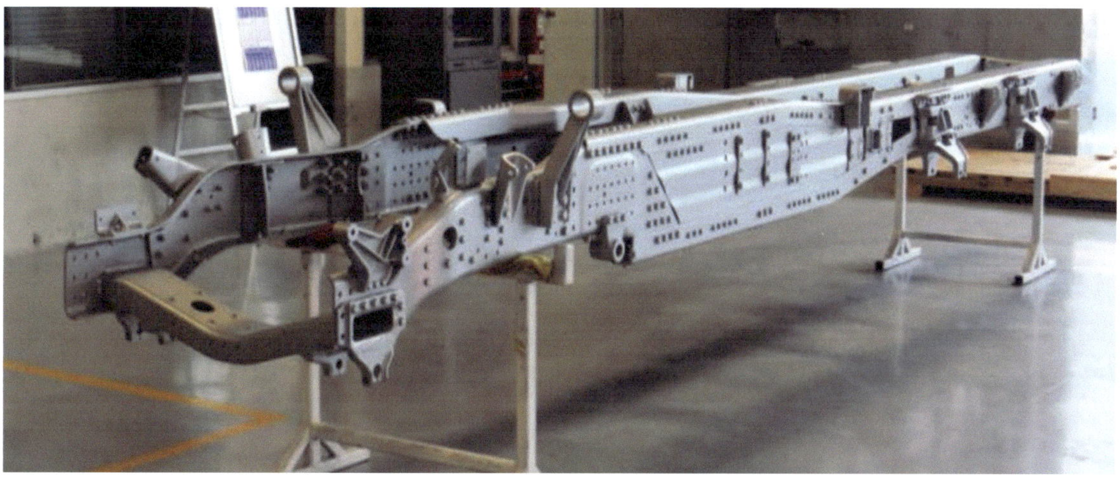

FIGURE 8-3 Prototype Volvo aluminum tractor frame. SOURCE: Amar (2014).

load profile, it was possible to establish rough requirements for energy storage capacity and potential fuel savings.

The requirements established for the APU included 10,000 Btu for cooling, 10,000 BTU for heating, flexibility to operate during driving and when parked in hotel mode, and the ability to operate from a battery pack or from shore power. A supplier has been selected, and the design direction was decided in early 2013. The first prototype system was bench tested in late 2013, and the first chassis installation took place early in 2014 to identify any changes necessary to the design prior to the final assembly.

The application of intelligent controls includes both a more fuel-efficient "look-ahead" GPS-type cruise control and the management of power-consuming auxiliaries. In cruise, the vehicle will legally accelerate on downgrades but will hold a gear and reduce speed slightly while cresting a hill. Auxiliaries will be engaged during downhill operation to maximize the use of available vehicle kinetic energy. Modeling work done by Volvo predicts that these intelligent controls will provide the following fuel economy improvements: rolling terrain 3 to 5 percent, and hilly terrain 5 to 8 percent. The development of the Phase 2 power train and vehicle is progressing. A technology package to enable the 11 L engine to meet the 50 percent BTE target at cruise has been defined, and performance development work is under way. Volvo told the committee that it plans to have the final demonstration vehicle completed and begin testing it in the fourth quarter of 2015.

Volvo has listed some issues and barriers that need to be dealt with before certain SuperTruck technologies can go into production (Gibble, 2014):

- Cost-effective and timely evaluation of advanced components and configurations;
- Added weight, packaging difficulty, and complexity of certain technologies;
- Reduced aftertreatment efficiency at low exhaust temperatures (a natural result of engine efficiency improvement and vehicle power demand reduction);
- Integration of interdependent technologies; and
- Operational effectiveness and end-user acceptance of advanced concepts.

Volvo's 55 percent BTE effort is discussed in Chapter 3.

Volvo Program Management

Volvo, like the other original equipment manufacturers (OEMs), also noted how SuperTruck allowed much more extensive technology development and learning from system integration. A large integration program like SuperTruck allows:

- A wide scope of technologies to be evaluated,
- A range of technologies from short term to exotic to be evaluated, and
- Vehicle-level targets drive more innovation than component-level targets.

Volvo noted that a number of graduate students have been developed as a result of SuperTruck funding, and several of them are now working in the industry, making use of the ideas they worked on under the program. Volvo expects some of the features developed under the SuperTruck program to be in production soon. The committee was shown a 2016 model tractor, which incorporates some aerodynamic features first developed for SuperTruck. Volvo had some suggestions for potential follow-on projects:

- Platooning,[7]
- Regional haul, and
- Vocational trucks

The discussion around regional haul focused on one main question: Is the regional haul duty cycle different enough from the long-haul cycle to drive different technical solutions? This is a question that could itself form the basis of a research project. One difficulty with a vocational truck project is that there are so many types of vocational operations to choose from, not one of which accounts for a really large portion of medium- and heavy-duty vehicle fuel consumption. Some care would be needed to define a vocational project that would develop technologies applicable across a fairly wide range of vocational applications.

NAVISTAR PROJECT

The Navistar SuperTruck project is unique among the projects in one way: the Navistar project was initiated in the fourth quarter of 2010 but put on hold in the third quarter of 2012. Navistar resumed work on November 1, 2014, after a 2-year pause. As a result, this project will be completed well after the other SuperTruck projects. The current schedule calls for completion at the end of 2016. Navistar provided a listing of project partners and suppliers in 2011, but in its November 2014 presentation to the committee, it stated that it had reduced the list of partners from 8 to 2 (Zukouski, 2014, slide 2) without identifying them to the committee. The Navistar presentation to the committee did list Wabash as the trailer partner, Michelin as the tire partner, and Lawrence Livermore National Lab as an aerodynamics partner. A revised list of partners and suppliers has not been provided to the committee. Navistar presented their plans and results to date to the committee on November 18, 2014 (Zukouski, 2014). Navistar will retain the baseline 13 L engine, with extensive changes to reduce friction, reduce heat loss, improve combustion efficiency, and reduce parasitic

[7] See Chapter 9 for further discussion of platooning.

losses. Work is under way to improve the efficiency of the air handling system (EGR, turbocharger, ports, and valve events). Navistar has not defined if or by how much the SuperTruck power and torque curves might vary from the baseline engine.

Navistar's November 18 presentation shows that conventional diesel combustion was retained, but with revisions to the combustion system (compression ratio, piston bowl, injector specification, and calibration) (Zukouski, 2014). Several turbocharger efficiency options will be evaluated, including ball bearings, e-boost (an electric motor/generator that can either help spool up the turbocharger when more boost is required, or extract electrical energy from the exhaust when boost is adequate), and turbocompound. However, initial results with turbocompounding show a net loss at the cruise point (Zukouski, 2014, slide 14). Changes were made to reduce cylinder kit friction. A variable-speed water pump and variable-displacement oil pump are used to reduce parasitic power, and low-viscosity oil is employed at a higher than normal operating temperature in an effort to reduce friction. The aftertreatment system was upgraded to allow for higher engine-out NO_x and lower back pressure. The engine was calibrated to produce higher engine-out NO_x than the baseline engine. In response to a question during the public session of the November 18 committee meeting, Navistar stated that engine-out NO_x would increase to 6 g/ hp-hr.

The original Navistar plan called for a 360 kW, 700 V series hybrid system. This system proved to be very heavy, expensive, and complex for a modest fuel saving performance, so it was dropped from the project. Navistar is still considering a 48 V mild hybrid system that would recuperate some braking energy and allow for electrification of accessories. The system includes a 48 V motor/generator, a supercapacitor, an electric A/C compressor, and a nickel-zinc (NiZn) battery (Zukouski, 2014, slide 25).

Navistar plans to use a direct drive automated manual transmission combined with a prototype axle ratio of 1.91:1 (Zukouski, 2014, slide 8). This would provide a cruise rpm of just under 1,050 rpm at 65 mph in a direct drive top gear. An alternative ratio of 2.05:1 would allow a cruise speed of about 1,125 rpm with a direct-drive top gear. Navistar projects the fuel savings from a direct top gear to be 1.5 percent, with additional fuel savings coming from the low cruise rpm.

Navistar is considering a couple of aerodynamic technologies that are not in the other SuperTruck team plans. One is active ride height and pitch control, which is the subject of an invention disclosure. Another item is an active trailer gap/flow control device. The list of controls features such as smart cruise control, and parasitic power demand reduction features is similar to that of the other SuperTruck teams. One possibly unique feature is "intelligent air control with dryer," which is not further described (Zukouski, 2014, slide 31).

Navistar's weight reduction plans include a large weight reduction effort on the trailer, where a 3,700 lb weight reduction is projected (Zukouski, 2014, slide 29). Trailer features include composite nose, sides, roof, rear door, and skirts, along with a 1.125 in. thick aluminum composite floor, aluminum cross members, light weight landing gear, and a new boat tail. The tractor weight reduction is targeted at 3,250 pounds, including:

- Composite cab (240 lb),
- MMC brakes (tractor and trailer, 600 lb),
- Light driveshaft and axles (190 lb),
- Aluminum wheel hubs and bearings (150 lb),
- Aluminum engine and transmission mounts (50 lb),
- Aluminum intensive rear suspension (370 lb),
- Thin wall, advanced material, and downsized fuel tanks (500 lb),
- Aluminum wide base single wheels (350 lb), and
- Michelin wide-base single tires (350 lb).

Navistar Project Results

Table 8-11 lists the results achieved by the Navistar team as of April 2015. Keep in mind that work has just resumed, after a 2-yr hold.

Navistar's 55 percent BTE effort is discussed in Chapter 3.

Navistar Program Management

Because Navistar has been out of the SuperTruck program until recently, the committee is not in a position to review program management.

OVERALL SUPERTRUCK PROGRAM REVIEW

The consensus of the committee is that the SuperTruck projects have provided a significant advancement in the state of the art. By combining a large number of technologies into complete, running vehicles, many useful results have been obtained, such as

- Some entirely new technologies have been developed and implemented;
- Some existing technologies have had their performance improved;
- Technology combinations that have never been tried before were evaluated;
- Cost/benefit information on many technologies has been developed, although most of this information will remain proprietary to the companies involved;
- Participating companies have improved their fuel efficiency simulation and test capability;
- Participating companies have already selected certain technologies for either continued development or production implementation; and
- Importantly, participating companies have the information to decide which technologies are not worth further development.

TABLE 8-11 Navistar Project Results to Date[a]

Target	Status	Complete?
50% engine BTE at cruise	47.4% BTE engine only	Late 2015
50% freight efficiency increase on a long- haul drive cycle (33% reduction in LSFC)	Demo vehicle is being designed	Late 2016
68% freight efficiency increase on a 24 hr duty cycle (40.5% reduction in LSFC)	Demo vehicle is being designed	Late 2016
55% engine BTE	Dual fuel work under way at ANL	Late 2016

[a] The results presented in this table are based on data collected by the committee prior to May 2015. At the DOE Annual Merit Review in June 2015, Russ Zukouski of Navistar presented the following results: Engine BTE of 48.3% demonstrated, and 50% planned in early 2016 (Zukouski, 2015). The remaining plans presented in the 2015 AMR match the table above. However, the committee was not in a position to review these results from the June 2015 AMR.

The two teams that have finished their vehicle demonstrations (Cummins-Peterbilt and Daimler/Detroit) have produced results substantially better than the targets called for in the SuperTruck contracts. Their results are very impressive and represent substantial achievements by the two teams. Both teams just cleared the 50 percent engine BTE target, but they went well beyond the goals for overall freight efficiency improvement. It is clear to the committee that competition among the teams drove them to go well beyond the contract requirements. The remaining two teams are expected to try to match or beat the results already posted.

The program could be criticized on the grounds that the four SuperTruck teams have in many cases converged on similar technical solutions. However, there are enough differences between the four trucks to provide alternative approaches to many technologies. For example, there are hybrid and nonhybrid vehicles, and there are single and dual clutch transmissions. There is also some variation in the approach to aerodynamics. One team chose an integrated tractor-trailer approach, where the vehicle must be used as a set. Other teams chose an independent approach, where the SuperTruck tractor could pull a standard trailer, or the SuperTruck trailer could go behind a current tractor. Several other features could be listed that are unique to one or two of the teams.

Another difficulty with the SuperTruck program is the lack of a common point for comparing the results of the four SuperTruck teams. Each team chose its own baseline (a 2009 model engine and vehicle, not all of which were identically equipped) and its own test routes (similar in concept, but never identical). As a result, if two teams both state that they achieved an x percent reduction in fuel consumption, this does not necessarily mean that they ended up with identical fuel consumption. Their baselines and test routes differ, so x improvement for one team cannot be directly compared to x for another team. The same issue applies to results such as x miles per gallon or y gallons per 1,000 ton-miles. Since these results were obtained on different test routes, they are not directly comparable. Unfortunately, there are no plans to test either the baseline vehicles or the final demonstration vehicles under comparable conditions.

Another issue that the committee found in the SuperTruck program is that teams are allowed to take any net reductions in empty vehicle weight and increase the payload to maintain the same operating weight. For vehicles that run at maximum legal weight (80,000 lb in most cases), this is an appropriate approach. However, many vehicles "cube out," i.e., run with the trailer full at a gross vehicle weight (GVW) of less than 80,000 lb. To account for this, the SuperTruck program specified a 65,000 lb test weight. To allow the teams to take credit for increased payload at a 65,000 lb GVW artificially inflates the apparent benefits of the weight reductions achieved by the SuperTruck teams. A more realistic accounting for the benefit of weight reductions would factor in the limited percentage of vehicles that can actually turn an empty weight reduction into increased payload.

DOE has selected the working GVW for 21CTP at 65,000 lb, which is near the average GVW value but not necessarily representative of typical truck operating weights. In practice, trucks tend to run near 80,000 lb when full (with exceptions), and then near 34,000 lb when running empty. This is shown by weigh-in-motion (WIM) data collected by DOT, as shown in Figure 5-4 (Quinley, 2010). The figure shows that 65,000 lb is actually a rather unusual operating condition.

It is also clear that a 65,000 lb truck is an inefficient vehicle by international standards in terms of cargo mass capacity (the amount of payload weight that can be carried, which in turn affects LSFC). This makes a 65,000 lb truck inconsistent with the essential requirements of an exemplar future truck. The consequences of low-cargo-weight limited vehicles could be significant, because more efficient vehicles would require fewer trips for a given freight task, thereby reducing fuel use and emissions by virtue of the lower number of trips required (exposure). Weight limits would also influence crash frequency, given that they are related to exposure (Woodrooffe, 2001; Montufar et al., 2007). A further possible unintended safety consequence of choosing the lower GVW target of 65,000 lb is that engine downsizing and power reduction become viable strategies for reducing fuel consumption, which can have an impact on vehicle speed performance in the fully loaded condition, particularly on grades. The resulting speed differentials relative to other vehicle classes may constitute a safety hazard. It appears that the four SuperTrucks in the current program have enough power to comply with the contract requirement calling for comparable performance, with the possible exception of speed on a long grade in the case of the Daimler/Detroit and Volvo trucks.

Safety systems development and evaluation were not part of the contractual requirements of the SuperTruck program, so there has been no reported safety development activity.

The committee did not observe any apparent safety issues with the SuperTruck vehicles. One technology that will require some safety related review as development continues is the waste heat recovery systems. These systems often use a flammable working fluid under considerable pressure, so provisions to mitigate fire risk must be in place.

Appropriate Federal Role

The SuperTruck program covered a range of technologies from relatively straightforward items that can be quickly implemented to high-complexity, high-risk, long-term technologies such as WHR systems. It could be argued that the government role does not extend to the short-term, low-risk technologies, but never before have so many fuel-saving technologies been brought together into actual vehicles for on-road testing. The committee finds that the industry on its own would never have put together such an extensive vehicle integration project. The high level of learning and advancement of the state of the art that came from the SuperTruck program is exemplary for the sort of results that government-sponsored R&D is meant to achieve.

SuperTruck Future

The committee believes that there are roles for a range of R&D projects under the 21st Century Truck Partnership. These could range from gathering and analyzing data from field operations, to the development of specific technologies, to large technology integration programs like SuperTruck. The clear benefits of the SuperTruck programs mean that follow-on projects would be welcome. However, budget realities mean that it will not be possible to fund four projects of such a magnitude at the same time in the foreseeable future. An alternative approach would be to competitively fund one large program every year or two. Potential future topics are discussed in the Program Management sections of this chapter and in the Findings and Recommendations section.

One of the ideas for a follow-on project—a SuperTrailer program—is brought up in Recommendation 8-2. The idea for such a program stems from the following circumstance: Trailer manufacturers have little engineering capability in aerodynamics despite the fact that trailer aerodynamics accounts for a significant portion of the fuel savings for all of the SuperTruck programs. Because trailer makers have limited capability, truck OEMs and specialist suppliers took the lead in developing SuperTruck trailer aerodynamic features. A SuperTrailer program could help to grow an engineering capability among trailer manufacturers.

In considering possible future vehicle integration programs, DOE and 21CTP need to consider a number of issues, such as

- Does the regional haul duty cycle differ enough from long haul that it would result in a significantly different technical solution from the current SuperTruck program?
- Can a vocational SuperTruck project be defined that covers enough vocational applications to represent a worthwhile portion of total medium- and heavy-duty fuel consumption?
- Would a SuperTrailer program have sufficient research content to justify government funding?
- Should a project target 55 percent BTE at the best point, or should some other metric be devised that better represents actual on-road engine efficiency?

At the February 17, 2015, committee meeting, Ken Howden of DOE explained that the DOE is planning to start a SuperTruck 2 project in 2015. Under the current plan, there will be a single winning team, and the goal will be a 100 percent improvement in freight efficiency over a 24-hour duty cycle (50 percent LSFC reduction). This project is meant to build on the existing accomplishments of the SuperTruck teams. The project may also involve consideration of regional haul applications. This presentation was made before Daimler/Detroit announced its 115 percent freight efficiency improvement over a 24-hour cycle, so the goals may be revised.

21CTP Partnership Responses to NRC Phase 2 Recommendations

Recommendation 3-2 in the NRC Phase 2 report (NRC, 2012) asked DOE to ensure that the 50 percent engine BTE requirement gets a sufficient share of the SuperTruck funding and that previous DOE-funded research be utilized to give a good chance of success. The Partnership replied that this was indeed the case, and that the contractors working on the SuperTruck programs had already participated in a number of DOE-funded research projects related to engine efficiency, with considerable success. Given the success to date of the three SuperTruck teams who have been working continuously on the 50 percent BTE goal since 2011, the committee agrees that the Partnership has done a good job of promoting development of 50 percent BTE engines. Two SuperTruck teams have met the 50 percent BTE goal, and the other two teams have plausible plans in place to meet the goal.

The NRC Phase 2 Recommendation 5-1 suggested that the Partnership consider setting a stretch goal of a 40 percent reduction in aerodynamic drag, compared to the existing goal of 30 percent. The Partnership responded that it would consider adjusting its goals,based on results obtained from the SuperTruck program. Cummins-Peterbilt claims a 46 percent C_d reduction from 2009,[8] while Daimler/Detroit Diesel has shown 39 percent C_d reduction in scale-model wind tunnel testing (Singh, 2014) and Volvo has demonstrated 30 percent C_d reduction but plans to reach over 40 percent.[9] It now appears that all four SuperTruck programs will probably approach or even exceed the 40 percent target suggested in the NRC Phase 2 report.

The NRC Phase 2 Finding 6-3 and Recommendation 6-3 concluded that the Delphi solid oxide fuel cell APU prototype was, after 10 years of DOE funding and development, far from meeting performance criteria suitable for truck applications. A reassessment of the solid oxide fuel cell program was called for to determine whether it made sense to continue development. This recommendation was not taken up by the Partnership, but after another year of continuing development issues, the two SuperTruck teams using the solid oxide fuel cell dropped it. Delphi has since discontinued its effort to develop the solid oxide fuel cell APU.

The NRC Phase 2 Recommendation 8-1 asked that LSFC be used as the metric for the SuperTruck program (gallons per 1,000 ton-miles). The Partnership responded that the fuel economy metrics of ton-miles per gallon had been part of the initial solicitation and could not be changed. However, the Partnership promised that it would "present the results for SuperTruck ... in terms of reductions in fuel consumption ... wherever possible." In reviewing reports on SuperTruck activity, such as the Annual Merit Reviews and presentations made to the committee, the committee finds no indication that the Partnership has actually started to provide results in terms of LSFC. Indeed, the LSFC values in this report had to be calculated by the committee from fuel economy data provided by the SuperTruck teams.

The NRC Phase 2 Recommendation 8-2 asked that the SuperTruck contractors agree on "at least one common vehicle duty cycle that will be used to compare the performance of all three [now four] SuperTruck vehicles." The Partnership responded that different OEMs were developing with different customers in mind, and that using a common test protocol for all the vehicles would be prohibitively expensive. The committee finds the first objection to be questionable, given that all four SuperTruck projects are aimed at long-haul vehicles with identical payloads and performance targets. However, while getting all four vehicles together for joint testing (or at least the three vehicles likely to be available soon) would admittedly be quite expensive, getting accurate data for comparison does require running vehicles over the same route at the same time. The committee believes that a comparison test under identical conditions would be very useful, and suggests that additional funding be found to support such a test.

FINDINGS AND RECOMMENDATIONS

Finding 8-1. Overall, the committee finds the SuperTruck program to be a great success, and that the system integration aspect of SuperTruck was a key to the program's success. The SuperTruck program drove technology development at a faster pace than industry would have achieved on its own. SuperTruck teams used the program to do the following:

- Increase both test and analysis capabilities, and improve the correlation between test and analysis;
- Use simulation results to drive improved experimental techniques, and use experimental results to help improve simulation techniques;
- Integrate combinations of technologies that had never been tested on a complete vehicle before;
- Learn about opportunities, issues, and trade-offs with fuel saving technologies in real-world vehicle testing; and
- Understand the challenges that must be overcome in order to make certain technologies cost effective.

Finding 8-2. The Cummins-Peterbilt and Daimler SuperTruck teams have met the goal of an engine with 50 percent brake thermal efficiency (BTE) at the cruise power point, and the other two teams are working to meet this goal. The Cummins and Daimler teams have also exceeded by a wide margin the goal of a 50 percent increase in freight efficiency (33 percent reduction in LSFC) over a long-haul drive cycle. The other two teams are also working to meet or exceed the program goal in 2015 (Volvo) and early 2016 (Navistar).

Finding 8-3. The Cummins-Peterbilt SuperTruck team has comfortably exceeded a self-imposed goal of a 68 percent increase in freight efficiency (40.5 percent reduction in LSFC) over a 24-hr long-haul duty cycle. They achieved an 86 percent increase in freight efficiency (46 percent reduction in LSFC). The Daimler Trucks North America team demonstrated a 115 percent increase in freight efficiency (53.5 percent reduction in LSFC) on a different 24-hour duty cycle. This 24-hour goal does not apply to the Volvo program, and Navistar's status is yet to be determined.

Finding 8-4. Fuel-saving technologies such as extensive aerodynamic features, a WHR system, and an APU to handle hotel loads all add substantial weight (3,850 pounds in the

[8] E. Koeberlein, Cummins-Peterbilt, "Cummins SuperTruck Program: Technology and System Level Demonstration of Highly Efficient and Clean, Diesel Powered Class 8 Trucks," presentation to the committee, May 14, 2014.

[9] A. Greszler, Volvo Group Truck Technology "SuperTruck: Development and Demonstration of a Fuel-Efficient Class 8 Highway Vehicle," presentation to the committee, May 15, 2014.

Cummins-Peterbilt truck). This weight penalty represents a significant challenge for improved efficiency trucks, and project teams had to work very hard and implement some very expensive weight reduction features to achieve an overall vehicle empty weight reduction.

Finding 8-5. All project teams report that the bulk of the aerodynamic improvements achieved in SuperTruck projects result from features added to the trailer, despite the fact that most of the engineering effort went into tractor improvements. This highlights the critical role of the trailer in achieving real-world fuel savings.

Finding 8-6. Using the results available to date, about 26 to 44 percent of the total vehicle fuel savings are due to engine efficiency improvements, while about 56 to 74 percent are due to vehicle power demand reduction. In the Cummins-Peterbilt project, 42 percent of fuel savings are due to the engine and WHR, 14 percent to tractor aerodynamics, 28 percent to trailer aerodynamics, and 15 percent to tire and driveline improvements. In the Daimler SuperTruck, engine improvements account for 26 percent of the total fuel savings while 74 percent is a result of vehicle power demand reductions, including the effect of the hybrid system.

Finding 8-7. SuperTruck project results show a limited potential impact on long-haul duty cycles for hybrid systems using currently available technology. Much of the benefit of a hybrid system can be captured with much less expensive and not as heavy alternatives, such as a GPS-based cruise control that uses the vehicle as a kinetic energy storage device. Microhybrid systems (smart control of auxiliary power demand, possibly combined with limited energy storage to handle auxiliary and/or hotel loads) may be a more promising hybrid approach for long-haul trucks.

Finding 8-8. Additional component-level R&D is required to generate new technologies that could enable a future SuperTruck program to exceed the results achieved by the current program. Promising areas of research include the engine and power train system, controls features, and the trailer and its integration with the tractor.

Finding 8-9. The SuperTruck vehicles incorporate technologies with a wide range of production readiness: Some will be going into production soon; some will never become cost-effective with technology that is now known. The outstanding fuel savings achieved in this program thus need to be treated carefully. Actual production vehicles achieving SuperTruck fuel savings may not be cost-effective for several decades, unless fuel costs increase substantially.

Finding 8-10. The SuperTruck contract goals required testing at 65,000 lb gross combined weight (GCW) in recognition of the fact that many trucks cube out rather than gross out (fill the trailer with cargo without reaching the weight limit). The goals also allow teams that achieve an empty weight reduction to add freight to maintain the 65,000 lb GCW. This gives a very large benefit in terms of freight efficiency for any weight reduction achieved. Since an operator that cubes out would not be able to take advantage of a weight reduction in practice, the project goals tend to overemphasize the benefit of vehicle weight reduction.

Finding 8-11. The SuperTruck program allowed each OEM to select different 2009 baseline vehicles and test routes, so the results are not directly comparable. This limits the ability to compare the results of the four vehicles.

Finding 8-12. Although it did not conduct a detailed safety analysis, the committee believes that it is unlikely that most of the efficiency technologies under consideration in the SuperTruck program will have a negative impact on safety. However, due to elevated temperatures and pressures and potentially flammable working fluids, the safety aspects of WHR systems will need to be considered during their development.

Finding 8-13. DOE is still using fuel economy (FE) in miles/gallon and freight efficiency in ton-miles/gallon for their fuel use metric, while NHTSA regulations that were published 5 years ago use fuel consumption (FC) in gallons/100 miles and load specific fuel consumption (LSFC) in gallons/1,000 ton miles. When experimental or modeling studies of percent improvement for technologies are calculated, a 50 percent increase in FE results in a 33 percent reduction in FC. FC and LSFC are the correct metrics to use since they are used in the regulations and they also multiply directly by miles driven to get fuel usage while FE and freight efficiency are inversely related to miles driven to get fuel usage. This nonlinear relationship is harder to understand without doing a calculation.

Recommendation 8-1. The SuperTruck demonstration vehicles represent a very large investment. DOE should consider ways of extracting additional research results from this investment by using the trucks that have been built to evaluate additional technologies. Some possibilities include:

- Evaluation of additional technologies, such as microhybrid;
- Comparison of SuperTrucks on identical test cycles, with additional work to help understand any differences in performance;
- Vehicle evaluation of hardware resulting from future system or subsystem research projects; and
- Exploration of a range of routes and payloads to determine the sensitivity of technologies to various applications.

Recommendation 8-2. Because of the great value demonstrated by the SuperTruck program, DOE should maintain at least one vehicle integration project at any given time. Owing to likely funding limitations, however, it will probably not be possible to have three or four similar projects running. A range of integration projects are possible, such as:

- A regional haul SuperTruck;
- A heavy-duty vocational SuperTruck (refuse, dump, etc.);
- A SuperTrailer program (to help trailer manufacturers build engineering capability); and
- A delivery truck of Class 3, 4, 5, or 6.

Recommendation 8-3. Any future system integration program with more than one team should entail performance testing on identical duty cycles, so that differences in the performance of specific technologies can be better understood. The vehicle should also maintain the acceleration and speed-on-grade performance of the baseline truck.

Recommendation 8-4. Future complete vehicle programs should account for the benefit of weight reductions in an appropriate way (at 80,000 lb only for a tractor-trailer), taking into account that because only a portion of a vehicle fleet runs at the legal weight limit, only that portion of the fleet could carry additional cargo if vehicle empty weight is reduced.

Recommendation 8-5. It is important for the 21CTP, probably through DOT, to monitor and analyze in detail the technologies implemented in the SuperTruck projects to verify that they do not have a negative effect on safety, since one or more of these technologies may be considered for future production vehicles.

Recommendation 8-6. DOE should use FC and LSFC in its studies in order to be consistent with EPA/NHTSA regulations and to provide in the literature the percent improvements in terms that relate to the metrics used in the regulations. Also, DOE should take the lead in changing the culture so that FC and LSFC metrics become accepted by industry.

REFERENCES

Amar, P. 2013. 2013 FY Progress Report for Vehicle Systems Simulation and Testing: Energy Efficiency and Renewable Energy, Vehicle Technologies Office.

Amar, P. 2014. Development and Demonstration of a Fuel-Efficient Class 8 Highway Vehicle: Vehicle Systems. Volvo SuperTruck. DOE Annual Merit Review VSS081, Washington, D.C., June 19.

Amar, P. 2015. Volvo SuperTruck. DOE Annual Merit Review VSS081, Arlington, Va., June 11.

Anderson, D. 2014. Vehicle Technologies Office Vehicle and Systems Simulation and Testing. 21st Century Truck Partnership. Presentation to NRC Committee on Review of the 21st Century Truck Partnership, Phase 3, Washington, D.C., September 3.

Damon, K. 2014. DOE SuperTruck Program: Technology and System Level Demonstration of Highly Efficient and Clean, Diesel Powered Class 8 Trucks. Peterbilt Motors Company. DOE Annual Merit Review ARRAVT081, Washington, D.C., June 19.

DOE VTO. (Department of Energy Vehicle Technologies Office). 2013. Volvo 2013 FY Progress Report for Vehicle Systems Simulation and Testing. Energy Efficiency and Renewable Energy Vehicle Technologies Office. U.S. Department of Energy.

Fleet Owner Magazine. 2010. Cummins Picks VanDyne SuperTurbo for SuperTruck Project. Highbeam Business, August 17.

Gibble, J. 2014. Powertrain Technologies for Efficiency Improvement. Volvo SuperTruck. DOE Annual Merit Review ACE060, Washington, D.C., June 20.

Gibble, J. 2015. Volvo SuperTruck–Powertrain Technologies for Efficiency Improvement. DOE Annual Merit Review ACE060, Arlington, Va., June 12.

Koeberlein, E. 2014. Cummins SuperTruck Program: Technology and System Level Demonstration of Highly Efficient and Clean, Diesel Powered Class 8 Trucks. Cummins-Peterbilt. DOE Annual Merit Review ACE057, Washington, D.C., June 20.

Koeberlein, E. 2015. Cummins SuperTruck Program Technology and System Level Demonstration of Highly Efficient and Clean, Diesel Powered Class 8 Trucks. Cummins-Peterbilt. DOE Annual Merit Review ACE057, Arlington, Va., June 12.

Montufar, J., J. Regehr, and G. Rempel. 2007. Long Combination Vehicle (LCV) Safety Performance in Alberta 1999–2005. Alberta, Canada: Alberta Transportation. http://www.transportation.alberta.ca/Content/docType61/production/LCVFinalReport2005.pdf.

NRC (National Research Council). 2008. *Review of the 21st Century Truck Partnership*. Washington, D.C.: The National Academies Press.

NRC. 2010. *Technologies and Approaches to Reducing the Fuel Consumption of Medium- and Heavy-Duty Vehicles*. Washington, D.C.: The National Academies Press.

NRC. 2012. *Review of the 21st Century Truck Partnership, Second Report*. Washington, D.C.: The National Academies Press.

Quinley, R. 2010. WIM Data Analyst's Manual, FHWA Report IF-10-018, June 2018. http://www.fhwa.dot.gov/pavement/wim/pubs/if10018/if10018.pdf.

Rotz, D. 2014. Super Truck Program: Vehicle Project Review. Daimler Truck North America LLC. DOE Annual Merit Review ARRAVT080, Washington, D.C., June 19.

Singh, S. 2014. SuperTruck Program: Engine Project Review. DOE Annual Merit Review, Washington, D.C., June 20.

Singh, S. 2015. SuperTruck Program: Engine Project Review. DOE Annual Merit Review ACE058, Arlington, Va., June 12.

Volvo Trucks. n.d. Volvo I-Shift. Volvo Group North America. http://www.volvotrucks.com/SiteCollectionDocuments/VTNA_Tree/ILF/Products/Powertrain/i-shift_brochure_050510.pdf.

Woodrooffe, J. 2001. Long Combination Vehicle (LCV) Safety Performance In Alberta 1995–1998. Alberta, Canada: Alberta Transportation. http://www.transportation.alberta.ca/Content/docType61/production/LCVSafetyPerformanceReport.pdf

Woodrooffe, J., and D. Blower. 2013. Heavy Truck Crashworthiness: Injury Mechanisms and Countermeasures to Improve Occupant Safety. UMTRI Report 2013-41.

Zukouski, R. 2014. SuperTruck–Development and Demonstration of a Fuel-Efficient Class 8 Tractor and Trailer. Navistar SuperTruck. Presentation to NRC Committee on Review of the 21st Century Truck Partnership, Phase 3, Washington, D.C., November 18.

Zukouski, R. 2015. SuperTruck–Development and Demonstration of a Fuel-Efficient Class 8 Tractor and Trailer, Engine Systems. DOE Annual Merit Review ACE059, Arlington, Va., June 12.

9

Efficient Operations

INTRODUCTION

There are many available technologies that can be applied to reduce the fuel consumption of trucks, but there are other ways of saving fuel that do not require any changes to vehicle or engine technologies. The overall fuel consumption of a truck fleet can be influenced significantly by the ways in which the vehicles are operated and maintained. Factors such as how close to full payload the trucks operate, whether they run on the most efficient routes, and even driver training can play a role in determining overall fleet fuel consumption. In addition, regulations can directly affect fuel consumption by constraining or promoting technology implementation and efficient operations. The infrastructure on which trucks operate also affects fuel consumption, through factors such as speed fluctuation and congestion. Electronic features can be added to the truck to modify the performance of the engine or vehicle in ways that can save fuel. All of these possibilities fall under the category of "efficient operations." The U.S. Department of Energy (DOE) and the U.S. Department of Transportation (DOT) proposed efficient operations as a new area for work under the 21st Century Truck Partnership (21CTP) in 2011. At that time no programs or work had been initiated in the 21CTP in this new area. The 21CTP's proposal for work on efficient operations was initially laid out in a February 2011 draft white paper, "Reducing Fuel Consumption in U.S. Trucking—A DOE-DOT Joint Study and Whitepaper" (DOE-DOT, 2011). In this draft, the two agencies explored opportunities to improve the efficiency of trucking operations. The paper focused on two areas:

- Opportunities for joint research and development (R&D) effort between the DOE and the DOT and
- Opportunities for modifying regulations (primarily DOT regulations) in ways that could permit more efficient operations.

The draft white paper became available just as the National Research Council (NRC) Phase 2 report (NRC, 2012) was being prepared. The Phase 2 committee devoted Chapter 9 of that report to a review of the draft white paper, as well as suggestions for improvement. Chapter 9 of the NRC Phase 2 report provided information on numerous approaches to making operations more efficient:

- Vehicle maintenance;
- Optimization of packaging for goods to be shipped, to increase the number of units per truckload;
- Load management optimization;[1]
- Route optimization;
- Supply chain optimization to limit the amount of shipping required;
- Infrastructure improvements;
- Application of Intelligent Transportation Systems (ITS);
- Expanding driver training to include fuel efficiency;
- Applying driver management controls features; and
- Reconsideration of regulatory constraints.

The NRC Phase 2 report recommended a number of changes to the draft white paper, to broaden its scope and include more areas of potential cooperation among the agencies. For example, research on ITS usually does not take freight efficiency into account, although this can be a significant benefit of ITS applications. The Phase 2 report recommended consideration of high-productivity vehicles, with higher weight limits and/or larger cubic capacity. The MAP-21 project on truck size and weight is now exploring some of these options. Finally, the Phase 2 report recommended that specific productivity goals be set in the final version of the white paper.

[1] Load management refers to efforts by trucking companies to ensure that trucks run as close to full payload as possible over the shortest distance needed to make deliveries.

In February 2013, the Partnership issued an updated roadmap and technical white papers (21CTP, 2013). Goals were stated for the areas of interest for the Partnership. Many of the approaches to making operations more efficient identified in the Phase 2 report and listed above were included in the white paper. Areas not addressed in the white paper include vehicle maintenance, optimization of packaging, load management optimization, route optimization, and infrastructure improvements. The goals from the February 2013 white paper for efficient operations are repeated below, along with the understood progress toward each goal.

GOALS AND STATUS

Goal: Minimizing Impact of Driver Behavior for Optimal Acceleration

Develop and demonstrate technologies that minimize the impact of driver behavior for optimal acceleration efficiency by automatically controlling vehicle accelerations at a level for which the engine operates in its most efficient operational state for the current environment. Driver feedback information devices can also be implemented as a retrofit option for existing vehicles.

Status

This goal has been met. Various vehicle acceleration control products are available today in the U.S. heavy-duty industry. One such product from Cummins is SmartTorque2 and Vehicle Acceleration Management (VAM). SmartTorque2 automatically senses vehicle weight, grade, and operating gear and then selects the optimum torque for the best fuel consumption and performance in every gear. VAM is a unique electronic feature that is enabled on Cummins ISX15 ST2 engines in the SmartAdvantage Powertrain. It controls the acceleration rate from the launch of the vehicle, maintaining more consistent acceleration for a more efficient transition through the gears. Cummins claims that VAM results in smoother acceleration and reduced driveline wear, while improving fuel consumption (Cummins, Inc., 2013).

Industry has introduced products that go beyond acceleration control to improve fuel consumption. These are referred to as predictive adaptive cruise control. Simply stated, these products use knowledge of the vertical terrain via loaded or learned maps and the Global Positioning System (GPS) to control the engine and transmission to optimize the use of vehicle kinetic energy. This approach is effective at improving fuel consumption in rolling or hilly terrains. Such products include Volvo's I-See, currently available in Europe, and Daimler's new Intelligent Powertrain Management (IPM). Daimler Trucks North America announced on December 5, 2014, that IPM will be standard on all of Detroit Diesel Corporations's DT12 automated manual transmissions paired with any heavy-duty Detroit Diesel engine, beginning in March 2015.[2]

Predictive adaptive cruise control technology is being used in all the SuperTruck projects as well. Additional features are being developed under SuperTruck, such as intelligent control of auxiliaries.

ITS technologies that reduce traffic congestion and improve traffic flow, as well as technologies such as adaptive cruise control, have good potential to reduce fleet fuel consumption.

Goal: Tools to Estimate Fuel Savings of Advanced Technologies

Develop simple tools for the trucking industry that will provide estimates of the fuel savings potential of advanced efficiency technologies and technology combinations depending on specific usage information of a particular fleet (measured drive cycle data). The tools will provide cost and benefit analyses for the selection of technologies on a case-by-case basis when representative drive cycles for an individual fleet or owner-operator are available (and recommendations to the fleet for obtaining the drive cycles can be provided).

Status

The committee is not aware of any activity toward this goal under the Partnership. The committee is aware that all the major OEMs have very sophisticated tools that can take duty-cycle data and payload data and use it to optimize a vehicle specification for a given customer. These tools are widely used to optimize vehicle specifications for customers. As a result of this situation, it may not make sense for government agencies to invest in the development of additional tools.

Goal: ITS/Connected Vehicle Technologies to Reduce Fuel Consumption

Conduct a study to identify proposed ITS/connected vehicle technologies that offer significant fuel savings and quantify the reduction in fuel consumption for technologies that offer the greatest benefits. Select one technology, evaluate the benefits for fuel consumption as a function of market penetration and identify the infrastructure needs and costs for deployment of the technology to a level at which the benefits of vehicle-to-vehicle (V2V) and vehicle-to-infrastructure (V2I) networking are realized.

[2] Detroit Diesel Corporation is a subsidiary of Daimler Trucks North America.

Status

Vehicle Platooning[3]

The National Renewable Energy Laboratory (NREL) and Peloton have conducted controlled testing to evaluate the fuel economy improvement potential provided by heavy truck "platooning" (DOE Project VSS001; Walkowicz et al., 2014). SAE J1321 Type II fuel economy tests were performed under controlled track conditions. Two platooned SmartWay tractors were used: 2011 Peterbilt 386 line haul sleepers with Cummins ISX 450 engines. The two trailers were 53-ft van bodies with side skirts. The testing was conducted at 55, 65, and 70 miles per hour (mph), with vehicle gaps from 20 to 75 ft and gross vehicle weights (GVWs) of 65,000 and 80,000 lb. Fuel savings were seen with both the lead and following tractor-trailers (truck). The best "team" fuel saving was 6.4 percent at 65,000 lb, 55 mph, and 30 ft following distance. The lead truck saw increasing benefit with closer following distances at all speeds. The savings ranged from 1.7 to 5.3 percent at 65,000 lb. The trailing vehicle saw savings of 2.8 to 9.7 percent at 65,000 lb. Savings on the trailing truck were reduced at the closer following distances due to a higher percentage of engine fan-on time. The authors comment that in order to maximize the savings for the trailing truck, the following distance should be adjusted based on coolant temperature to minimize the engine fan-on duty cycle.

It can be concluded that line-haul fuel savings are possible through platooning. Such close following distances for heavy trucks raise safety concerns so that additional study and testing are required to address this concern. V2V communication should greatly improve the safety aspects of platooning.

Look-Ahead Driver Feedback and Power Train Management

The objective of this project is to develop and demonstrate on real vehicles a driver assistance technology to reduce commercial fleet average fuel consumption by at least 2 percent (DOE Project VSS087; Verma, 2014). In this project, information from various sources, including radar, V2I, V2V, GPS, the vehicle data bus and a three-dimensional digital map are fed to an "intelligent driver assistance system." The system, through recognition of the environment and driver behavior, will estimate optimal fuel consumption behavior. A combination of power train control and advisory feedback provided to the driver via a human-to-machine interface (HMI) will maximize fuel savings with minimal distraction. The HMI algorithm and hardware have been developed and built. The data acquisition system is integrated and the entire system has been installed and validated on a prototype vehicle. The project has partnered with Con-Way to install and evaluate the system on two pilot vehicles. Preparation of the pilot vehicles is in progress.

Connected Vehicle Program

The Connected Vehicle program at DOT (ITS Joint Program Office) has several aspects that touch on 21CTP interests. A prime example is the Crash Avoidance Metrics Partnership (CAMP), the "connected vehicles and infrastructure" pilot project recently concluded in Ann Arbor. This 2012-2014 field project followed 16 heavy trucks and 3 buses fitted with safety warning equipment to avoid crashes. No fuel efficiency or derivative data were collected in Phase I of that pilot project. However, future embodiments of the research will be conducted under the University of Michigan's Mobility Transformation Center, and the scope will be broadened to include DOE input on desired measures of performance. It appears that the scope of the Connected Vehicle Program could be broadened to include commercial vehicle fuel consumption.

Goal: A Real-World Test Corridor to Improve Vehicle Operations

Establish a real-world test corridor for commercial vehicles focused on improving commercial vehicle operations, including fuel efficiency. The test corridor could have infrastructure content such as DSRC (dedicated short range communication) and Wi-Max technologies, to provide an environment compatible with future V2I communications. The concept would involve one or more fleets enabled for DSRC/Wi-Max capability and outfitted with various applications designed for improved efficiency of commercial vehicles.

Status

The committee is not aware of any Partnership activity addressing this goal.

Goal: Regulatory Changes to Replace Mirrors with Cameras

- Explore regulatory changes to permit the replacement of body-mounted mirrors with a camera-based system and quantify the fuel saving benefits associated with such a change.
- Quantify the fuel consumption penalty imposed by mirror regulations on highway-based commercial vehicle operations.
- Develop safety and robustness requirements for camera-based systems and conduct human factors research to develop and demonstrate equivalent safety of a camera-based system.

[3] Platooning refers to a convoy of two or more trucks linked electronically with an active driver in each.

- Assess procedural requirements for implementing the necessary regulatory changes and quantify the efforts required to modify regulations to permit camera-based systems in place of mirrors for Class 8 long-haul vehicles.

Status

The committee is not aware of any Partnership activity with this as a goal. A report by the Truck Manufacturers Association (NETL, 2007) shows that replacing mirrors with cameras has the potential to reduce fuel consumption at high-speed cruise by up to 3 percent, depending on the type of mirrors being replaced. The truck size and weight study being conducted under MAP-21 is examining triple trailer configurations (FHWA, 2014). The study will address fuel savings, safety, infrastructure, and potential freight diversion.

Goal: Use of Long Combination Vehicles

- Demonstrate the fuel savings benefits and develop policy guidelines for extending the use of long combination vehicles (LCVs), particularly triple-trailer units.
- Complete long-term in-fleet measurements to quantify the fuel savings of triple-trailer combination vehicles in comparison with single- and double-trailer operations in the same fleet, on a load specific fuel consumption basis (LSFC).
- Conduct an analysis to quantify the fuel savings if triple trailers are permitted on all interstate highways in the United States.

Status

The committee is not aware of any active programs with this goal under the Partnership.

Goal: Improved Supply Chain Management Strategies

- Promote improved supply chain management strategies in the commercial freight industry with an objective to increase the loads carried per truck and reduce vehicle miles traveled (VMT).
- Conduct a study to identify fleet best practices for supply chain management, and quantify the fuel savings that are achieved with efficient fleet operations vs. operations of fleets that do not have streamlined supply chains.

Status

The committee is not aware of active programs toward this goal under the Partnership. Since there is considerable commercial incentive to improve supply chain management, there is a lot of activity by the trucking industry on this topic. As such, it is not clear that this is an appropriate area for government-sponsored research.

RESPONSE TO RECOMMENDATIONS FROM THE NRC PHASE 2 REVIEW

In this section the recommendations made by the committee in the NRC Phase 2 report and the Partnership's responses are addressed.

NRC Phase 2 Recommendation 9-1. As suggested in the draft white paper on efficient operations, the DOE and DOT, in cooperation with the EPA and other agencies, should conduct joint research on efficient operations and should cooperate as appropriate on any regulations that affect fuel use and safety.

21CTP Response: The Partnership is pleased that the NRC panel recognizes the significance of efficient operations and supports the objectives of this new white paper. The 21CTP concurs with the recommendation for joint DOE and DOT research in efficient operations, and members of the Partnership are committed to supporting this effort. The Partnership is also interested in identifying opportunities for streamlining regulations that can improve operational efficiencies to achieve maximum benefit while maintaining or improving safety and other areas of operation.

Committee Comment on Response to 9-1

This response is quite positive, and various goals were established in the Partnership's February 2013 white paper.

***

NRC Phase 2 Recommendation 9-2. The available data show that trailer aerodynamic-improvement features and rolling resistance contribute significantly to overall vehicle fuel consumption. Therefore, the DOE and DOT should look in detail at options for trailer improvement.

21CTP Response: The Partnership agrees with the recommendation and supports research to quantify the fuel savings benefits of trailer technologies in addition to tractor-based technology improvements. A systems approach that considers all components of the vehicle is expected to provide the greatest benefits, and the Partnership understands the gains that can be achieved from trailer technologies and promoting their use and further development is a worthwhile pursuit. The DOE and DOT are committed to better understanding the real world fuel savings offered by advanced technologies for both tractors and trailers. For example, the DOE conducted a long-term in-fleet study in which new generation wide base single (NGWBS) tire fuel efficiency benefits were evaluated on both tractors and trailers. Such studies are very helpful in promoting the benefits of these technologies, and the Partnership supports the continuation of this type of research.

Committee Comment on Response to 9-2

The response gently sidesteps the actual recommendation regarding trailer aerodynamic and rolling resistance features. However, the SuperTruck participants are seriously pursuing improvements in trailer aerodynamics and are demonstrating its beneficial effects on fuel consumption.

***

NRC Phase 2 Recommendation 9-3. Traditionally, ITSs have been viewed as a way of improving safety. As suggested in the draft white paper on efficient operations, the DOT and DOE should conduct additional research and development devoted to exploiting the potential for reduced fuel consumption.

Partnership Response: The Partnership agrees with the NRC panel's finding and recommendation for further ITS-based R&D aimed at reducing fuel consumption. The 21CTP welcomes the opportunity to help advance this exciting new technology area to extend fuel savings even beyond what is possible with traditional engine and vehicle technologies. ITS technology is very unique in that it requires a strong infrastructure and participation of numerous vehicles before the full benefits can be realized, and the Partnership feels that its structure as a broad private-public partnership offers a unique opportunity to coordinate the development of critical infrastructure and standards while simultaneously deploying new technologies into the vehicle market.

Committee Comment on Response to 9-3

The Partnership's response is quite positive but lacking in any specific goals or plans.

***

NRC Phase 2 Recommendation 9-4. The DOE and DOT should work with the trucking industry to take advantage of the ideas, data, and experience that the industry can provide to accelerate efficiency improvements and to avoid unintended negative outcomes of efforts to improve trucking efficiency.

Partnership Response: The Partnership concurs with this finding and recommendation. Any efforts aimed at improved logistics management and trucking operations must be performed in collaboration with the trucking industry to ensure that best practices are not violated and any new proposed solutions can be effectively implemented with minimal or no negative consequences on fleet operations, safety or road damage.

Committee Comment on Response to 9-4

The Partnership's response is quite positive but lacking in any specific goals or plans.

***

NRC Phase 2 Recommendation 9-5. The DOT and DOE should look at the full range of high productivity vehicles in use in some U.S. states and around the world and review the literature available on the safety and fuel-saving performance of these vehicles. The assessment should take into consideration that the higher productivity of these vehicles can also be used to justify the implementation of additional safety technologies.

Partnership Response: The NRC panel pointed out a number of important additional points concerning high productivity vehicle use that were not highlighted in the draft white paper on efficient operations. A more thorough literature review in the white paper is appropriate for this topic and is planned for the next version of the white paper.

Committee Comment on Response to 9-5

The Partnership's response is quite positive, and while it provides a plan to address the issue, a more thorough literature review was not included in the February 2013 white paper. The white paper did establish a goal concerning the use of long combination vehicles (LCVs).

This recommendation is partially addressed by the Federal Truck Size and Weight study under Map-21. Provisions in MAP-21 require DOT to conduct a study addressing safety risks, infrastructure impacts, and the effect of enforcement for trucks operating at or within federal truck size and weight limits in contrast to more productive trucks legally operating in excess of federal limits. The fuel savings potential for these higher productivity vehicles is to be assessed, together with estimates of freight diversion from other modes that may occur as a result of their introduction. The study report was due to Congress by November 15, 2014, but is now anticipated in 2015.

***

NRC Phase 2 Recommendation 9-6. The DOT and DOE, in discussion with the Congress, should consider the recommendations of the Transportation Research Board regarding the establishment of a Commercial Traffic Effects Institute or a similar approach.

21CTP Response: The Partnership fully agrees with these findings and recommendations, including the consideration of recommendations made in TRB Special Report 267 (Transportation Research Board 2002). In particular the establishment of a Commercial Traffic Effects Institute.

Committee Comment on Response to 9-6

The Partnership's response is quite positive but lacking in any specific plan.

***

NRC Phase 2 Recommendation 9-7. Specific goals for efficient operations should be developed, with strong consideration given to exploiting the potential for intelligent transportation systems to reduce fuel consumption. In addition, priorities should be set for the R&D, testing, and data collection needed to analyze the benefits, drawbacks, and potential unintended consequences of removing barriers, including regulatory barriers, to the application of fuel-saving features. The draft white paper on efficient operations should be rewritten to take the findings and recommendations of the committee into account. The 21CTP partners, trucking fleets, and major suppliers should be involved in setting goals and research priorities.

Partnership Response: Although the draft white paper on efficient operations available at the time of the NRC's panel review did not include goals on this topic, the Partnership has added a set of specific goals that are consistent with this recommendation. A further rewrite of the draft white paper is also planned that will address the panel's recommendations. The Partnership is aware that many of the approaches proposed for efficient operations involve multiple complexities and agrees that detailed studied are needed to assess the benefits, drawbacks, and potential unintended consequences of removing barriers for efficient operations.

Committee Comment on Response to 9-7

As these goals are established they should have numerical targets and should be prioritized.

***

NRC Phase 2 Recommendation 9-8. The DOE and DOT should study the potential fuel savings from efficient operations in more detail, including a review of cost-effectiveness and ease of implementation. Once this information is available, goals, targets, and timetables for fuel savings from efficient operations should be established. Programs should then be developed and implemented to realize the available fuel savings.

21CTP Response: The Partnership agrees that research is needed to quantify the benefits as well as the costs and challenges of implementation associated with the proposed methods for efficient operations. Ultimately, the end goal is to implement these approaches for which the benefits clearly justify the costs, and the Partnership concurs with the NRC panel's recommended course of action to arrive at this objective.

Committee Comment on Response to 9-8

In order to set specific goals and R&D priorities, it will be necessary to do some up-front research. The committee's recommended project would quantify the potential benefits of various technologies on efficient operations, determine which ones are appropriate targets for government-funded research, and identify those technologies with the highest potential benefit.

FINDINGS AND RECOMMENDATIONS

Finding 9-1. The committee supports the effort by the Partnership to revise the Efficient Operations section of the February 2013 white paper to include specific goals for Efficient Operations. Unfortunately, activity and progress toward articulating these goals are limited.

Recommendation 9-1. In light of the limited activity in Efficient Operations, the Partnership should revisit, revise, and prioritize the goals to better reflect the areas where government-sponsored research can lead to significant fuel consumption improvements. A preliminary study may be needed to help set appropriate goals and priorities. 21CTP should revisit goals in two areas in particular: high productivity vehicles and expanded use of ITS to reduce fuel consumption.

Finding 9-2. As noted in Finding 9-3 of the NRC Phase 2 report, ITS technology holds considerable potential for improving fuel consumption in commercial vehicles.

Finding 9-3. The DOT's Connected Vehicle Program phase 1 pilot program did not collect data on fuel efficiency or related derivative data.

Recommendation 9-2. DOT should expand the scope of its Connected Vehicle Program to measure the effects of various technology implementations on fleet fuel consumption. Once the potential for fuel savings is clarified, fuel savings targets should be set for future projects.

Finding 9-4. The close following distances that are required for vehicle platooning with heavy trucks raise safety concerns.

Recommendation 9-3. 21CTP should conduct additional study and testing to address any potential safety concerns associated with platooning of heavy trucks.

REFERENCES

21CTP. 2013. *21CTP Roadmap and Technical White Papers* (February). Washington, D.C.: Office of Vehicle Technologies. http://energy.gov/eere/vehicles/downloads/roadmap-and-technical-white-papers-21st-century-truck-partnership. Accessed October 10, 2014.

Cummins, Inc. 2013. "ISX15 SmartAdvantage Powertrain – HD Truck (2013)," Cumminsengines.com. http://cumminsengines.com/smartadvantage#overview. Accessed September 1, 2015.

DOE-DOT (Department of Energy-Department of Transportation). 2011. Reducing Fuel Consumption in U.S. Trucking–A DOE-DOT Joint Study and White Paper. 21st Century Truck Partnership Efficient Operations White Paper Draft of February 15, 2011. Washington, D.C.

FHWA (Federal Highway Administration). 2014. MAP-21 Comprehensive Truck Size and Weight Limits Study. http://www.ops.fhwa.dot.gov/freight/SW/map21tswstudy/index.htm. Accessed April 24, 2015.

NETL (National Energy Technology Laboratory). 2007. Test, Evaluation, and Demonstration of Practical Devices/Systems to Reduce Aerodynamic Drag of Tractor/Semitrailer Combination Unit Trucks. Prepared for the National Energy Technology Laboratory by the Truck Manufacturer's Association, Contract DE-FC26-04NT42117, April.

National Research Council (NRC). 2012. *Review of the 21st Century Truck Partnership, Second Report.* Washington, D.C.: The National Academies Press.

Verma, R. 2014. Look-ahead Driver Feedback and Powertrain Management. Eaton Corporation. DOE Annual Merit Review VSS087. Washington, D.C., June 19.

Walkowicz, K., M. Lammert, A. Ragatz, A. Duran, and M. Simpson. 2014. Medium and Heavy Duty Vehicle Field Evaluations. National Renewable Energy Laboratory. DOE Annual Merit Review VSS001. Washington, D.C., June 17.

Appendixes

Appendix A

Biographical Sketches of Committee Members

John H. Johnson (*Chair*) is a presidential professor emeritus in the Department of Mechanical Engineering-Engineering Mechanics at Michigan Technological University (MTU) and a fellow of the Society of Automotive Engineers (SAE) and the American Society of Mechanical Engineers (ASME). His experience spans a wide range of analysis and experimental work related to advanced engine concepts, diesel and other internal engine emissions studies, fuel systems, and engine simulation. He was previously project engineer at the U.S. Army Tank Automotive Center and chief engineer in applied engine research at the International Harvester Company before joining the MTU mechanical engineering faculty. He served as chairman of the MTU mechanical engineering and engineering mechanics department from 1986 to 1993. He has served on many committees related to engine technology, engine emissions, and health effects—for example, committees of the SAE, the National Research Council (NRC), the Combustion Institute, the Health Effects Institute, and the Environmental Protection Agency—and consults to a number of government and private sector institutions. In particular, he served on many NRC committees, including the Committee on Fuel Economy of Automobiles and Light Trucks, the Committee on Advanced Automotive Technologies Plan, the Committee on the Impact and Effectiveness of Corporate Average Fuel Economy (CAFE) Standards, and the Committee to Assess Fuel Economy for Medium and Heavy-Duty Vehicles. He chaired the NRC Committee on Review of DOE's Office of Heavy Vehicle Technologies, the NRC Committee on Review of the 21st Century Truck Partnership, Phase 1, and the NRC Committee on Review of the 21st Century Truck Partnership, Phase 2. Dr. Johnson received from SAE the Horning Memorial Award, Colwell Merit Award (two), McFarland Award, Myers Award for Outstanding Student Paper, the Franz Pischinger Powertrain Innovation Award, and from ASME the Honda Medal and the Internal Combustion Engine Award. He received his Ph.D. in mechanical engineering from the University of Wisconsin.

Julie Chen is a professor of mechanical engineering and vice provost for research at the University of Massachusetts Lowell (UML). She was one of three founding codirectors of the UML Nanomanufacturing Center of Excellence and is also codirector of the Advanced Composite Materials and Textile Research Laboratory. From 2002 until 2004, Dr. Chen served as a program director for materials processing and nanomanufacturing at the National Science Foundation. She has been a NASA-Langley Summer Faculty fellow and an invited participant on three occasions in the National Academy of Engineering's (NAE's) Frontiers of Engineering Program. Dr. Chen has more than 25 years of experience in the mechanical behavior and deformation of fiber structures, fiber assemblies, and composite materials, with an emphasis on composites processing and nanomanufacturing. She recently served as a member of the NRC Committee on Benchmarking the Technology and Application of Lightweighting, which wrote the report *Application of Lightweighting Technology to Military Vehicles, Vessels and Aircraft*; the NRC Panel on Review of Manufacturing-Related Programs at the National Institute of Standards and Technology; and the NRC Panel on Air and Ground Vehicles Technology. Dr. Chen has co-organized several national and international symposia and workshops on composites manufacturing, including a National Science Foundation (NSF) composites sheet forming workshop, which led to an international benchmarking effort and the ASC International Symposium on Affordable Composites Manufacturing. Dr. Chen served as the technical program chair for the 2010 ASME International Mechanical Engineers Congress and as the ASME Materials Division chair. Dr. Chen holds B.S., M.S., and Ph.D. degrees in mechanical engineering, all from the Massachusetts Institute of Technology (MIT).

David E. Foster, the Phil and Jean Myers Professor Emeritus of Mechanical Engineering, received his B.S. and M.S. degrees in mechanical engineering from the University of Wisconsin-Madison in 1973 and 1975 respectively. He received his Ph.D. in mechanical engineering in 1979 from MIT. He was a faculty member at the University of Wisconsin (UW) after completing his Ph.D. He is an active member of the Engine Research Center (ERC), which he served as director from 1994 through 1999 and from September 2008 through December 2011. He was also the founding codirector of the General Motors–ERC–collaborative research laboratory, from its inception in 2002 until he retired in July 2012. Dr. Foster is a registered professional engineer in the State of Wisconsin and has won departmental, engineering society, and university awards for his classroom teaching. He was a member of NRC's standing committee to review the Partnership for a New Generation of Vehicles for 6 years, and has served on the Committee to Assess Fuel Economy Technologies of Medium- and Heavy-Duty Vehicles, the NRC committee to review the DOE FreedomCAR and Fuels Partnership Program, the 21st Century Truck Review, and USDRIVE program review. He has been the recipient of the Academic Contribution Award from JSAE, the UW Engineering Byron Bird Excellence in Research Publication Award, the ASME Honda Gold Medal for outstanding contributions in the field of personal transportation, the 2011 SAE Horning Award, and is a fellow of SAE.

Thomas M. Jahns (NAE), Grainger Professor of Power Electronics and Electric Machines at the University of Wisconsin-Madison, has been a driving force behind the development of high-performance permanent magnet (PM) synchronous machine drives, distinguished by magnets in their spinning rotors. Since early in his professional career at General Electric, Dr. Jahns has made important technical contributions leading to pioneering applications of PM drives in machine tools, home appliances, aerospace actuators, and electric vehicles. Drawing on these principles, nearly all hybrid and battery-electric passenger vehicles in high-volume commercial production today have adopted PM synchronous machines for their electric propulsion systems. A fellow of the Institute of Electrical and Electronics Engineers (IEEE), Dr. Jahns's many honors include the 2005 IEEE Nikola Tesla Technical Field Award that recognizes the significance of his PM machine contributions. He has served as president of the IEEE Power Electronics Society and as Division II director on the IEEE board of directors. Both the IEEE Industry Applications Society and the IEEE Power Electronics Society have recognized him as a Distinguished Lecturer. He has served on a number of NRC committees, including the Committee on Review of the 21st Century Truck Partnership, Phase 1; the Review for the Intelligent Vehicle Initiative, Phase 1, and the Committee on Advanced Automotive Technologies Plan. He earned his Ph.D. in electrical engineering from MIT.

Timothy V. Johnson is director for emerging regulations and technologies at Corning Incorporated. He tracks emerging mobile emissions regulations and technologies and helps develop strategic positioning for new products. He has been with Corning for 28 years, and has 18 years in his current position. He is a three-time recipient of the Lloyd L. Withrow Distinguished Speaker Awards from SAE, and in 2008 was named SAE fellow. He also received California's 2009 Haagen-Smit Clean Air Award. Dr. Johnson is active in various advisory committees with government agencies, universities, and private organizations and is a frequent speaker at international technical conferences. He is on the editorial board of three leading engine journals. He earned B.S. and M.S. degrees in engineering from the University of Minnesota in 1978 and 1979 and received a Doctor of Science from MIT in 1987.

Paul Menig is CEO of Tech-I-M, a consultancy. Previously he was employed by Freightliner, where he was responsible for daily production problems, field problems, custom work orders, and advanced engineering for electrical and electronic items such as engines, transmissions, brakes, and safety devices. Mr. Menig joined Daimler Trucks North America in July of 1994 and initially led the development of electronics for the new Freightliner Century Class truck product line. Before joining Freightliner, Mr. Menig spent 7 years with Eaton Truck Components, leading a team of as many as 65 people in the development of electronic products for automated mechanical transmissions, brakes, and tire pressure control. These activities included some worldwide responsibility and coordination with engineering in Europe and joint venture development with Japanese companies. Before that, Mr. Menig worked for the industrial automation part of Eaton known as Cutler-Hammer. During those 8 years he lead teams working on sensors, factory communications, programmable and motion controllers and vision inspection equipment. Earlier, Mr. Menig worked 5 years for General Electric in the areas of medical equipment for hospitals, remotely guided military vehicles (smart bombs), and charge- coupled device imagers and signal processors. He is currently cochair of the Future Truck/Far Horizon committee of the Technology and Maintenance Council of the American Trucking Associations and also serves on the NRC Committee on Approaches and Technologies for Reducing the Fuel Consumption of Medium- and Heavy-Duty Vehicles. Mr. Menig graduated from MIT in 1976 with a bachelor's degree in electrical engineering. He participated in the ABC program of General Electric, completing the A and B portions. Master's degree work in electrical engineering was completed with the exception of a thesis at Marquette University. In addition, Mr. Menig has participated in numerous training programs such as total quality management, software development, strategic planning, finance for the nonfinancial manager, ISO 9000, and vehicle dynamics.

James W. Morris is retired from his position as director, advanced engineering, Volvo Powertrain North America (VPTNA). Previous positions with Mack Trucks and Volvo

Powertrain included director, Product Development Laboratories and Engineering Services; chief engineer for the Mack ETECH Engine and Engineering Services; Mack engineering liaison to the Atlantis Project, a joint program between Renault and Mack to design and develop a new 13 L engine; chief engineer 10 L and 16 L product engineering; manager of emissions control; manager of engine development; and a number of the positions related to the development of not only engines but also electronic control engine products, transmissions, structures, and fuel systems. At Mack Trucks and Volvo he gained extensive expertise in development, design, validation, and production support of new technologies, especially for engines, drivelines, exhaust emissions, engine maintenance, and cost reduction, as well as knowledge of the needs and requirements of the trucking industry. He also served as a consultant from September 2010 to June 2012 for the Advanced Combustion Group of Volvo Power Train. He has a BSME from Pennsylvania State University.

Thomas E. Reinhart is an institute engineer in the Department of Engine Design and Development, which is part of the Division of Emissions and Vehicle Research at Southwest Research Institute. His previous positions include Cummins Inc., 1980-2000; Roush Industries, Inc., 2001-2004, and Visteon Corporation, 2004-2005. He leads projects in engine design, performance and emissions development, and gasoline and diesel engine noise, vibration, and harshness (NVH) improvement. Since 2007, he has led a number of projects to investigate technologies for improved engine, power train, and vehicle fuel efficiency and GHG reduction, focused on medium- and heavy-duty vehicles. Currently, Mr. Reinhart is leading projects for the National Highway Traffic Safety Administration (NHTSA) and the National Academies of Sciences, Engineering, and Medicine to evaluate the costs and benefits of various fuel efficiency technologies that could be applied to comply with future truck fuel efficiency regulations. Mr. Reinhart has served on two previous NRC committees: Technologies and Approaches to Reducing the Fuel Consumption of Medium- and Heavy-Duty Vehicles and Review of the 21st Century Truck Partnership, Phase 2. He is a member of SAE, the Institute of Noise Control Engineering (INCE), where he served on the board of directors from 2008 through 2011, and the International Institute of Acoustics and Vibration (IIAV). He has a B.S. and an M.S. in mechanical engineering from Purdue University.

Bernard Robertson (NAE) is president of BIR1, LLC, an engineering consultancy specializing in transportation and energy matters that he founded in January 2004 on his retirement from DaimlerChrysler Corporation. During the latter part of his 38-year career in the automotive industry, Mr. Robertson was elected an officer of Chrysler Corporation in February 1992. He was appointed senior vice president coincident with the merger of Chrysler Corporation and Daimler-Benz AG in November 1998 and was named senior vice president of engineering technologies and regulatory affairs in January 2001. In his last position, he led the Liberty and Technical Affairs Research group; advanced technology management and FreedomCAR activities; and hybrid electric, battery electric, fuel cell, and military vehicle development. In addition, he was responsible for regulatory analysis and compliance for safety and emissions. He is a member of the NAE, a fellow of the Institute of Mechanical Engineers (U.K.), a Chartered Engineer (U.K.), and a fellow of SAE. He has served on a number of NRC committees, including the Committee on Review of the Research Program of the U.S. DRIVE Partnership and the Committee on Review of the 21st Century Truck Partnership, Phase 2. Mr. Robertson holds an M.B.A. degree from Michigan State University, a master's degree in automotive engineering from the Chrysler Institute, and a master's degree in mechanical sciences from Cambridge University, England.

Subhash C. Singhal (NAE) is Battelle fellow emeritus, Pacific Northwest National Laboratory (PNNL). At PNNL he worked in the Energy Science and Technology Directorate after having worked at Siemens Power Generation (formerly Westinghouse Electric Corporation) for over 29 years. At PNNL, Dr. Singhal provided senior technical, managerial, and commercialization leadership to the laboratory's extensive fuel cell program. At Siemens Westinghouse, he conducted and/or managed major research, development, and demonstration programs in advanced materials for various energy conversion systems, including steam and gas turbines, coal gasification, and fuel cells. He was manager of fuel cell technology there, and was responsible for the development of high-temperature solid oxide fuel cells (SOFCs) for stationary power generation. In this role, he led an internationally recognized group in SOFC technology and brought this technology from a laboratory curiosity of a few watts to fully integrated, 200 kW power generation systems. He has authored 100 scientific publications, edited 17 books, received 13 patents, and given numerous plenary, keynote, and other invited presentations worldwide. Dr. Singhal is a member of the National Academy of Engineering; a fellow of four professional societies (American Ceramic Society, The Electrochemical Society, ASM International, and the American Association for the Advancement of Science [AAAS]); a senior member of the Mineral, Metals & Materials Society (TMS), received the Electrochemical Society's Outstanding Achievement Award in High Temperature Materials in 1994, and continues as the chairman of the Society's International Symposium on Solid Oxide Fuel Cells. He served as president of the International Society for Solid State Ionics during 2003-2005. He received the American Ceramic Society's Edward Orton Jr. Memorial Award in 2001; an Invited Professorship Award from the Japan Ministry of Science, Education and Culture in 2002; and the Christian Friedrich Schoenbein Gold Medal from the European Fuel Cell Forum in 2006. He serves on the editorial boards of Elsevier's *Journal of Power Sources* and the *Fuel Cell Virtual Journal* and is an associate editor of ASME's *Journal of Fuel Cell Science and Technology.* He has also served on many national and international advisory panels including those of the National Materials Advisory Board of the National Research Council, National Science Foundation, Materials Properties Council, U.S. Department of Energy, NATO Advanced Study Institutes and NATO Science for Peace Programs, United Nations Development Program (UNDP), United Nations Industrial Development Organization (UNIDO), International Energy Agency (IEA), and the European Commission. He has a B.S. in physics, chemistry, and mathematics from Agra University; a B.E. in metallurgy from the Indian Institute of Science; a Ph.D. in materials engineering and science from the University of Pennsylvania; and an M.B.A. in technology management from the University of Pittsburgh.

James A. (Jim) Spearot is currently president of his own consulting company, Mountain Ridgeline Consulting, LLC. His consulting efforts focus on transportation energy and automotive fuel and lubricant issues as they affect emissions and fuel efficiency. In 2009, Dr. Spearot retired from General Motors Research and Development Center, where he was director of the Chemical and Environmental Sciences Laboratory, whose mission was to develop cost-effective environmental strategies and systems for GM's products and processes. Additionally, Dr. Spearot served as chief scientist for GM's Public Policy Center, lead executive for research programs in Russia and CIS countries, and manager of GM's Hydrogen Storage Innovation Program. Dr. Spearot began his GM career in 1972 as an assistant senior research engineer in the Fuels and Lubricants Department. He was appointed department head of Fuels and Lubricants in 1992 and director of the Chemical and Environmental Sciences Laboratory in 1998. He is a member of several organizations: SAE, the Society of Rheology, the Society of Tribologists and Lubrication Engineers, and the American Institute of Chemical Engineers. He is a former chairman of the SAE Fuels and Lubricants Division and a former chairman of the Coordinating Research Council (CRC). He has served as chairman of the Fuels Working Group of the U.S. Council for Automotive Research (USCAR) and the USCAR Environmental and Hydrogen Technical Leadership Councils. His professional honors include an ASTM Award for Excellence in 1990; the Arch T. Colwell Merit Award from SAE in 1987; and the Award for Research on Automotive Lubricants, also from the SAE, in 1987. He is a fellow member of the SAE and has received a Lifetime Achievement award from USCAR. He holds a B.S. in chemical engineering from Syracuse University and master's and doctorate degrees, also in chemical engineering, from the University of Delaware.

Kathleen C. Taylor (NAE) is retired director of the Materials and Processes Laboratory at General Motors Research and Development and Planning Center. She was simultaneously chief scientist for General Motors of Canada, Ltd. in Oshawa, Ontario. Earlier Dr. Taylor was department head for physics and physical chemistry and department head for environmental sciences. Currently Dr. Taylor is a member of the DOE Hydrogen Technology Advisory Committee. Dr. Taylor was awarded the Garvan Medal from the American Chemical Society. She is a member of NAE, a fellow of SAE International and the American Academy of Arts and Sciences, and a foreign fellow of the Indian National Academy of Engineering. She has been president of the Materials Research Society and chair of the board of directors of the Gordon Research Conferences. She has served on many NRC committees, including the Committee on Review of the U.S. DRIVE Research Program, Phase 4, and the Review of the 21st Century Truck, Phase 2, and was a member of the Board on Energy and Environmental Systems. She has expertise in R&D management, fuel cells, batteries, catalysis, exhaust emission control, and automotive materials. She received an A.B. in chemistry from Douglass College and a Ph.D. in physical chemistry from Northwestern University.

John Woodrooffe heads Transportation Safety Analytics and is director of the Commercial Vehicle Research and Policy program at the University of Michigan Transportation Institute (UMTRI). He is responsible for the Center for National Truck and Bus Statistics, which conducts nationwide surveys of trucks involved in fatal accidents (TIFA) and buses involved in fatal accidents (BIFA), and for the Statistical Analysis Group, which performs analytical modeling and conducts research to advance statistical methods for road and vehicle safety analysis. He is an international expert on policy and safety evaluation of large vehicles, including stability and control, accident reconstruction, vehicle productivity, fuel use, and environmental impact. He has participated in many large international technical projects and has been a member of vehicle-related Organisation for Economic Development (OECD) technical expert working groups, most recently the OECD/JTRC project entitled "Heavy Vehicles: Regulatory, Operational and Productivity Improvements." This Paris-based international task force examined regulatory concepts and future truck technology for sustainable road transport. Before joining UMTRI, Mr. Woodrooffe founded the Road Vehicle Research Program at the National Research Council of Canada and developed it into a successful, internationally active heavy truck research laboratory. He served on the NRC Committee on Fuel Economy of Medium- and Heavy-Duty Vehicles, Phase 1, and is currently serving on the Committee on Fuel Economy of Medium- and Heavy-Duty Vehicles, Phase 2, and the Committee on Motor Vehicle Size and Weight. He was a consultant to Australia's National Road Transport Commission for a unique 3-year performance-based standards development project that produced a new performance-based regulatory system for large vehicle combinations. Mr. Woodrooffe holds master's and bachelor's degrees in mechanical engineering from the University of Ottawa.

Appendix B

Committee Meetings and Presentations

MAY 14-15, 2014, WASHINGTON, D.C.

Overview of DOE's Vehicle Technologies Office R&D
Patrick Davis, DOE Office of Vehicle Technologies

Overview of 21st Century Truck Partnership
Ken Howden, DOE Office of Vehicle Technologies

Overview of U.S. Army (NAC/TARDEC) Activities
David Gorsich, U.S. Army Tank Automotive Research, Development and Engineering Center (TARDEC)

Alternative Fuel Safety Projects at FMCSA
Brian Routhier, Federal Motor Carrier Safety Administration (FMCSA)

NHTSA Heavy Vehicle Safety Research
Alrik Svenson, National Highway Traffic Safety Administration (NHTSA)

Review of Previous NAS Review Recommendations, with Responses
Ken Howden, DOE Office of Vehicle Technologies

Cummins SuperTruck Team Presentation
David Koeberlein, Cummins Inc.

Daimler SuperTruck Team Presentation
Derek Rotz, Daimler Trucks North America

Volvo SuperTruck Team Presentation
Tony Greszler, Volvo Trucks North America

SEPTEMBER 3-4, 2014, WASHINGTON, D.C.

Introduction and Background on 21CTP Goals and Activities
Ken Howden, DOE Office of Vehicle Technologies

Overview of Advanced Combustion Engine R&D Program
Gurpreet Singh, DOE Office of Vehicle Technologies

Overview of Fuel and Lubricant Technologies R&D
Kevin Stork, DOE Office of Vehicle Technologies

Vehicle Systems Simulation and Testing
David Anderson, DOE Office of Vehicle Technologies

Electric Drive Technologies Overview
Steven Boyd, DOE Office of Vehicle Technologies

Energy Storage Overview
Brian Cunningham, DOE Office of Vehicle Technologies

Review of Materials Program
Jerry Gibbs, DOE Office of Vehicle Technologies

Update on Idling Reduction Activities
Glenn Keller, Argonne National Laboratory

Supporting Safe and Efficient Goods Movement on the Nation's Highways: An Overview of Research and Data Programs
Caitlin Hughes Rayman, Federal Highway Administration

Looking Ahead to the Next Phase of Heavy-Duty Greenhouse Gas and Fuel Efficiency Standards
Matthew Spears, Environmental Protection Agency

21st Century Truck Partnership Management Structures and Considerations
Ken Howden, DOE Office of Vehicle Technologies

NOVEMBER 18-19, 2014, WASHINGTON, D.C.

The Navistar SuperTruck Project: Development and Demonstration of a Fuel-Efficient Class 8 Tractor & Trailer Vehicle System
Russ Zukouski, Navistar

Oak Ridge National Laboratory (ORNL) Vehicle Systems Integration (VSI) Laboratory
David Smith, ORNL

Vehicle Technologies Office Connected and Automated Vehicles Overview
David Anderson, DOE Office of Vehicle Technologies

FMCSA Research and Technology
Luke Loy, Federal Motor Carrier Safety Administration

Technologies for MD/HD GHG and Fuel Efficiency
Tom Reinhart, Committee Member, Southwest Research Institute (SwRI)

Summary of ORNL Site Visit
Kathy Taylor, Committee Member

FEBRUARY 18-19, 2015, WASHINGTON, D.C.

Energy Efficiency and Renewable Energy FY 2016 Budget Request Presentation
Ken Howden, DOE Office of Vehicle Technologies

COMMITTEE SUBGROUP MEETINGS

Committee subgroups also made visits to Cummins Inc., Columbus, Indiana; Daimler Trucks North America, Portland, Oregon; ORNL, Oak Ridge, Tennessee; Detroit Diesel Corporation, Detroit, Michigan; U.S. Army TARDEC, Warren, Michigan; Volvo Trucks, Greensboro, North Carolina.

Appendix C

21CTP Responses to Findings and Recommendations from NRC Phase 2 Report

This document provides a compilation of the findings and recommendations from the National Academy of Sciences *Review of the 21st Century Truck Partnership, Second Report*, published in June 2012. This document also provides the 21st Century Truck Partnership's responses to these findings and recommendations, organized in groups by report section. Within each section, findings and responses have been grouped together (and responded to) by topic.

NOTE: Findings/Recommendations marked with ❶ were highlighted in the executive summary of the 2012 report and are considered of particular importance.

OVERALL FINDINGS

Subject	NAS Findings and Recommendations	Partnership Response
O-1 **Overall Observations: 21CTP Overview**	*FINDING S-1.* The key benefit of the 21st Century Truck Partnership is the coordination of research programs directed toward the goal of reducing fuel usage and emissions while increasing the safety of heavy-duty vehicles. Federal involvement is bringing stakeholders to the table and accelerating the pace of technological development. Given the federal regulatory requirements to reduce emissions and fuel consumption, it seems the sharing of research and development (R&D) costs between the government and U.S. manufacturers of trucks and buses or heavy-duty vehicle components is appropriate to develop new technologies. Thus, the 21CTP is providing access to the extraordinary expertise and equipment in federal laboratories, in addition to seed funding that draws financial commitment from the companies to push forward in new technology areas. The Partnership provides the United States with a forum in which the member agencies, in combination with industry, academia, and federal laboratories, can better coordinate their programs. The steady decline in research funding from FY 2003 through FY 2007 was threatening the attainment of program goals. The actual funding and need for R&D are discussed in Chapter 1. The funding level in the years prior to the availability of funding through the American Recovery and Reinvestment Act of 2009 (ARRA) was not in proportion to the importance of the goal of reducing the fuel consumption of heavy-duty vehicles and providing advanced technology for the industry to meet the 2014-2018 and later fuel consumption regulations. The ARRA funds provided by Congress in 2009-2010 have significantly enhanced the ability of the Partnership to meet and demonstrate the goals for reducing fuel-consumption and improving safety in prototype vehicles. *RECOMMENDATION S-1.* The 21CTP should be continued to help meet the nation's goal of reduced fuel consumption in the transportation sector. In addition, the Partnership needs to review whether additional partners—such as major truck and component manufacturers that are not currently members—that could contribute to the R&D program should be recruited. Research funding should be commensurate with well-formulated goals that are strategic to reducing the fuel consumption of heavy-duty vehicles while improving safety, and all projects should be prioritized so that the 21CTP R&D program can be implemented within the available budget.	The Partnership concurs with the recommendation for continuation of our efforts, and remains committed to reducing fuel use in medium and heavy truck transportation in line with our stated mission and goals. The Partnership appreciates that the NAS panel is taking note of the enhanced activities toward meeting our goals, especially the new SuperTruck initiatives to enhance Class 8 truck efficiency from a whole-vehicle perspective. The Partnership concurs that a review of additional partner participation should be undertaken. Through the SuperTruck projects, the Partnership has already been making contact with organizations beyond our existing partners, especially in the tire and trailer industries. In addition, the Partnership is beginning to build relationships with organizations by inviting them to present a technical discussion at full Partnership conference calls. The Partnership has also developed specific procedures for adding new members to the group, including the potential for adding associate members as appropriate. The Partnership concurs that well-formulated goals are critical to establishing research priorities and demonstrating success, and will continue to work collaboratively to revise the Partnership's goals as such revisions prove necessary. We will work to encourage the federal government agencies to prioritize heavy vehicle research efforts that are in line with these goals.

Subject	NAS Findings and Recommendations	Partnership Response
O-2 **Overall Observations: 21CTP Goals, and Response to Recommendations**	*FINDING S-2.* The 21CTP leadership responded substantively to most of the recommendations of the National Research Council's (NRC) Phase 1 review, which helped to contribute to the improved program that was the subject of this Phase 2 review. The committee commends the leadership of the Partnership for this effort. *RECOMMENDATION S-2:* The 21CTP program goals should continue to be established, reviewed, updated, related to available funding, and clearly stated in measurable engineering terms. The white papers defining the various technical areas of R&D should be reviewed and revised, as appropriate, periodically and prior to any future NRC review of the 21CTP. Given the "virtual" nature of the Partnership among 4 agencies and 15 industrial partners, the projects that are considered to be part of 21CTP should be better defined and, if part of the Partnership, indicated by a 21CTP notation in any 21CTP documentation.	The Partnership appreciates the recognition for its efforts in responding to the panel's Phase 1 review, and appreciates the opportunity to interact with the panel to gain this valuable feedback. The Partnership concurs that the 21CTP goals should remain relevant through periodic review and update. To that end, we will be conducting periodic reviews of the goals and the white papers, and revising them as needed to reflect current technical needs, funding availability, and Partnership direction. To the maximum extent possible, the Partnership will ensure that goals are relevant, related to available funding, and measurable. The frequency of such reviews has not yet been determined, but a yearly review has been suggested as appropriate. The Partnership concurs that it is important to have a clear definition of the projects considered to be part of 21CTP. For this reason, the Partnership is reviewing the potential to restart the project inventory activity that had been historically conducted to catalog 21CTP projects. In this way, the Partnership will be more readily able to define the list of projects relevant to 21CTP goals, and characterize the investments being made in technology research and development.

INTRODUCTION AND BACKGROUND

This section of the NAS report contained no findings or recommendations.

Management Strategy and Priority Setting

Subject	NAS Findings and Recommendations	Partnership Response
M-1 **Management: "Virtual Organization" and Management Structure**	*Finding 2-1.* The 21CTP is a virtual organization facilitating communication among four agencies, government laboratories, and industry, but it has no direct control over research activities or funding across the agencies or by its industry partners. The committee continues to believe that the lack of single-point 21CTP authority is far from optimal, although it recognizes that this is necessary because of the various congressional committees that the agencies report to and that provide their budgets. *Recommendation 2-1.* The Department of Energy (DOE) is urged to continue to improve the functioning of the 21CTP "virtual" management structure in every way possible. Such improved functioning would include strengthening interagency collaboration (particularly that involving the Environmental Protection Agency [EPA] and the Department of Defense [DOD]) and documenting and publishing specific 21CTP activity within all four agencies. (NOTE: Subsequent to the committee's review of 21CTP programs, the DOE and the DOD entered into the Advanced Vehicle Power Technology Alliance (AVPTA) partnership on July 18, 2011.)	The Partnership believes its informal virtual organization allows us to respond quickly to new initiatives and market changes, and is an advantage of the Partnership. We are consistently seeking ways to improve the operation of its virtual management structure to build on its strengths. We are exploring new communication methods for our members, such as a significantly revised internal 21CTP website. This new website, built on a SharePoint platform, is facilitating greater interaction and collaboration among government and industry partners through new tools and techniques available through SharePoint. The website was deployed in spring 2012. In addition, the Partnership has continued its efforts to strengthen interagency partnerships. The new Advanced Vehicle Power Technology Alliance (AVPTA) between DOD and DOE will assist in this regard by establishing clear research links between these agencies, specifically in the vehicle technologies area. DOE and DOT are maintaining and expanding their information exchange related to heavy truck fuel efficiency regulations, and the relationship of those regulations with 21CTP research efforts. DOE is working with EPA on a collaborative research program related to hydraulic hybrids, and maintains an information connection with the SmartWay Transport Partnership (partly through efforts with the DOE Clean Cities program). The Partnership plans to continue its schedule of meetings and conference calls to maintain connections among partners and to facilitate information exchange. The Partnership has encouraged the development of meetings at national laboratory sites whose expertise is relevant to 21CTP partner goals and objectives, and several new initiatives have resulted from these meetings.

Subject	NAS Findings and Recommendations	Partnership Response
M-2 Management: Need for Annual Reporting	*Finding 2-2.* The EPA, DOD, and Department of Transportation (DOT) did not have a well-defined list of the projects and associated budgets that were included under the 21CTP umbrella. This stems in part from the virtual nature of the Partnership and partly, particularly within DOE, from the natural overlap in activities on batteries, hybrids, materials, and other areas between the activities for light-duty vehicles and the 21CTP. Many of these activities are reviewed at the annual DOE Merit Review and at Directions in Engine-Efficiency and Emissions Research (DEER) conferences, and the new SuperTruck projects include an annual reporting requirement, but there is no dedicated report for the 21CTP. *Recommendation 2-2.* The DOE should issue a brief annual report documenting the specific projects within the 21CTP and the progress made. The annual report should provide references to published technical reports from the involved agencies. This would especially help outside groups, future review committees, the Congress, and others to understand the structure, activities, and progress of the Partnership.	The Partnership concurs that no dedicated 21CTP report is available to outline progress made on 21CTP-specific projects. We do note, however, that detailed information on DOE-funded projects is made publicly available through the Vehicle Technologies Program website, in annual progress reports and reviews at conferences such as the merit review and DEER. Duplication of these information dissemination efforts would not be the best use of 21CTP resources. The Partnership will, however, consider the development of an annual report that would document 21CTP project progress and be complementary to existing report products.

Engine Systems and Fuels

Subject	NAS Findings and Recommendations	Partnership Response
E-1 **Engine Technology: 50% Thermal Efficiency Goals**	*Finding 3-1.* The committee reviewed nine diesel engine programs that were funded at a total of more than $100 million by DOE and industry and that included the High Efficiency Clean Combustion (HECC) program, the Waste Heat Recovery (WHR) program, and others. Some programs met or exceeded their goals, for example achieving a 10.2 percent improvement in brake thermal efficiency (BTE) versus a 10 percent goal, whereas others did not quite meet the goals of 5 percent or 10 percent improvement in BTE. By combining HECC and WHR, each demonstrating greater than 10 percent improvement in BTE, together with other technologies, it should be possible to improve BTE by 20 percent to achieve the original DOE target of 50 percent peak BTE. However, the DOE target of 50 percent peak BTE was not met by the original goal of 2010. *Finding 3-1a.* The DOE has shifted the original target of 50 percent peak brake thermal efficiency by 2010 to a new target of 50 percent BTE at an operating point representative of vehicle load during highway cruise operation. This makes the efficiency target more difficult to meet and may require complex and expensive technology that extends beyond the technologies demonstrated on engines to date. These technologies will not necessarily be production-feasible or cost-effective. *(No Recommendations Provided)*	To respond to the recommendations provided in the first NAS review of 21CTP, DOE changed the operating point for the 50 percent thermal efficiency target within SuperTruck to be more representative of actual vehicle operation. Specifically, the operating point has been shifted from peak BTE to a load and speed point representative of 65 mph steady state operation (see Finding 3-6 from the first NAS review, and responses). The SuperTruck teams will be demonstrating achievement of the 50 percent BTE goal in actual engine operation as part of their development work. The Partnership agrees that achievement of this target at the 65-mph speed-load point will be challenging, and that new technologies may be required. DOE addresses the question of production feasibility of technologies through the use of 50 percent cost-shared research. The 50 percent investment of company funding to supplement the DOE funding ensures that the participating companies have a financial stake in the research. This in turn encourages participating companies to focus on research directions that will provide a more immediate payback for their investment through production hardware.
E-2 **Engine Technology: 55% Thermal Efficiency Goals**	*Finding 3-2.* The DOE-funded research in advanced engine combustion at the national laboratories, in industry, and at universities is well managed and addresses important aspects for achieving an integration of advanced combustion processes that should be important enablers for achieving the 55 percent BTE goal as well as providing ongoing improvements. There also appears to be good interaction between the researchers performing the work and the industry stakeholders. Efforts to achieve 55 percent BTE are going to require complex and expensive technologies. It will be some time before it becomes clear whether there is a production-feasible and cost-effective way to achieve the 55 percent BTE target. The committee believes that this target carries considerable risk, even at the test cell demonstration stage. *Recommendation 3-1.* The 21CTP fundamental research program should continue to provide important enablers for the 55 percent BTE goal, and DOE should continue to look for leverage opportunities with other government- and industry-funded projects.	The Partnership agrees with the need to continue research toward the 55 percent thermal efficiency goal, and has included this as a research goal for the SuperTruck partners (with technology scoping toward this goal being the major activity). The Partnership will continue to look for new opportunities to work together: one possible new collaborative arena is the recently announced partnership between DOE and the U.S. Army (the Advanced Vehicle Power and Technology Alliance). DOE is working with the U.S. Army to identify areas of common interest that could result in collaborative research efforts.

Subject	NAS Findings and Recommendations	Partnership Response
E-3 **Engine Technology: Future R&D-SuperTruck**	*Finding 3-3.* Future engine R&D for Goal 1, develop and demonstrate 50 percent BTE at over-the- road cruise conditions by 2015, and for Goal 2, research and develop technology pathways to achieve a stretch goal of 55 percent BTE in a 2010 emissions-compliant engine system in the laboratory by 2015, will be carried out under the SuperTruck program. The engine programs outlined by the three SuperTruck project teams appear to be comprehensive and are expected to achieve the 50 percent BTE goal, although there is risk in being able to achieve the goal at a cruise condition with the significantly reduced power demand level of the SuperTruck. Developing engine technology pathways to achieve the stretch goal of 55 percent BTE in an engine in a laboratory by 2015 is considered very high risk, but might be achievable. *Recommendation 3-2.* The DOE should ensure that the engine R&D for the goal of 50 percent BTE at over-the-road cruise conditions and the stretch goal of 55 percent BTE in an engine in a laboratory that will now be carried out under the SuperTruck program receive the appropriate share of the SuperTruck funding and benefit extensively from the DOE-funded research programs in advanced engine combustion.	Funding for advanced combustion engine research is coordinated and focused, and SuperTruck is a high priority within this research. Even though DOE funding is subject to annual appropriations, funding will be available to complete the engine R&D efforts contained in the SuperTruck teams. The teams are also receiving funds from other parts of program (including the Vehicle Systems and Materials sub-programs). Participating SuperTruck companies are also involved in the rest of the VTP R&D program (the advanced combustion MOU, the advanced engine crosscut team, and the Annual Merit Review), and are thus made aware of the DOE-funded advanced engine combustion programs. DOE's Annual Merit Review included the SuperTruck team members as active participants, and presented the entire research portfolio to them. This ensures that SuperTruck teams are aware of the portfolio and can harvest breakthrough results for their use.
E-4 **Engine Technology: Alcohol-Fueled Engines (EPA)**	*Finding 3-4.* The EPA has demonstrated that optimized E85 alcohol-fueled engines using conventional three-way catalysts for meeting 2010 emissions standards can achieve current diesel levels of BTE that can potentially provide engine technology suitable for both conventional and hybrid vehicles for the medium-duty fleet truck market. *(No Recommendations Provided)*	The Partnership appreciates the NRC panel's observations on the potential benefits of optimized E85 engines for efficiency and emissions, and the work that EPA has done to explore this technology. Additional work on E85-optimized engines was performed by DOE through three industry projects with Ford, Delphi, and Bosch. These successful projects, which ran from late 2007 to late 2010, sought to reduce fuel consumption of engines operating on E85 while meeting all prevailing emission standards.
E-5 **Engine Technology: HCCI (EPA)**	*Finding 3-5.* The EPA has developed an HCCI engine that operates in the HCCI mode at all times using low-pressure, port fuel injectors suited to the unique operating conditions of a series hydraulic hybrid vehicle. The unique operating conditions include a narrow range of operation at the best BTE condition for each engine speed, with only slow transient response times for changes in power demands. At these unique operating conditions, NOx and PM are below the levels required by the 2010 emissions standards without aftertreatment; HC and CO emissions are controlled with oxidation catalysts. *(No Recommendations Provided)*	The Partnership appreciates the NRC panel's highlight of EPA's HCCI engine research to support its hydraulic hybrid development efforts. The Partnership plans to continue work on HCCI engines for efficiency and emissions improvements, subject to funding availability.

Subject	NAS Findings and Recommendations	Partnership Response
E-6 **Engine Technology: Complementary DOE and DOD Programs**	*Finding 3-6.* The DOD has engine programs that are cooperative between industry and universities and have goals of improved BTE and other goals more specific to the Army. *Recommendation 3-3.* The DOD and the DOE should increase their awareness of one another's programs and look for opportunities to share technologies on areas of joint interest, such as thermal efficiency. One way to encourage interaction is for the DOE to invite DOD program participants to present their findings at the DEER (Diesel Engine-Efficiency and Emissions Research) Conference.	In 2011, DOE and the U.S. Army announced the formation of a research collaboration, the Advanced Vehicle Power and Technology Alliance. DOE is working with the U.S. Army to identify areas of common interest that could result in collaborative research efforts. This partnership should enhance the interaction between these federal departments: some areas of collaboration have already been identified. The U.S. Army also participates in meetings of the Diesel Crosscut Team and the light-duty USCAR partnership with DOE and industry partners. Incorporation of DOD presentations at the yearly DEER meeting will also be considered: DOD has presented papers at DEER in the past, and DOE's role as the chair for the meeting will ensure that DOD can have access to presenter slots as needed.
E-7 **Fuels Technology: Petroleum-Based Fuels**	*Finding 3-7.* In spite of efforts to reduce the fuel consumption of light-duty and heavy-duty vehicles and to develop biomass-derived fuels (an effort which, except for corn-based ethanol, has not progressed as much as had been expected), petroleum will remain the primary source of light-duty and heavy-duty vehicle fuel for many years to come. Whereas future U.S. gasoline demand is expected to be flat for the next 20 years, diesel fuel demand is expected to grow, necessitating changes in refinery operations. *Recommendation 3-4.* The DOE should reinstate its program for advanced petroleum-derived fuels (they will be transportation's primary fuels for many years to come) with the objective of maximizing the efficiency of their use.	The new, consolidated line incorporates the activities of both previous lines. Advanced petroleum based fuels are already the subject of a large portion of the projects supported under the new line.
E-8 **Fuels Technology: Lubricant Technology**	*Finding 3-8.* The DOE recognizes the importance of reducing truck powertrain friction and the need for improved lubricants that reduce fuel consumption. *Recommendation 3-5.* The DOE must work closely with industry in exploring improved lubricants that reduce fuel consumption, especially with regard to using such lubricants in existing truck engines and transmissions.	The lubricants activity is relatively new, but DOE has always strived to work with vehicle and engine OEMs, as well as oil and additive companies. DOE is currently partnered directly with vehicle OEMs, such as Ford and GM, on projects looking at next generation oils. DOE also has partnerships on projects with engine manufacturers, such as Cummins, to look at advanced engine oil additives. The program also interacts with OEMs to develop lower-friction engine components through participation in the MIT Lubrication in Internal Combustion Engines Consortium, which includes Daimler, Volkswagen, Volvo, Toyota, PSA, Renault, and Mahle. DOE intends to continue and expand these collaborations in the future. It is also important to note lubricants will likely never drive major decisions at either engine companies or oil companies; therefore, a government role is essential in assuring this social good, i.e., a 2% increase in fuel economy.

Subject	NAS Findings and Recommendations	Partnership Response
E-9 **Fuels Technology: Program Goals**	*Finding 3-9.* The DOE established three different sets of goals for the fuels program from 2008 to 2011, which made an assessment of progress against the goals difficult. In total, little progress has been made toward the achievement of these DOE goals, which were not specified goals of the 21CTP. *Recommendation 3-6.* The DOE fuel goals should be re-evaluated in line with the FY 2012 budget and the recommendations of this report. Specific plans for achieving these goals should be established.	We are continually open to reevaluation of our goals in light of budget changes. Recent budgets have been volatile, which complicates the effort -- e.g., between the FY12 Omnibus appropriation and the FY13 marks there has been a greater-than-40% cut -- but we will continue to reevaluate as appropriate.
E-10 **Aftertreatment: Program Activities**	*Finding 3-10.* The research agenda of the 21CTP is focused on improving the NOx reduction performance of selective catalytic reduction (SCR) and lean- NOx-trap systems, improving the efficiency of and reducing the fuel consumption associated with particulate matter (PM) filter regeneration, and improving the ability to model aftertreatment systems. The DOE Cross-cut Lean Exhaust Emissions Reductions Simulations (CLEERS) program does a good job of coordinating the aftertreatment research programs within the 21CTP and disseminating the results to the technical community at large. *Finding 3-11.* The demands on the aftertreatment system and its performance are intimately linked to the combustion process taking place within the cylinder. Consequently, the aftertreatment system must be developed and its performance evaluated in conjunction with the combustion system. The 21CTP realizes this, and its new goals for the aftertreatment program specifically state this. *Recommendation 3-7.* The aftertreatment program within the 21CTP should be continued, and DOE should continue to support the activities of CLEERS that interface with the activities of the aftertreatment technical community at large.	The Partnership agrees with this assessment to continue the aftertreatment programs. Combustion and aftertreatment activities are continuing under the SuperTruck projects, which are looking to achieve stretch efficiency goals while meeting current stringent emission standards: this produces a need for continuing aftertreatment research.
E-11 **Aftertreatment: Particle Number Emissions Studies**	*Finding 3-12.* Particulate size distribution is not a problem with current diesel-type combustion using DPFs. However, as new combustion processes, possibly using different fuels ranging from petroleum-derived fuels to bio-fuels and synthetics, are integrated into future engine operating maps, it is important to assess particulate size distribution characteristics if particulate filter designs are changed or if DPFs are not used. *Recommendation 3-8.* In light of the progress being made with new combustion technologies, which show potential for very low cylinder-out NOx and particulate emissions, the 21CTP should incorporate studies of particulate number emissions into their research portfolio.	The Partnership is aware of the evolving interest in particulate number regulation (number of particles and size distribution), especially in Europe. We are currently measuring these parameters in several projects with the national laboratories, universities, and industry.
E-12 **Health Impacts**	*Finding 3-13.* The Advanced Collaborative Emissions Study (ACES), the Collaborative Lubricating Oil Study on Emissions (CLOSE), and the project on Measurement and Characterization of Unregulated Emissions from Advanced Technologies are comprehensive and cooperative projects that are investigating important issues related to potential heavy-duty diesel engine health effects. Based on the activities reported, the committee finds a high degree of collaboration among government agencies, national laboratories, and industry stakeholders. *Recommendation 3-9.* The DOE should continue funding the Advanced Collaborative Emissions Study, the Collaborative Lubricating Oil Study on Emissions, and the project on Measurement and Characterization of Unregulated Emissions from Advanced Technologies until results are finalized and reported for all three studies.	The Advanced Collaborative Emissions Study will continue for FY 2013, and the other named projects have achieved their objectives.

Subject	NAS Findings and Recommendations	Partnership Response
E-13 **Propulsion Materials**	*Finding 3-14.* The propulsion materials program is addressing a broad range of materials issues associated with heavy-truck propulsion systems. Many of the initiatives are funded as cooperative R&D agreements (CRADAs) with significant industry cost sharing, showing strong support by industry for this area of work. *Recommendation 3-10.* The DOE should fund programs in the areas outlined in its "21st Century Truck Partnership White Paper on Engines and Fuels" (February 25, 2011) in the section "Approach to Reaching Goals" covering materials R&D for valve trains, major engine components, air-handling systems (turbochargers and exhaust gas recirculation [EGR] systems), and exhaust manifold sealing materials.	The Partnership agrees with this finding. The U.S. Department of Energy continues to fund research in materials that will enable improved efficiency in HD engines and after treatment devices.
E-14 **High-Temperature Materials Laboratory**	*Finding 3-15.* The HTML continues to be a valuable resource for materials research for the 21CTP, providing specialized and in many cases unique instrumentation and professional expertise. The expertise of those who oversee the laboratory, and therefore the value of HTML to all users, is enhanced by the participation of the HTML staff in the research. *Recommendation 3-11.* The DOE should continue to provide 21CTP researchers and other potential users access to HTML, and it should make every effort to maintain support for HTML and to maintain the cutting-edge capability of the facility. Moreover, DOE should provide sufficient funding for HTML, and for the research specialists who oversee and operate the facility, to enable continued research collaboration with the academic community, other government laboratories, and industry. In particular, HTML support should not be reduced to a level that allows only maintenance of the equipment for paying users.	The Partnership agrees with this finding, stating that the HTML is a valuable resource to 21CTP materials researchers. The prioritization of funding for DOE programs resides with Congressional budget authority and is beyond the scope of the 21CTP partnership.

Medium and Heavy-Duty Hybrid Vehicles

Subject	NAS Findings and Recommendations	Partnership Response
H-1 **Hybrid and Battery Goals**	*Finding 4-1.* Although 2012 has been established as the deadline for 21CTP Goals 1 and 2 for hybrid vehicle technology, it is unlikely that these will be met, as there has been no funding for either goal. However, with regard to Goal 2, the DOE Office of Vehicle Technologies' battery R&D program is supporting a large number of programs addressing issues ranging from fundamental materials research through battery development and testing. Significant progress has been made in developing domestic manufacturing facilities for battery materials production and recycling, cell production, and pack assembly. Although the applicability of these programs to heavy-duty applications was not provided to the committee, the committee believes that these developments are supportive of the needs of medium- and heavy-duty hybrid applications. *(No Recommendations Provided)*	The Partnership agrees with this finding. The DOE Vehicle Technologies Program's hybrid and electric system R&D program is supporting a large number of projects on development of advanced batteries, power electronics, and electric machines. Significant progress has been made in developing domestic manufacturing facilities for lead-acid and lithium-ion batteries, and for electric drive system components. These developments are directly applicable or broadly supportive of the diverse needs of the various highly differentiated medium- and heavy-duty hybrid vehicle mission profiles.
H-2 **Hybrid Goals**	*Finding 4-2.* The DOE did not receive any funding for heavy-duty hybrid R&D in FY 2007 through FY 2010. Consequently, no progress was reported toward the 21CTP's three heavy-duty hybrid goals, primarily focused on R&D, for achieving 15 years of design life, achieving cost goals for drive-unit systems and energy storage systems, and achieving a 60 percent improvement in fuel economy (38 percent reduction in fuel consumption). During this period, the DOE made progress in developing heavy-duty hybrid simulations and models and conducting fleet testing and evaluations of heavy-duty hybrid vehicles. *Recommendation 4-1.* The DOE should provide an up-to-date status with respect to the heavy-duty hybrid goals. The DOE should partition the available hybrid funds between heavy-duty and light-duty hybrid R&D technology to promote the R&D required for the development of heavy-duty hybrid technologies, since heavy-duty hybrid requirements are significantly different from light-duty requirements.	DOE does not have specific hybrid goals for light-duty hybrids. Research and Development (R&D) and corresponding goals are for component technologies (e.g. batteries, electric motors, etc...). These technologies and the R&D advances should be scalable across vehicle weight classes in many cases.
H-3 **ARRA-Transportation Electrification**	*Finding 4-3.* More than 1,800 commercial vehicles are being funded through the ARRA by the DOE to aid in the development and demonstration of plug-in hybrid electric vehicles and battery electric vehicles in fleets. One of the objectives of this program is to develop U.S. manufacturing capacity for all-electric drive components (energy storage, drive motors, power electronics, etc.). However, in at least one of these projects, the battery cells are being manufactured outside the United States. *Recommendation 4-2.* The DOE should determine what is needed for the battery cells and other electric drive components in the ARRA-Transportation Electrification programs aimed at development and manufacturing in the United States, as specified in the objectives of these programs.	The objective of the ARRA Transportation Electrification grants are to demonstrate, collect data, and evaluate potential grid impacts of electric-drive vehicles that are ultimately produced in the United States. While DOE encourages domestic sourcing of components used in the vehicles, there is no requirement that the components be manufactured in the United States.

Subject	NAS Findings and Recommendations	Partnership Response
H-4 **Hybrid Emissions Certification**	*Finding 4-4.* The EPA and DOT's National Highway Traffic Safety Administration (NHTSA) issued their final rules on September 15, 2011, for "Greenhouse Gas Emissions Standards and Fuel Efficiency Standards for Medium- and Heavy-Duty Engines and Vehicles." Although these standards contain test procedures for determining fuel consumption for heavy-duty hybrid trucks, a manufacturer still needs a certificate of conformity showing that a vehicle's internal combustion engine meets the EPA criteria emission standards for heavy-duty engines (a procedure that does not recognize hybrid heavy-duty trucks). The California Air Resources Board (CARB) is currently drafting vehicle-level test procedures for heavy-duty hybrid vehicles. *Recommendation 4-3.* As partners of the 21CTP, EPA and DOT's NHTSA should work with CARB to develop test procedures for the certification process for criteria emissions so that the emissions benefits of hybridization will be recognized, allowing the reduction in size or simplification of the emission control system of hybrid heavy-duty vehicles to be realized.	DOE agrees that the proposed test procedure development should be performed by EPA and DOT's NHTSA.
H-5 **Hybrid Business Case/Break-even Time**	*Finding 4-5.* The 21CTP acknowledges that current heavy-duty hybrid vehicle break-even times, without subsidies, based on current costs and fuel consumption improvements, are typically twice as long as the 5 years that fleets normally require for a return on investment on new hardware for cost savings. Heavy-duty hybrid components tend to be costly since they are not designed or optimized for the application and are produced in low volumes. Fuel-economy improvements of heavy-duty hybrid vehicles have not achieved the 60 percent improvement goal (38 percent reduction in fuel consumption). *Recommendation 4-4.* Dual paths should be pursued to achieve a break-even time of 5 years for heavy-duty hybrid vehicles. First, the DOE should use its vehicle simulation tools to determine the advanced technologies needed to meet the goal of 60 percent improvement in fuel economy (38 percent reduction in fuel consumption), from the current status of 20 to 40 percent improvement (17 to 29 percent reduction in fuel consumption) and initiate R&D programs to develop these technologies. Second, manufacturers should be encouraged to explore modular, flexible designs, which could yield higher production volumes and thus achieve significant reductions in capital costs of hybrid systems.	DOE is prepared to assist industry in these types of studies. DOE does not plan to conduct or initiate hybrid centric R&D programs. DOE's focus is on electric-drive component R&D to develop technologies that can be integrated by manufacturers into advanced technology vehicles.
H-6 **Revised Hybrid Goals**	*Finding 4-6.* Six new stretch technical goals have been established by the 21CTP for heavy-duty hybrid vehicles. The committee agrees with the 21CTP that these are indeed stretch goals. Specific plans for achieving these new goals, some of which were carried over from the previous three goals that had been set for hybrids, were not provided to the committee. Nor was the rationale provided for these new goals, although they are appropriately focused on fuel consumption reductions, cost reduction, and a 15-year design life for the technologies. They appear to be reasonable technical goals. The cost and design life objectives in the previous goals had been identified earlier by the 21CTP as being necessary for achieving commercially viable heavy-duty hybrid vehicles. It is expected that a significant budget would be required through the target dates specified in the new goals, and a significant increase from the zero budget for heavy-duty hybrid R&D over the past 3 years would be required. *Recommendation 4-5.* The 21CTP should establish plans and develop realistic budgets for accomplishing the six new stretch goals for heavy-duty hybrid vehicles in accordance with the committee's findings, explain the rationale behind the new goals, and provide the current status of the applicable technology for each of the goals so that the magnitude of the tasks for each can be assessed.	The Partnership concurs that planning for these updated goals is critical: the Partnership industry and government members will be working as a team to conduct these planning efforts and identify the appropriate parameters for successful achievement of the goals, subject to available funding. Ongoing research results will inform goal revisions. Two of the SuperTruck teams are developing and integrating full hybrid systems into Class 8 vehicles. In addition, ORNL will be installing and testing a full heavy-duty hybrid system in a dedicated test cell. 21CTP will use these project findings to revise goals as appropriate.

Vehicle Power Demands

Subject	NAS Findings and Recommendations	Partnership Response
VPD-1 **Vehicle Aerodynamics**	*Finding 5-1.* Aerodynamic improvement studies need to become increasingly integrated, as individual component improvements are typically not additive. Appropriately, the perspective of the 21CTP for the SuperTruck projects is to utilize a vehicle systems approach for the validation of research and development results. *Finding 5-2.* The aerodynamic test procedures may not be sufficiently precise and only wind tunnel testing accounts for important yaw effects, so that competitive pressures discourage truck-tractor manufacturers from publishing Cd figures. Recommendation 5- 15 from the National Research Council's 2010 report entitled **Technologies and Approaches to Reducing the Fuel Consumption of Medium- and Heavy-Duty Vehicles** provided good suggestions for standardizing Cd reporting. *Finding 5-3.* The proposed EPA/NHTSA greenhouse gas emissions standards rule chose not to regulate trailer operational efficiency. Regardless of the reasons, this seems a significant omission, because both trailer aerodynamic devices and low-rolling-resistance tires that are currently production-available can provide an immediate, combined fuel consumption reduction of about 13 percent (compared to the rule's baselines). *Finding 5-4.* Aerodynamic design packages are expected to improve tractor-trailer fuel consumption by 19 percent at 65 mph when fully developed in the 2015-2020 time period. This reduction corresponds to a Cd reduction of nearly 40 percent (from the newly adopted 0.69 Cd baseline). *Recommendation 5-1.* The Partnership should consider setting an aerodynamic drag stretch goal of 40 percent instead of 30 percent.	The Partnership is aware of the NRC's recent work on heavy truck fuel consumption for EPA and NHTSA and the results of that work. The Partnership's work acknowledges the importance of trailers to the operational efficiency of the vehicle, and has made efforts to include trailer efficiency considerations in its SuperTruck research activities, from a vehicle systems perspective. The Partnership periodically reviews its goals and objectives to ensure they are in alignment with current technology progress and government agency research plans. SuperTruck research results will help inform future aerodynamic goal revisions. As information about the technology status of the aerodynamics work within SuperTruck becomes available, the Partnership will re-examine its goals for aerodynamics and adjust as necessary to provide the appropriate stretch targets.
VPD-2 **Wide-Base Single Tires – Rolling Resistance Goal**	*Finding 5-5.* Next-generation wide-base single tires (NGWBSTs) can provide a combination tractor-trailer with an immediate 10.5 percent fuel-consumption reduction and up to a 15 percent reduction in the next 5 years, but many fleets do not yet embrace the technology. *Recommendation 5-2.* The DOE should set the goal for reduced rolling resistance for the tires of the combination tractor-van trailer, rather than for the tractor drive wheels only, since improved-performance trailer tires are equally important to realizing the full benefit of reduced rolling resistance designs. This benefit can be achieved by combining the EPA base values for steer and drive tires in the EPA/NHTSA GHG rule, with an assumed trailer tire Crr value of about 0.0072.	The Partnership concurs that a systems view of tire rolling resistance (including both tractor and trailer tires) is important to realizing the benefits of these tire technologies, and will take this into consideration when reviewing and revising Partnership goals. DOE, as a member of the Partnership, has initiated three tire technology projects in FY2012 (cross-cutting between light duty and heavy duty vehicles) that target 2% fuel consumption reduction for the full vehicle from rolling resistance improvements and automatic tire inflation.
VPD-3 **Wide-Base Single Tires - Retrofits**	*Finding 5-6.* Carriers need to follow carefully the recommendations of axle manufacturers for replacing dual tires with single-wide tires to ensure that the integrity of the load system is not compromised. *Recommendation 5-3.* The 21CTP should consider producing a comprehensive summary that can be updated giving the prescriptions and precautions that carriers should consider when retrofitting NGWBSTs onto original equipment axles fitted with dual wheels and tires. This effect might best be managed in conjunction with the American Trucking Associations' (ATA's) Technology and Maintenance Council, which has drafted such a Recommended Practice and is a specialist in creating such directives for ATA membership (ATA, 2007).	The Partnership agrees that safety is extremely important when considering retrofits of NGWBS tires on existing trucks. The Partnership would encourage the use and promotion of Technology and Maintenance Council Recommended Practices to address this issue, and will consider addressing relevant safety concerns in the white papers and other 21CTP documentation addressing the use of NGWBS tires.

Subject	NAS Findings and Recommendations	Partnership Response
VPD-4 **Wide-Base Single Tires – Rolling Resistance Test Procedure**	*Finding 5-7.* There is no rolling resistance test procedure with inter-laboratory correlation universally employed as an industry standard. *Recommendation 5-4.* The 21CTP, strongly supported by DOT and EPA (the latter through its SmartWay program), should conduct an authoritative study of the several barriers (e.g., related to tread life, truck stability in blowouts, run-flat tires, and other topics) to the widespread carrier adoption of next generation wide base single (NGWBS) tires. The DOT should specifically support reduction of barriers to NGWBS tire acceptance by requiring the universal use by tire manufacturers of a rolling resistance test procedure like that in ISO (International Organization for Standardization) 28580, to ensure that comparative inter-laboratory data exist.	The Partnership agrees that identifying and addressing barriers to NGWBS tire acceptance are critical in expanding the use of this technology to improve truck efficiency. The Partnership will consider the possibility of conducting a study of barriers, subject to available resources. Truck tire manufacturers at present do not correlate rolling resistance measurements among one another to any large extent: this may be due to the fact that rolling resistance has not been a specification provided to tire manufacturers by the vehicle OEMs. (In the case of light-duty tires, the vehicle OEM considers tire rolling resistance to be a very important performance requirement.) This may change as new truck fuel consumption regulations are imposed, and the need for lower rolling resistance tires increases. It should be noted that the ISO 28580 standard calls for a reference laboratory, but this has not yet been identified. The Partnership agrees that lack of consistent rolling resistance measurement could be a barrier to increased acceptance of NGWBS tires, along with the lack of education for fleets and owner-operators on the benefits of low rolling resistance tires. Absent any requirements to provide rolling resistance information at the point of sale, this information is not generally available to the tire purchaser.
VPD-5 **Auxiliary Power Demands R&D**	*Finding 5-8.* The More Electric Truck may achieve about one-third of the auxiliaries' reduction goal for a loaded tractor-trailer. Better quantification is expected to result through two of the SuperTruck projects. *Recommendation 5-5.* The Partnership should renew R&D efforts to further reduce fuel consumption related to auxiliary power demands.	The partnership should monitor auxiliary load improvements resulting from the SuperTruck projects.
VPD-6 **Lightweight Materials**	*Finding 5-9.* Several projects that were carried out prior to 2007 have shown the potential for the reduction in weight of individual components and subsystems. However, to date there has been no integrated full vehicle project to show that the goal of reducing the weight of a Class 8 tractor-trailer by 3,400 lb can be achieved. Moreover, the NRC Phase 1 report had recommended that such a project, using prototype components, vehicle integration, and full-vehicle system analysis, should be carried out by industrial partners—led by original equipment manufacturers. The new SuperTruck program appears to be a response to this suggestion. *(No Recommendations Provided)*	The Partnership agrees with this finding. The current SuperTruck program is addressing the use of advanced materials to reduce the weight and improve the freight efficiency of class 8 heavy trucks.
VPD-7 **Thermal Management**	*Finding 5-10.* Heavy-duty truck thermal management objectives are growing in importance as new systems to improve both engine and truck efficiency, particularly waste heat recovery systems, become reality. These are accompanied by new heat management issues and are expected to be added to trucks in the current decade. *Recommendation 5-6.* The Partnership should continue priority support of nano-fluid and high-efficiency under-hood cooling systems, as well as review other potential technical concepts, and validate them as an integrated system.	DOE is planning to expand R&D on high efficiency HVAC systems. DOE agrees and is continuing support of nano-fluid and high-efficiency under-hood cooling systems. DOE will monitor other potential technology solutions to reach thermal management objectives.

Subject	NAS Findings and Recommendations	Partnership Response
VPD-8 **Driveline Power Demand**	*FINDING 5-11.* There is a need for an updated study of the current driveline power demand of 12 hp. Furthermore, to represent vehicle power demand power consumptions only, it is appropriate that the term "powertrain" be removed from the 21CTP Goal 5.b. statement. *RECOMMENDATION 5-7.* The term "powertrain" should be removed from the 21 CTP Goal 5.b statement. In addition, the Partnership should update its study on the driveline power demand of 12 hp.	The Partnership concurs: a subsequent revision to the Partnership's goal wording made after the completion of this review has removed the word "powertrain" from the subject goal, which will be published as part of the final white paper/roadmap document. The Partnership will review the current information on driveline power demand and consider updates to this study. The Partnership will review research results from the SuperTruck teams to gather current technology information for power demand, and revise assessments of power demand as appropriate.
VPD-9 **Lubricant Collaborations**	*FINDING 5-12.* There has been no apparent collaboration on lubricant projects between the DOE and OEM partners. *(NO RECOMMENDATIONS PROVIDED)*	As noted in the response to Recommendation 3-5, DOE works closely with a variety of OEM partners on lubricant projects, including both direct partnerships with vehicle and engine OEMs, and indirect partnerships through research consortia such as the MIT Lubrication in Internal Combustion Engines Consortium. While DOE has only worked directly with one 21CTP heavy-duty partner on lubricant issues (Cummins), results from the complete range of lubricant projects within VTP are made available to 21CTP partners through a variety of means. DOE is open to collaborative efforts on lubricant projects with 21CTP partners in the future.
VPD-10 **Overall Vehicle Power Demands Findings**	*FINDING 5-13.* Summarizing the committee's findings on vehicle power demands: Project prioritization by the 21CTP roughly follows the consumption ranking of the several heavy-duty truck operating loads in Table 5-1 (see Chapter 5 in the report) and technology risk. However, sometimes market forces provide considerable impetus for quite good development and implementation—for example, in tire rolling resistance and, to a lesser extent, trailer aerodynamic components. The DOE has identified a strong role in which technology development costs and risks are high, as in its vehicle systems simulation and testing activities for heavy-duty trucks. It has generally followed these principles, to address high costs and risks, in the vehicle power demand projects. The SuperTruck projects will provide a unique Partnership opportunity to provide both further high-risk technology results for certain vehicle power demand reductions and real-world validation of numerous integrated systems. *RECOMMENDATION 5-8.* Although it is tempting to assume that the SuperTruck projects will address all of the technologies required to reduce tractor-trailer fuel consumption, in practice many technologies may be left behind, particularly those that are not yet very mature. The Partnership should carefully review the technologies that have been identified and determine whether any technologies to reduce vehicle power demand are not being adequately addressed by the SuperTruck program. The DOE should define projects and find funding to support the development of technologies beyond the scope of SuperTruck.	The SuperTruck projects are designed to develop combinations of advanced technologies into a Class 8 platform that can be commercialized in the near-term. In order to ensure commercial viability, the technologies are chosen by each industry team and not dictated by DOE. Technical approaches for reducing petroleum consumption that are not addressed by the SuperTruck projects may be appropriate for investigation through other pathways that address longer term technology development.

Subject	NAS Findings and Recommendations	Partnership Response
IR-3 **Solid-Oxide Fuel Cells for Idle Reduction**	*FINDING 6-3.* The Delphi solid oxide fuel cell (SOFC) auxiliary power unit (APU) provides several advantages over diesel APUs, but it has significant issues in its current development status, including the following: low efficiency of 25 percent versus DOE's goal of 35 percent, and low demonstrated output power of 1.5 kW versus 3.0 kW believed sufficient by Delphi and 5 kW of typical diesel APUs; limited demonstrated durability; 2- to 5-hour warm-up time to the 750°C operating temperature; and the need to keep it operating at idle throughout the workday to maintain temperature. The 10-year funding for this program expires in 2011. *RECOMMENDATION 6-3.* The DOE should reassess the viability of the SOFC APU, particularly for application to the SuperTruck program, considering the following: (1) SOFC APU is still in the laboratory, (2) the low efficiency of 25 percent versus the DOE goal of 35 percent, (3) the low 1.5 kW output compared to the typical 5 kW diesel APUs, (4) the disadvantages associated with the requirement for continuous operation at 750°C, and (5) the expiration of funding from the DOE Office of Fossil Energy and EERE Fuel Cell Technologies Program of the DOE Office of Energy Efficiency and Renewable Energy after 10 years of development. The DOE should coordinate more closely with DOD in its fuel cell APU developments to ensure that the best technology is being pursued for the 21CTP's Goal 7 in the engine idle reduction focus area; that goal relates to the development and demonstration of viable fuel cell APU systems for military and other users (see Chapter 6 for the full text of Goal 7). (This recommendation is a follow-on to Recommendation 6-8 in the NRC Phase 1 report.)	The information presented during the 21CTP NAS review was based upon one of our initial A-Level prototype units. Since then, we have made significant progress and are now assembling our B-Level prototype units. These units should be capable of demonstrating the targeted goal of 35% efficiency and output of 3kW. On the SOFC stack, Delphi has completed more than 10,000 hours of durability testing. Additionally, we have accumulated thousands of hours of on-truck, real-work application data. We are scheduled to deliver a B-Level unit during Q1 '12 to a national fleet for use on one of their regular in-service long haul trucks. Currently, our start-up time is ~2 hours. The 5-hour example reported on represented a given demonstration. Our goal is to be at operating temperature in under 1-hour. Current costs reflect laboratory built prototype units. Delphi is investing in production intent tooling to drive down overall unit cost. Funding to date has allowed Delphi, as well as other fuel cell developers, to move their products from concept design to real-world demonstrations. Congress has recently reinstated funding for SECA and other fuel cell programs. Delphi will use the re-funded SECA program to further improve the power output and durability of its SOFC stack.
IR-4 **Relation of Idle Reduction Plans to Fuel Efficiency Standards**	*FINDING 6-4.* Idle reduction technologies could provide 6 percent reduction in overall fuel consumption for Class 8 long-haul trucks with sleeper cabs, which is nearly 30 percent of the 20 percent reduction in the fuel consumption required to meet the EPA/NHTSA proposed 2017 fuel consumption standards. *RECOMMENDATION 6-4.* The 21CTP should review and potentially revise its idle reduction plans and goals in view of the fact that the proposed 2017 fuel efficiency standards provide an incentive for the adoption of idle reduction technologies as a means for achieving these standards for Class 8 long-haul trucks with sleeper cabs.	The 21CTP agrees that the EPA's rulemaking to establish fuel efficiency standards for heavy-duty truck fleets provides an incentive to look beyond Class 8 long-haul trucks with sleeper cabs to other types of trucks for additional opportunities to apply idle reduction technologies. We feel that a substantial improvement to the idle reduction goal would include support to establish a program to address the fuel wasted in work day idling of all types of vocational trucks.

Subject	NAS Findings and Recommendations	Partnership Response
IR-5 **Idle Reduction Goals and Objectives**	*FINDING 6-5.* In February 2011, the 21CTP deleted the quantification of the overall goal to reduce fuel use and emissions produced by idling engines. The 21CTP issued five new goals for idle reduction and designated the goals that had been in place through 2010 as "action items." The new goals are generally not supported by the "action items." A separate budget for idle reduction for FY 2012 has not been proposed, although idle reduction will be addressed by the SuperTruck program. The 21CTP has stated that, "without funding dedicated to this effort [the idle reduction goals], it is quite difficult, if not impossible, to accomplish these goals" (DOE, 2011). *RECOMMENDATION 6-5.* The 21CTP should revise its new idle reduction goals to include metrics, funding, and timing for the overall goal of reducing fuel use and emissions produced by idling engines. The associated "action items" should be supportive of these goals.	The 21CTP agrees with the NAS recommendation that the inclusion of a progressive and measurable program for idle reduction goals development is needed along with the year-to-year funding necessary to develop data to enable such an approach.

SAFETY

Subject	NAS Findings and Recommendations	Partnership Response
S-1 **Safety Goals**	*FINDING 7-1.* The wording of 21CTP Safety Goals 1 and 2 as now written might be subject to misinterpretation by some as allowing the compromise of safety in the effort to improve fuel consumption. *RECOMMENDATION 7-1.* The Partnership should review the wording of its safety goals and consider rewording them so as to unambiguously state that safety will not be compromised in reducing fuel consumption.	The Partnership will review wording of safety goals to ensure appropriate emphasis is placed on safety—and that safety is not compromised in achieving fuel efficiency goals.
S-2 **Crash Avoidance**	*FINDING 7-2.* Vehicle crashworthiness and occupant protection systems have seen extensive deployment, have contributed greatly to improved highway safety, and have achieved extensive North American fleet penetration. The next important step is to prevent crashes altogether. *RECOMMENDATION 7-2.* The committee supports the emphasis that the DOT and the 21CTP are giving to crash-avoidance technologies and recommends that crash-avoidance technologies continue to be given high priority and technical support.	The Partnership agrees with the committee's observations and recommendations.
S-3 **Review of Safety Goals – Relation to TRB Study**	*FINDING 7-3.* The DOT has met its heavy-truck safety goals for the past 4 years. However, the committee observes that the Transportation Research Board's (TRB's) 2010 study Achieving Traffic Safety Goals in the United States: Lessons from Other Nations has shown that other nations have established more aggressive initiatives and goals with impressive results, and those results suggest that even greater improvement in highway safety is possible in the United States. The committee also notes that overall improvements in highway safety also yield improvements in heavy-duty truck safety, as most heavy-duty truck fatal accidents involve a light-duty vehicle. *RECOMMENDATION 7-3.* The DOT should evaluate the conclusions and recommendations of the TRB study Achieving Traffic Safety Goals in the United States: Lessons from Other Nations of highway safety in other nations, and consider the possibility of establishing more aggressive initiatives and goals for highway safety in general. The DOT should also consider establishing more aggressive goals for heavy-duty truck safety.	DOT will review the TRB study (Achieving Traffic Safety Goals in the United States: Lessons from Other Nations). DOT regularly re-evaluates its safety goals each year, and will take into consideration information from this study, as well as the special circumstances impacting traffic safety in United States.
S-4 **Fuel Consumption Impact of Safety Improvements**	*FINDING 7-4.* Some of the potential safety improvements considered by the committee may have negligible impact on fuel consumption and, in some cases, appear to have positive implications. However, further study of the potential highway safety impact of high productivity vehicles is warranted. *(NO RECOMMENDATIONS PROVIDED)*	USDOT will launch a major study of this issue based on direction given in MAP-21; specifically, Section 32801 requires completing a "Comprehensive Truck Size and Weight Limits Study". The scope of this study can be found in the authorizing legislation.

SuperTruck Projects

Subject	NAS Findings and Recommendations	Partnership Response
ST-1 **SuperTruck Overview**	*Finding 8-1.* The three SuperTruck projects will be the flagship projects under the 21CTP for FY 2011 through FY 2014; the goals are in concert with recommendations made in the 2008 NRC Phase 1 report. A large portion of the DOE 21CTP budget will be devoted to these three projects. Each SuperTruck project integrates a wide range of technologies into a single demonstration vehicle (engine, waste heat recovery, driveline, rolling resistance, tractor and trailer aerodynamics, idle reduction, weight reduction technologies, etc.), and the contractors are pursuing sufficiently different technical paths to avoid excessive duplication of effort. The results will help determine which fuel-saving technologies are ready and cost-effective for original equipment manufacturer (OEM)-level product development programs. *(No Recommendations Provided)*	The Partnership concurs with this finding, and appreciates the panel's observations about the alignment of goals with their previous recommendations. VTP is placing a high priority on completing these SuperTruck projects and achieving the goals set forth for them.
ST-2 **SuperTruck Goals**	*Finding 8-2.* Rather than have a number of targets for each subsystem, the SuperTruck projects have only two types of goals: one for the engine and one for overall vehicle fuel efficiency. This approach reflects the EPA/National Highway Traffic Safety Administration (NHTSA) approach to heavy-duty fuel efficiency regulations. Each project team is allowed to select a set of technologies that meet the project goals. The engine goal of 50 percent BTE for the demonstration vehicle appears to be feasible, although there is risk in being able to achieve it at a cruise condition. The engine goal of 55 percent BTE demonstrated in a test cell is very high risk but might be achievable. The overall vehicle goal of a 33 percent reduction in load-specific fuel consumption appears to be feasible. *(No Recommendations Provided)*	The SuperTruck projects are required to use modeling and analysis to identify the key pathways to the 55% efficiency goal, as outlined in the original solicitation (excerpted below). The teams are not required to demonstrate the 55% goal in a test cell engine, but can do so optionally. "As a separate and parallel effort, proposers shall identify, through modeling and analysis, key pathways to achieving our long-term goal of developing a 55% efficient (brake thermal efficiency) heavy-duty diesel engine. Critical components and/or systems needing specific development to achieve this goal should also be identified. This engine must be capable of meeting 2010 emission standards, and be commercially viable." As the NAS panel notes, the SuperTruck teams are required to demonstrate the 50% efficiency goal in actual engine hardware.
ST-3 **Fuel Consumption versus Fuel Economy**	*Finding 8-3.* Unfortunately, the SuperTruck program expresses vehicle efficiency targets in terms of fuel economy rather than fuel consumption. The vehicle target is stated as a 50 percent improvement in fuel economy rather than as a 33 percent reduction in fuel consumption. This can lead to confusion regarding the actual benefits of the program. *Recommendation 8-1.* The DOE should state the SuperTruck program vehicle efficiency goals in terms of load-specific fuel consumption and track progress on this basis—that is gallons per 1,000 ton-miles, which is the metric used in the EPA/NHTSA fuel consumption regulations.	The Partnership acknowledges the NAS panel's concerns, but notes that the 50 percent improvement in freight fuel efficiency goals for the SuperTruck program (expressed as ton miles per gallon) were established as part of the original solicitation, which cannot be altered at this stage. The Partnership will present the results of analysis efforts regarding projected program benefits for SuperTruck and other VTP initiatives in terms of reductions in fuel consumption (typically total gallons displaced) wherever possible. To the extent possible, reporting on SuperTruck benefits will be reported both in terms of freight fuel efficiency (ton miles per gallon) and fuel consumption (gallons per thousand ton miles).

Subject	NAS Findings and Recommendations	Partnership Response
ST-4 **SuperTruck Plans**	*FINDING 8-4.* The committee believes that the SuperTruck project teams have developed plans that address the needs of the SuperTruck program and that have a reasonable chance for success. The keys to success include proper implementation of the plans along with the flexibility to adapt to new information and intermediate results during the course of the project. *(NO RECOMMENDATIONS PROVIDED)*	The Partnership concurs with this finding about the importance of planning and implementation of the plans to the success of SuperTruck.
ST-5 **Test Cycles**	*FINDING 8-5.* The SuperTruck projects allow each team to design its own test duty cycle(s) within certain constraints. One negative consequence of this approach is that the three trucks may never be tested using a common cycle for comparison. *RECOMMENDATION 8-2.* The DOE and the SuperTruck contractors should agree on at least one common vehicle duty cycle that will be used to compare the performance of all three SuperTruck vehicles. In addition, fuel consumption improvements should be calculated on the basis of the EPA/NHTSA fuel consumption regulations.	The goal of the SuperTruck project was to develop efficiency technologies that would improve Class 8 truck efficiency, and do so from a systems viewpoint to optimize these technologies to fit customer needs. The project was not intended to focus simply on meeting the EPA/NHTSA fuel consumption regulations: rather, it was intended to be an examination of the future technology possibilities for greatly improving the efficiency of Class 8 trucks to go beyond the regulations. The Partnership notes that the concept of imposing a common duty cycle among the SuperTruck teams is a good engineering idea, but does not necessarily fit with how the participating SuperTruck vehicle OEMs develop vehicles for their customers. The OEMs have differing customer bases with differing duty cycle needs, and the vehicles are tailored to accommodate these needs. In addition, the OEMs address different market segments, so a truck optimized for a single duty cycle may not represent all customers. It would certainly be possible to test all SuperTruck prototypes on all the duty cycles identified by the teams, but this testing would be extensive and likely cost prohibitive. The allowance for team-driven duty cycles was the best tradeoff that DOE could make at the time of the RFP release.
ST-6 **Scope of SuperTruck – Vehicle Demonstration**	*FINDING 8-6.* The SuperTruck projects go beyond the scope of previous 21CTP projects. Instead of relying entirely on simulations and laboratory testing, each of these projects will result in a drivable truck. The committee believes that it is important to take technologies that have been developed to date and implement them in a real vehicle. Often, the application of new technologies in real-world applications yields unexpected results, and these results must be explored before any new technology can be considered ready for production implementation. *(NO RECOMMENDATIONS PROVIDED)*	The Partnership concurs with this finding, and also believes that it is critical to implement efficiency technologies in real vehicle applications to demonstrate their effectiveness and encourage their ultimate market uptake.

Subject	NAS Findings and Recommendations	Partnership Response
EO-4 **Driver Management**	*Finding 9-4.* Driver-management features must be carefully researched and developed in cooperation with vehicle manufacturers and operators. There are important concerns with driver-management features that need to be addressed, regarding unintended consequences stemming from allowing the vehicle (or its controller) to ignore or modify driver input. Consideration must also be given to identifying the types of intervention that drivers would accept. *(No Recommendations Provided)*	Driver-management features such as progressive shift, road-speed governors and smart cruise control systems provide a technical means to achieve more efficient driving by influencing driver behavior with imposed speed or operational controls. The 21CTP agrees that such systems can provide significant fuel saving benefits and that further research in this area is needed. Of course, it is understood that the driver's ability to control a vehicle is critical to highway safety, and the Partnership strongly agrees that development of driver management/feedback systems should be done with careful consideration of any potential negative consequences to safety or other aspects of vehicle operation.
EO-5 **Trucking Efficiency – Work with Trucking Companies**	*Finding 9-5.* Trucking companies already have very strong economic incentives to improve operational efficiency and average load factors. As a result, they are making significant investment in logistics technology. In addition, shippers have an economic incentive to reduce the size and weight of packaging materials. The trucking industry is a valuable source of ideas, data, and experience regarding efficiency, and the industry can help agencies avoid unintended negative consequences of efforts to improve efficiency. *Recommendation 9-4.* The DOE and DOT should work with the trucking industry to take advantage of the ideas, data, and experience that the industry can provide to accelerate efficiency improvements and to avoid unintended negative outcomes of efforts to improve trucking efficiency.	The Partnership concurs with this finding and recommendation. Any efforts aimed at improved logistics management and trucking operations must be performed in collaboration with the trucking industry to ensure that best practices are not violated and any newly proposed solutions can be effectively implemented with minimal or no negative consequences on fleet operations, safety, or road damage.

Subject	NAS Findings and Recommendations	Partnership Response
EO-6 **High-Productivity Vehicles**	*FINDING 9-6.* High-productivity vehicles, known as HPVs or LCVs, as currently configured and using current technology, can reduce fuel consumption by up to 28 percent. In addition, HPVs can reduce greenhouse gas emissions, truck vehicle miles traveled, congestion, shipper costs, truck highway accidents, road damage, and truck driver shortages. *FINDING 9-7.* High-productivity vehicles have proven to be a highly controversial and emotional topic. Some U.S. states, as well as countries including Canada, Australia, and the Scandinavian countries, have extensive experience with HPV operations and safety performance. Operational limitations and equipment policy used for decades in Canada have significantly increased safety for HPVs compared with that of more conventional tractor-trailers. In 2002, the NRC's Transportation Research Board proposed a process, to be led by a congressionally chartered Commercial Traffic Effects Institute, to make decisions regarding a number of critical and historically controversial issues that effectively have prevented the growth of HPV use for nearly three decades. As far as the committee can determine, no action on the CTEI recommendation has been considered by Congress. *FINDING 9-8.* The draft white paper on efficient operations brings up the topic of high productivity vehicles and the possibility of raising weight and size limits to accommodate them. However, the white paper focuses narrowly on 6-axle tractor-trailer combinations with weights up to 100,000 lb (45.5 metric tons) and does not address other options that increase volumetric freight capacity or that allow weights beyond 100,000 lb . *FINDING 9-9.* The committee finds the case for fuel savings of HPVs compelling, and the case for improved safety of HPVs compared to that of standard 5-axle semi-tractor trucks is also strong. *RECOMMENDATION 9-5.* The DOT and DOE should look at the full range of high productivity vehicles in use in some U.S. states and around the world and review the literature available on the safety and fuel-saving performance of these vehicles. The assessment should take into consideration that the higher productivity of these vehicles can also be used to justify the implementation of additional safety technologies. *RECOMMENDATION 9-6.* The DOT and DOE, in discussion with the Congress, should consider the recommendations of the Transportation Research Board regarding the establishment of a Commercial Traffic Effects Institute or a similar approach.	The NRC panel pointed out a number of important additional points concerning high productivity vehicle use that were not highlighted in the draft white paper on efficient operations. A more thorough literature review in the white paper is appropriate for this topic and is planned for the next version of the white paper. The Partnership fully agrees with these findings and the recommendations, including the consideration of recommendations made in TRB Special Report 267 (Transportation Research Board, 2002), in particular the establishment of a Commercial Traffic Effects Institute. DOT has recently supported several studies in which safety, road damage and fuel savings have been investigated. Further analysis of HPV operation data both in the U.S. and abroad is warranted by the DOE and DOT to build further support for moving forward with changes that would allow more extensive and unified HPV operation throughout the United States.
EO-7 **Goals for Efficient Operations**	❶*FINDING 9-10.* The DOE-DOT draft white paper on efficient operations in its current form does not include any goals that could be used to prioritize and drive R&D efforts on efficient operations. ❶*RECOMMENDATION 9-7.* Specific goals for efficient operations should be developed, with strong consideration given to exploiting the potential for intelligent transportation systems (ITS) to reduce fuel consumption. In addition, priorities should be set for the R&D, testing, and data collection needed to analyze the benefits, drawbacks, and potential unintended consequences of removing barriers, including regulatory barriers, to the application of fuel-saving features. The draft white paper on efficient operations should be rewritten to take the findings and recommendations of the committee into account. The 21CTP partners, trucking fleets, and major suppliers should be involved in setting goals and research priorities.	Although the draft white paper on efficient operations available at the time of the NRC panel's review did not include goals on this topic, the Partnership has added a set of specific goals that are consistent with this recommendation. A further rewrite of the draft white paper is also planned that will address the panel's recommendations. The Partnership is aware that many of the approaches proposed for efficient operations involve multiple complexities and agrees that detailed studies are needed to assess the benefits, drawbacks, and potential unintended consequences of removing barriers for efficient operations.

Subject	NAS Findings and Recommendations	Partnership Response
EO-8 **Fuel Savings Potential for Efficient Operations**	*FINDING 9-11:* There is a need for a more detailed evaluation of the large potential for fuel savings from efficient operations than is provided in the existing DOE-DOT draft white paper of February 25, 2011. This more detailed study can be used to set goals, targets, and timetables for fuel savings from efficient operations. *RECOMMENDATION 9-8:* The DOE and DOT should study the potential fuel savings from efficient operations in more detail, including a review of cost-effectiveness and ease of implementation. Once this information is available, goals, targets, and timetables for fuel savings from efficient operations should be established. Programs should then be developed and implemented to realize the available fuel savings.	The Partnership agrees that research is needed to quantify the benefits as well as the costs and challenges of implementation associated with the proposed methods for efficient operations. Ultimately, the end goal is to implement those approaches for which the benefits clearly justify the costs, and the Partnership concurs with the NRC panel's recommended course of action to arrive at this objective.

Appendix D

21CTP Project Inventory and Summary of 21CTP Goals

The following project list showing 21st Century Truck Partnership activities was submitted to the committee by the 21st Century Truck Partnership on December 29, 2014, along with the summary of 21CTP goals listed at the end.

21CTP Project Inventory

Agency	Technology Area	Sub-grouping	Internal Project Title	Public Review Project Title	Project ID	Recipient	Principal Investigator	Lab/Industry/Univ Funding	ARRA Funding	2012 Funding	2013 Funding	2014 Funding	Notes
DOD	Energy Storage		Computer-Aided Engineering for Military Batteries	Computer-Aided Engineering for Military Batteries	N/A	N/A	N/A	Lab	$ -	$ 500,000	$ 500,000	$ 500,000	To accelerate development and fielding safe, high-performance, and long-lasting lithium ion battery(LIBs) modules for military and commercial vehicle applications, the Army requires computer tools that accurately represent cell and module multi-physics phenomena occurring across a wide range of scale. TARDEC, in the Department of Army, is teaming with DOE's Vehicle Technologies Program to leverage the battery modeling and simulation efforts under the Computer Aided Engineering of Batteries (CAEBAT) activity. The objectives of this proposed project are to expand the cell modeling and battery architecture currently being funded by DOE and DOD to generate battery module and pack models and additional data for validation of the CAEMBAT tools.
DOD	Materials Technology		Multi-Material Joining	Multi-Material Joining	N/A	N/A	N/A	Lab	$ -		$ 5,000,000	$ 2,500,000	To overcome technology gaps (predictive tools and low cost processes) to enable joining of dissimilar materials for both light duty and heavy duty systems (Body-in-white, cab, chassis, and closures) as evidenced from the AVPTA materials workshop.
DOD	Fuels Technologies		Fuel Bulk Modulus	Fuel Bulk Modulus	N/A	N/A	N/A	Lab	$ -	$ 551,000	$ 30,000	$ 10,000	Develop an improved test method to determine the bulk modulus of liquid transportation fuels • Bulk modulus of alternative liquid fuels can vary greatly from existing petroleum fuels affecting fuel's performance in engine • Current methodology is subjective and very complex with no standard apparatus
DOD	Advanced Combustion Engines		Thermoelectric Enabled Engine	Thermoelectric Enabled Engine	N/A	N/A	N/A	Lab	$ -	$ 3,175,000	$ 250,000	$ 350,000	To achieve improved efficiency and reduced emissions in advanced combustion engines for passenger and commercial vehicle applications through: 1) accelerated development of cost competitive advanced second generation thermoelectric devices for vehicle applications and 2) accelerated development of new or early stage enabling technologies needed to improve fuel efficiency, performance, and emissions in internal combustion engines.
DOD	Materials Technology		Light Weight Vehicle Structures	Light Weight Vehicle Structures	N/A	N/A	N/A	Lab	$ -	$ 2,000,000	$ 350,000	$ 350,000	Purpose: Demonstrate best practices in affordable, multi-material design for structures to reduce ground vehicle weight 1) Utilize automotive technology investments in vehicle structure light weighting, 2) • Demonstrate a weight savings of up to 1,500 lbs on a testable ground vehicle with the same level of protective performance. and 3) • Evaluate the current technical capability of the supply chain.
DOD	Electric Drive Technologies		Non-Rare Earth Materials for Motors	Non-Rare Earth Materials for Motors	N/A	N/A	N/A	Lab	$ -	$ 1,500,000	$ 500,000	$ 500,000	To develop and demonstrate two Integrated Starter-Generators (ISGs) without rare-earth permanent magnet materials suitable for onboard vehicle power generation in accordance with the Warfighter's requirements. Most high-performance electric machines in hybrid electric applications use rare-earth magnets (Neodymium & Samarium-Cobalt) to achieve the required power and torque densities. This effort will additionally explore alternate electric machine topologies to achieve its goals.
DOD	Fuels Technologies		Lubricant Formulations to Enhance Fuel Efficiency	Lubricant Formulations to Enhance Fuel Efficiency	N/A	N/A	N/A	Lab	$ -		$ 500,000	$ 500,000	The focus of this collaboration is to develop novel lubricant formulations that will improve fuel efficiency of legacy light- to heavy-duty vehicles by at least 2%.
DOD	Energy Storage		CAEBATT-II	CAEBATT-II	N/A	N/A	N/A	Lab	$ -		$ 500,000	$ 590,000	The objective of CAEBAT-II is to develop and validate an electrochemically-thermally coupled large-format Li-ion battery cell and pack model focusing largely on safety/abuse and degradation.

21CTP Project Inventory

Agency	Technology Area	Sub-grouping	Internal Project Title	Public Review Project Title	Project ID	Recipient	Principal Investigator	Lab/Industry/Univ Funding	ARRA Funding	2012 Funding	2013 Funding	2014 Funding	Notes
DOD	Materials Technology		Dissimilar Material Joining	Dissimilar Material Joining	N/A	N/A	N/A	Lab	$ -		$ 1,000,000		The Dissimilar Material Joining project investigates breakthrough ideas for joining dissimilar structural materials (Al, steel, Mg, Carbon-Fiber Composites), beyond such developed techniques such as fusion or friction welding, riveting, or ultrasonic joining.
DOD	Fuels Technologies		Ignition Models for Heavy Hydrocarbons Fuels	Ignition Models for Heavy Hydrocarbons Fuels	N/A	N/A	N/A	Lab	$ -	$ 200,000			To perform fundamental combustion measurements of two Tank Automotive Research, Development & Engineering Center (TARDEC) selected fuels at the Sandia National Lab (SNL) Combustion Research Facility. Such measurements include ignition delay, liquid length, and spray penetration.
DOD	Energy Storage		Beyond Li-Ion	Beyond Li-Ion	N/A	N/A	N/A	Lab	$ -			$ 1,140,000	
DOD	Vehicle Systems		Tire Efficiency	Tire Efficiency	N/A	N/A	N/A	Lab	$ -			$ 330,000	
DOD	Fuels Technologies		Advanced Lubricants	Advanced Lubricants	N/A	N/A	N/A	Lab	$ -			$ 900,000	
DOD	Electric Drive Technologies		Modeling and Optimization of Electrified Propulsion Systems	Modeling and Optimization of Electrified Propulsion Systems	N/A	N/A	N/A	Lab	$ -			$ 98,000	Part of list presented to NAS engine subgroup that visited TARDEC in November 2014
DOD	Advanced Combustion Engines		Combustion Behavior and Fuel Economy of Modern Heavy-Duty Diesel Engine using JP-8 and Alternative Fuels	Combustion Behavior and Fuel Economy of Modern Heavy-Duty Diesel Engine using JP-8 and Alternative Fuels	N/A	N/A	N/A	Lab	$ -			$ 121,000	Part of list presented to NAS engine subgroup that visited TARDEC in November 2014
DOD	Electric Drive Technologies		High Energy Density Asymmetric Capacitors	High Energy Density Asymmetric Capacitors	N/A	N/A	N/A	Lab	$ -			$ 99,000	Part of list presented to NAS engine subgroup that visited TARDEC in November 2014
DOD	Electric Drive Technologies		Powertrain Thermal Management - Integration & Control of a Hybrid Electric Vehicle Battery Pack, E-Motor Drive, and Internal Combustion Engine Multiple Loop Cooling System	Powertrain Thermal Management - Integration & Control of a Hybrid Electric Vehicle Battery Pack, E-Motor Drive, and Internal Combustion Engine Multiple Loop Cooling System	N/A	N/A	N/A	Lab	$ -			$ 50,000	Part of list presented to NAS engine subgroup that visited TARDEC in November 2014
DOD	Electric Drive Technologies		Advanced Models for Electric Machines	Advanced Models for Electric Machines	N/A	N/A	N/A	Lab	$ -			$ 78,000	Part of list presented to NAS engine subgroup that visited TARDEC in November 2014
DOD	Advanced Combustion Engines		Powertrain Thermal Management - Combined Experimental and Computational Study of Battery Cooling in Hybrid Electric Vehicles	Powertrain Thermal Management - Combined Experimental and Computational Study of Battery Cooling in Hybrid Electric Vehicles	N/A	N/A	N/A	Lab	$ -			$ 79,000	Part of list presented to NAS engine subgroup that visited TARDEC in November 2014
DOD	Energy Storage		Electro-Thermal Planar Dynamics and Control of Prismatic Li-Ion Cells	Electro-Thermal Planar Dynamics and Control of Prismatic Li-Ion Cells	N/A	N/A	N/A	Lab	$ -			$ 121,000	Part of list presented to NAS engine subgroup that visited TARDEC in November 2014
DOD	Fuels Technologies		Reaction Pathway and Elementary Ignition Behavior of Surrogates for JP-8 and Alternative JP-8 Fuels	Reaction Pathway and Elementary Ignition Behavior of Surrogates for JP-8 and Alternative JP-8 Fuels	N/A	N/A	N/A	Lab	$ -			$ 120,000	Part of list presented to NAS engine subgroup that visited TARDEC in November 2014
DOD	Advanced Combustion Engines		Simulation and Control of Combustion in Military Diesel Engines	Simulation and Control of Combustion in Military Diesel Engines	N/A	N/A	N/A	Lab	$ -			$ 140,000	Part of list presented to NAS engine subgroup that visited TARDEC in November 2014
DOD	Fuels Technologies		Validation of JP-8 Surrogates in an Optical Engine	Validation of JP-8 Surrogates in an Optical Engine	N/A	N/A	N/A	Lab	$ -			$ 123,000	Part of list presented to NAS engine subgroup that visited TARDEC in November 2014
DOD	Materials Technology		Nano-Materials Design for Enhanced Thermal and Mechanical Properties	Nano-Materials Design for Enhanced Thermal and Mechanical Properties	N/A	N/A	N/A	Lab	$ -			$ 104,000	Part of list presented to NAS engine subgroup that visited TARDEC in November 2014
DOD	Fuels Technologies		Bulk Modulus of Compressibility Measurements of Conventional and Alternative Military Fuels	Bulk Modulus of Compressibility Measurements of Conventional and Alternative Military Fuels	N/A	N/A	N/A	Lab	$ -			$ 110,000	Part of list presented to NAS engine subgroup that visited TARDEC in November 2014
DOD	Advanced Combustion Engines		Characterization and Warm-Up of Li-Ion Cells from Sub-Zero Temperatures	Characterization and Warm-Up of Li-Ion Cells from Sub-Zero Temperatures	N/A	N/A	N/A	Lab	$ -			$ 145,000	Part of list presented to NAS engine subgroup that visited TARDEC in November 2014
DOE	Advanced Combustion Engines	Combustion Research	Advanced Heavy-Duty LTC/Diesel Combustion Research	Heavy-Duty Low-Temperature and Diesel Combustion & Heavy-Duty Combustion Modeling	ACE001	Sandia National Laboratories (SNL)	Mark Musculus	Lab	$ -	$ 815,000	$ 805,000	$ 825,000	
DOE	Advanced Combustion Engines	Combustion Research	HCCI Engine Research (Medium-Duty, Dual-Engine Facility)	Low-Temperature Gasoline Combustion (LTGC) Research	ACE004	Sandia National Laboratories (SNL)	John E. Dec	Lab	$ -	$ 760,000	$ 740,000	$ 720,000	
DOE	Advanced Combustion Engines	Combustion Research	Spray Combustion Research for Enabling High-Efficiency Engines	Spray Combustion Cross-Cut Engine Research	ACE005	Sandia National Laboratories (SNL)	Lyle Pickett	Lab	$ -	$ 730,000	$ 740,000	$ 950,000	
DOE	Advanced Combustion Engines	Combustion Research	Large Eddy Simulation (LES) Applied to Advanced Engine Combustion Research	Large Eddy Simulation (LES) Applied to Low-Temperature and Diesel Engine Combustion Research	ACE007	Sandia National Laboratories (SNL)	Joe Oefelein	Lab	$ -	$ 500,000	$ 450,000	$ 200,000	
DOE	Advanced Combustion Engines	Combustion Research	Fuel Spray Research Using Advanced Photon Source	Fuel Injection and Spray Research Using X-Ray Diagnostics	ACE010	Argonne National Laboratory (ANL)	Christopher Powell	Lab	$ -	$ 1,100,000	$ 1,000,000	$ 850,000	
DOE	Advanced Combustion Engines	Combustion Research	Advanced Numerical Methods for Modeling Clean and Efficient Combustion Regimes	Model Development and Analysis of Clean & Efficient Engine Combustion	ACE012	Lawrence Livermore National Laboratory (LLNL)	Russell Whitesides	Lab	$ -	$ 520,000	$ 740,000	$ 475,000	
DOE	Advanced Combustion Engines	Combustion Research	Chemical Kinetics Research of HCCI Engines	Chemical Kinetic Models for Advanced Engine Combustion	ACE013	Lawrence Livermore National Laboratory (LLNL)	Bill Pitz	Lab	$ -	$ 620,000	$ 600,000	$ 550,000	

21CTP Project Inventory

Agency	Technology Area	Sub-grouping	Internal Project Title	Public Review Project Title	Project ID	Recipient	Principal Investigator	Lab/Industry/Univ Funding	ARRA Funding	2012 Funding	2013 Funding	2014 Funding	Notes
DOE	Advanced Combustion Engines	Combustion Research	KIVA Modeling to Support Combustion Research	2014 KIVA Development	ACE014	Los Alamos National Laboratory (LANL)	David Carrington	Lab	$ -	$ 720,000	$ 763,000	$ -	
DOE	Advanced Combustion Engines	Combustion Research	Stretch Efficiency --Thermodynamic Analysis of New Combustion Regimes	Stretch Efficiency for Combustion Engines: Exploiting New Combustion Regimes	ACE015	Oak Ridge National Laboratory (ORNL)	Stuart Daw	Lab	$ -	$ 350,000	$ 350,000	$ 300,000	
DOE	Advanced Combustion Engines	Combustion Research	Neutron Imaging of Advanced Transportation Technologies	Neutron Imaging of Advanced Transportation Technologies	ACE052	Oak Ridge National Laboratory (ORNL)	Todd Toops	Lab	$ -	$ 200,000	$ 200,000	$ 200,000	
DOE	Advanced Combustion Engines	Combustion Research	Collaborative Combustion Research with Basic Energy Sciences	Collaborative Combustion Research with BES	ACE054	Argonne National Laboratory (ANL)	Scott Goldsborough	Lab	$ -	$ 400,000	$ 320,000	$ 325,000	
DOE	Advanced Combustion Engines	Combustion Research	Spray-Combustion Modeling	Advancement in Fuel Spray and Combustion Modeling for Compression Ignition Engine Applications	ACE075	Argonne National Laboratory (ANL)	Sibendu Som	Lab	$ -	$ 350,000	$ 500,000	$ 350,000	
DOE	Advanced Combustion Engines	Combustion Research	Modeling of Clean and Efficient Combustion Regimes	Improved Solvers for Advanced Engine Combustion Simulation	ACE076	Lawrence Livermore National Laboratory (LLNL)	Matthew McNenly	Lab	$ -	$ 340,000	$ 340,000	$ 475,000	
DOE	Advanced Combustion Engines	Combustion Research	CRADA with Cummins on Reduction of Combustion Variation	Cummins ORNL\FEERC Combustion CRADA: Characterization & Reduction of Combustion Variations	ACE077	Oak Ridge National Laboratory (ORNL)	W.P. Partridge, S. Geckler	Lab	$ -	$ 300,000	$ 300,000	$ 300,000	
DOE	Advanced Combustion Engines	Emission Control	Coordination of Cross-Cut Lean Exhaust Emission Reduction Simulation (CLEERS) Project	Joint Development and Coordination of Emissions Control Data and Models (CLEERS Analysis and Coordination)	ACE022	Oak Ridge National Laboratory (ORNL)	Stuart Daw	Lab	$ -	$ 350,000	$ 712,000	$ 558,000	
DOE	Advanced Combustion Engines	Emission Control	CLEERS Diesel Soot Filter Characterization (Directed research with OEMs and Suppliers)	CLEERS Aftertreatment Modeling and Analysis	ACE023	Pacific Northwest National Laboratory (PNNL)	George Muntean	Lab	$ -	$ 750,000	$ 750,000	$ 750,000	
DOE	Advanced Combustion Engines	Emission Control	Enhanced High and Low Temperature Performance of NOx Reduction Materials	Enhanced High and Low Temperature Performance of NOx Reduction Materials	ACE026	Pacific Northwest National Laboratory (PNNL)	Chuck Peden	Lab	$ -	$ 300,000	$ 300,000	$ 300,000	
DOE	Advanced Combustion Engines	Emission Control	Experimental Studies for CPF and SCR Model, Control System, and OBD Development for Engines Using Diesel and Biodiesel Fuels	Experimental Studies for CPF and SCR Model, Control System, and OBD Development for Engines Using Diesel and Biodiesel Fuels	ACE028	Michigan Technological University	Gordon Parker	University	$ -	$ 607,000	$ -	$ -	Project completed in FY 2012
DOE	Advanced Combustion Engines	Emission Control	CRADA with Cummins on NOx Control and Measurement Technology for Diesel Engines	Cummins/ORNL-FEERC CRADA: NOx Control & Measurement Technology for Heavy-Duty Diesel Engines, Self-Diagnosing SmartCatalyst Systems	ACE032	Oak Ridge National Laboratory (ORNL)	Bill Partridge	Lab	$ -	$ 450,000	$ 595,000	$ 232,000	
DOE	Advanced Combustion Engines	Emission Control	CLEERS: Benchmark Kinetics for NOx Adsorbers and Catalyzed DPF	Development of Chemical Kinetic Models for Lean NOx Traps	ACE035	Sandia National Laboratories (SNL)	Richard Lawson	Lab	$ -	$ 200,000	$ -	$ -	Project completed in FY 2012
DOE	Advanced Combustion Engines	Emission Control	Advanced Collaborative Emissions Study (ACES)	Advanced Collaborative Emissions Study (ACES)	ACE044	Health Effects Institute	Dan Greenbaum	Other	$ -	$ 650,000	$ -	$ -	Project completed in FY 2013
DOE	Advanced Combustion Engines	Emission Control	Development of Radio Frequency Diesel Particulate Filter Sensor and Controls for Advanced Low-Pressure Drop Systems to Reduce Engine Fuel Consumption	Development of Radio Frequency Diesel Particulate Filter Sensor and Controls for Advanced Low-Pressure Drop Systems to Reduce Engine Fuel Consumption	ACE089	Filter Sensing Technologies	Alexander Sappok	Industry	$ -	$ 486,775	$ 1,553,109	$ -	
DOE	Advanced Combustion Engines	High Efficiency Engine Technologies	SuperTruck-Cummins	Cummins SuperTruck Program - Technology and System Level Demonstration of Highly Efficient and Clean, Diesel Powered Class 8 Trucks	ACE057	Cummins Inc.	David Koeberlein	Industry	$ -	$ -	$ -	$ -	Funding included with overall SuperTruck award in VSST listing
DOE	Advanced Combustion Engines	High Efficiency Engine Technologies	SuperTruck-Daimler/DDC	Recovery Act – Class 8 Truck Freight Efficiency Improvement Project	ACE058	Detroit Diesel	Derek Rotz/Sandeep Singh	Industry	$ -	$ -	$ -	$ -	Funding included with overall SuperTruck award in VSST listing
DOE	Advanced Combustion Engines	High Efficiency Engine Technologies	SuperTruck-Navistar	Supertruck - Development and Demonstration of a Fuel-Efficient Class 8 Tractor & Trailer	ACE059	Navistar	Russ Zukouski	Industry	$ -	$ 2,254,842	$ 19,355	$ 5,680,000	
DOE	Advanced Combustion Engines	High Efficiency Engine Technologies	SuperTruck-Volvo	Powertrain Technologies for Efficiency Improvement	ACE060	Volvo	Pascal Amar	Industry	$ -	$ 7,819,158	$ 2,089,543	$ 4,063,739	
DOE	Advanced Combustion Engines	High Efficiency Engine Technologies	Advanced Technology Light Automotive Systems (ATLAS): Cummins Next Generation Tier 2 Bin 2 Diesel Engine	ATP-LD: Cummins Next Generation Tier 2 Bin 2 Diesel Engine	ACE061	Cummins Inc.	Michael Ruth	Industry	$ -	$ 5,250,000	$ 5,008,419	$ -	
DOE	Advanced Combustion Engines	High Efficiency Engine Technologies	Robust Nitrogen Oxide/Ammonia Sensors for Vehicle Onboard Emissiosn Control	Robust Nitrogen Oxide/Ammonia Sensors for Vehicle Onboard Emissiosn Control	ACE079	Los Alamos National Laboratory (LANL)	Rangachary Mukundan	Lab	$ -	$ 350,000	$ 310,982	$ 389,018	
DOE	Advanced Combustion Engines	High Efficiency Engine Technologies	Heavy Duty Roots Expander Heat Energy Recovery	Heavy Duty Roots Expander Heat Energy Recovery	ACE088	Eaton Corporation	Swami Nathan Subramanian	Industry	$ -	$ 900,203	$ 1,031,882	$ 567,915	
DOE	Advanced Combustion Engines	High Efficiency Engine Technologies	ANL-EWO to Support SuperTruck Project	ANL-EWO to Support SuperTruck Project	N/A	Argonne National Laboratory (ANL)	Thomas Wallner	Lab	$ -	$ 1,251,000	$ 500,000	$ 406,000	

21CTP Project Inventory

Agency	Technology Area	Sub-grouping	Internal Project Title	Public Review Project Title	Project ID	Recipient	Principal Investigator	Lab/Industry/Univ Funding	ARRA Funding	2012 Funding	2013 Funding	2014 Funding	Notes
DOE	Fuels Technologies	Fuels Technologies	Reactivity Controlled Compression Ignition (RCCI) Combustion regimes	Demonstration/Development of Reactivity Controlled Compression Ignition (RCCI) Combustion for High Efficiency, Low Emissions Vehicle Applications	FT015	University of Wisconsin-Madison	Rolf Reitz	University	$ -	$ 500,000	$ 640,000	$ 360,000	Project end in FY 2014
DOE	Fuels Technologies	Fuels Technologies	High Compression Ratio Turbo Gasoline Engine Operation Using Alcohol Enhancement	High Compression Ratio Turbo Gasoline Engine Operation Using Alcohol Enhancement	FT016	Massachusetts Institute of Technology (MIT)	John Heywood	University	$ -	$ 408,000	$ 235,000	$ 320,000	
DOE	Fuels Technologies	Fuels Technologies	Fuel Properties to Enable Lifted-Flame Combustion	Fuel Properties to Enable Lifted-Flame Combustion	FT017	Ford Motor Company	Eric Kurtz	Industry	$ -	$ 436,904	$ 406,000	$ 694,000	
DOE	Fuels Technologies	Fuels Technologies	Advanced Nanolubricants for Improved Energy Efficiency and Reduced Emissions in Engines	Advanced Nanolubricants for Improved Energy Efficiency and Reduced Emissions in Engines	FT018	Argonne National Laboratory (ANL)	Ali Erdemir	Lab	$ -	$ -	$ 268,000	$ 267,000	
DOE	Fuels Technologies	Fuels Technologies	MIT lubes	Lubricant Formulations to Enhance Engine Efficiency (LFEEE) in Modern Internal Combustion Engines	FT019	Massachusetts Institute of Technology (MIT)	Wai Cheng	University	$ -	$ 630,000	$ 870,000	$ -	
DOE	Fuels Technologies	Fuels Technologies	Ford Lubes	Development of Modified PAG (polyalkylene glycol) High VI High Fuel Efficient Lubricant for LDV Applications	FT020	Ford Motor Company	Arup Gangopadhyay	Industry	$ -	$ 435,000	$ 327,916	$ 437,084	
DOE	Fuels Technologies	Fuels Technologies	Can hard coatings and lubricant anti-wear additives work together?	Can hard coatings and lubricant anti-wear additives work together?	FT021	Oak Ridge National Laboratory (ORNL)	Jun Qu	Lab	$ -	$ -	$ 250,000	$ 250,000	
DOE	Fuels Technologies	Fuels Technologies	CFD Simulations and Experiments to Determine the Feasibility of Various Alternate Fuels for Compression Ignition Engine Applications	CFD Simulations and Experiments to Determine the Feasibility of Various Alternate Fuels for Compression Ignition Engine Applications	FT022	Argonne National Laboratory (ANL)	Sibendu Som	Lab	$ -	$ 150,000	$ 150,000	$ -	
DOE	Fuels Technologies	Fuels Technologies	Natural Gas Engine Development with CEC and SCAQMD	Natural Gas Engine Development with CEC and SCAQMD	N/A	National Renewable Energy Laboratory (NREL)		Lab	$ -	$ -	$ -	$ -	Selections and funding were made for this RFP in 2010
DOE	Fuels Technologies	Fuels Technologies	PNNL Unconventional Fuels	PNNL Unconventional Fuels	N/A	Pacific Northwest National Laboratory (PNNL)	Tim Bays	Lab	$ -	$ 450,000	$ -	$ 220,000	
DOE	Fuels Technologies	Fuels Technologies	NWU/ANL Novel Lube Formulations	NWU/ANL Novel Lube Formulations	N/A	Northwestern University, Argonne National Laboratory (ANL)		University	$ -	$ -	$ -	$ 286,000	New, not reviewed at AMR
DOE	Fuels Technologies	Fuels Technologies	WTW analysis, refinery modeling of high-octane, NG pathways analysis, FT fuels, XTL fuels pathways.	WTW analysis, refinery modeling of high-octane, NG pathways analysis, FT fuels, XTL fuels pathways.	N/A	Argonne National Laboratory (ANL)	Michael Wang	Lab	$ -	$ 500,000	$ 500,000	$ 575,000	Not reviewed at AMR
DOE	Fuels Technologies	Fuels Technologies	Hyperbranched polymers as lubricants	Hyperbranched polymers as lubricants	N/A	Pacific Northwest National Laboratory (PNNL)		Lab	$ -	$ -	$ 519,375	$ 200,000	New, not reviewed at AMR
DOE	Materials Technology	Lightweight	Low Cost Carbon Fiber:Low cost carbon fiber	Low Cost Carbon Fiber Overview	LM002	Oak Ridge National Laboratory (ORNL)	Dave Warren	Lab	$ -	$ 1,750,000	$ 750,000	$ -	
DOE	Materials Technology	Lightweight	Carbon Fiber Technology Center Operating Funds	Carbon Fiber Technology Facility	LM003	Oak Ridge National Laboratory (ORNL)	Dave Warren	Lab	$ -	$ 1,000,000	$ 750,000	$ 750,000	50% relevance to heavy-duty
DOE	Materials Technology	Lightweight	Low Cost Carbon Fiber Composites for Lightweight Vehicle Parts	Low Cost Carbon Fiber Composites for Lightweight Vehicle Parts	LM047	Materials Innovation Technologies LLC	Mark Mauhar	Industry	$ -	$ 937,000	$ 500,000	$ -	50% relevance to heavy-duty
DOE	Materials Technology	Lightweight	Low Cost Carbon Fiber:Low cost carbon fiber	Development and Commercialization of a Novel Low-Cost Carbon Fiber	LM048	Zoltek	George Husman	Industry	$ -	$ 1,638,953	$ 700,000	$ -	Partially relevant to heavy-duty
DOE	Materials Technology	Lightweight	Aerodynamic Lightweight Cab Structure Components	Aerodynamic Lightweight Cab Structure Components	LM060	Pacific Northwest National Laboratory (PNNL)	Mark Smith	Lab	$ -	$ 365,000	$ 280,000	$ 65,000	2014 funding from AMR 2014
DOE	Materials Technology	Lightweight	Automotive Metals: Properties and Manufacturing	Improving Fatigue Performance of AHSS Welds	LM062	Oak Ridge National Laboratory (ORNL)	C. David Warren	Lab	$ -	$ 355,000	$ 125,000	$ 150,000	50% relevance to heavy-duty
DOE	Materials Technology	Lightweight	Polymer Composites: Polymer Composites	Engineering Property Prediction Tools for Tailored Polymer Composite Structures	LM068	Pacific Northwest National Laboratory (PNNL)	Ba Nghiep Nguyen	Lab	$ -	$ 132,040	$ -	$ -	Project end in FY 2012
DOE	Materials Technology	Lightweight	GATE AWARD light weight center	GATE Center of Excellence at UAB for Lightweight Materials and Manufacturing for Automotive, Truck and Mass Transit	LM081	University of Alabama at Birmingham	Uday Vaidya	University	$ -	$ 125,000	$ 60,000	$ -	50% relevance to heavy-duty
DOE	Materials Technology	Propulsion	HD -Cast Fe Alloys for High PCP Engines	HD -Cast Fe Alloys for High PCP Engines	N/A	National Energy Technology Laboratory (NETL)		Lab	$ -	$ -	$ 3,477,000	$ -	Funding = $2,179,020 PM / $1,298,110 LM, fully funded in FY13, 3 year project
DOE	Materials Technology	Propulsion	Tailored Materials for Improved Internal Combustion	Tailored Materials for Advanced CIDI Engines (through FY13) / Novel Manufacturing Technologies for High Power Induction and Permanent Magnet Electric Motors (FY 14)	PM004	Pacific Northwest National Laboratory (PNNL)	Glenn Grant	Lab	$ -	$ 350,000	$ 300,000	$ 225,000	
DOE	Materials Technology	Propulsion	NOX sensor development (Electrochemical Nox Sensor Devel)	NOx Sensor Development	PM005	Lawrence Livermore National Laboratory (LLNL)	Leta Woo	Lab	$ -	$ 400,000	$ 511,482	$ -	Project end in FY 2013
DOE	Materials Technology	Propulsion	Titanium Friction and wear	Friction and Wear Enhancement of Titanium Alloy Engine Components	PM007	Oak Ridge National Laboratory (ORNL)	Peter Blau	Lab	$ -	$ 125,000	$ -	$ -	Project end in FY 2012
DOE	Materials Technology	Propulsion	Materials Issues Associated with EGR Systems	Materials Issues Associated with EGR Systems	PM009	Oak Ridge National Laboratory (ORNL)	Michael Lance	Lab	$ -	$ 360,000	$ 170,000	$ 184,000	
DOE	Materials Technology	Propulsion	Durability Diesel Engine Partic. Filters	Durability of Diesel Engine Particulate Filters	PM010	Oak Ridge National Laboratory (ORNL)	Thomas Watkins	Lab	$ -	$ 350,000	$ 170,000	$ 265,000	

21CTP Project Inventory

Agency	Technology Area	Sub-grouping	Internal Project Title	Public Review Project Title	Project ID	Recipient	Principal Investigator	Lab/Industry/Univ Funding	ARRA Funding	2012 Funding	2013 Funding	2014 Funding	Notes
DOE	Materials Technology	Propulsion	Catalysts via First Principles	Catalysts via First Principles	PM011	Oak Ridge National Laboratory (ORNL)	Chaitanya K. Narula	Lab	$ -	$ 300,000	$ 300,000	$ -	Project end in FY 2012
DOE	Materials Technology	Propulsion	Thermoelectric Mechanical Reliability	Thermoelectric Mechanical Reliability	PM012	Oak Ridge National Laboratory (ORNL)	Andrew Wereszczak	Lab	$ -	$ 375,000	$ 340,000	$ -	Project end in FY 2014
DOE	Materials Technology	Propulsion	Thermoelectric Theory and Structure	Thermoelectrics Theory and Structure	PM013	Oak Ridge National Laboratory (ORNL)	David J. Singh	Lab	$ -	$ 375,000	$ 340,000	$ -	Project end in FY 2014
DOE	Materials Technology	Propulsion	Materials for HCCI Engine	Materials for HCCI Engines	PM018	Oak Ridge National Laboratory (ORNL)	Govindarajan Muralidharan	Lab	$ -	$ 225,000	$ -	$ -	Project end in FY 2012
DOE	Materials Technology	Propulsion	Compact Potentiometric NOx Sensor	Compact Potentiometric NOx Sensor	PM023	Argonne National Laboratory (ANL)	Dileep Singh	Lab	$ -	$ 60,000	$ 75,000	$ -	Project end in FY 2012
DOE	Materials Technology	Propulsion	Characterization of Catalyst Microstructures	Catalyst Characterization	PM028	Oak Ridge National Laboratory (ORNL)	Thomas Watkins	Lab	$ -	$ 325,000	$ -	$ -	Project end in FY 2012
DOE	Materials Technology	Propulsion	Materials for Advanced Turbocharger Designs	Materials for Advanced Turbocharger Designs	PM038	Oak Ridge National Laboratory (ORNL)	Phil Maziasz	Lab	$ -	$ 300,000	$ -	$ 250,000	
DOE	Materials Technology	Propulsion	Electrically Regenerated DPF Material	Electrically-Assisted Diesel Particulate Filter Regeneration	PM041	Oak Ridge National Laboratory (ORNL)	Michael Lance	Lab	$ -	$ 250,000	$ -	$ -	Project end in FY 2012
DOE	Materials Technology	Propulsion	High-Temperature Aluminum Alloys	High-Temperature Aluminum Alloys	PM044	Pacific Northwest National Laboratory (PNNL)	Stan Pitman	Lab	$ -	$ 395,000	$ 300,000	$ 125,000	
DOE	Materials Technology	Propulsion	Catalyst Characterization and Deactivation Mechanisms	Catalyst Characterization and Deactivation Mechanisms	PM049	Oak Ridge National Laboratory (ORNL)	Thomas Watkins	Lab	$ -	$ 200,000	$ 130,000	$ 75,000	
DOE	Materials Technology	Propulsion	Design Optimization of Piezoceramic Multilayer Actuators for Heavy Duty Diesel Engine Fuel Injectors	Design Optimization of Piezoceramic Multilayer Actuators for Heavy Duty Diesel Engine Fuel Injectors	PM051	Oak Ridge National Laboratory (ORNL)	H.T. Lin	Lab	$ -	$ 300,000	$ 190,000	$ 175,000	
DOE	Materials Technology	Propulsion	Friction and Wear	Friction Reduction through Surface Modification	PM052	Oak Ridge National Laboratory (ORNL)	Jun Qu	Lab	$ -	$ -	$ 260,000	$ 150,000	Project end in FY 2014
DOE	Materials Technology	Propulsion	High Temperature Materials for High Efficiency Engines	High Temperature Materials for High Efficiency Engines	PM053	Oak Ridge National Laboratory (ORNL)	Govindarajan Muralidharan	Lab	$ -	$ -	$ 200,000	$ -	
DOE	Materials Technology	Propulsion	Biofuels impact on DPF durability	Biofuels Impact on Aftertreatment Devices	PM055	Oak Ridge National Laboratory (ORNL)	Michael Lance	Lab	$ -	$ 300,000	$ 150,000	$ 135,000	
DOE	Materials Technology	Propulsion	Applied ICME For New Propulsion Materials	Applied ICME For New Propulsion Materials	PM057	Oak Ridge National Laboratory (ORNL)	David Singh	Lab	$ -	$ -	$ 68,711	$ 825,176	
DOE	Materials Technology	Propulsion	HD - High Performance Cast Steels for Crankshafts (CAT/GM)	HD - High Performance Cast Steels for Crankshafts (CAT/GM)	PM058 (ANL) / PM059 (Cat)	National Energy Technology Laboratory (NETL)	John Hryn (ANL) / Richard Huff (Cat)	Lab	$ -	$ -	$ 2,100,000	$ -	Fully funded in FY 13, 3 year project, Caterpillar and GM
DOE	Vehicle Systems Simulation and Testing	Component and Systems Evaluations	The Meritor Dual Mode Hybrid Powertrain CRADA	The ArvinMeritor Dual Mode Hybrid Powertrain (DMHP): Opportunities and Potential for Systems Optimization	VSS062	Oak Ridge National Laboratory (ORNL)	David Smith	Lab	$ -	$ 500,000	$ 500,000	$ -	Project end in FY 2013
DOE	Vehicle Systems Simulation and Testing	Industry Awards	Zero Emission Cargo Transport (Phase 2)	TBD	N/A	TBD	TBD	Industry	$ -	$ -	$ -	$ 4,000,000	Phase 2 awards to be announced by September 2014
DOE	Vehicle Systems Simulation and Testing	Industry Awards	SuperTruck (non-ARRA)	SuperTruck – Development and Demonstration of a Fuel-Efficient Class 8 Tractor & Trailer	VSS064	Navistar	Dale Oehlerking	Industry	$ -	$ 2,600,000	$ -	$ 3,000,000	
DOE	Vehicle Systems Simulation and Testing	Industry Awards	SuperTruck (non-ARRA)	Super Truck – Development and Demonstration of a Fuel Efficient Class 8 Highway Vehicle (Volvo), Tractor & Trailer (Navistar)	VSS081	Volvo	Pascal Amar	Industry	$ -	$ 4,400,000	$ 3,780,000	$ -	
DOE	Vehicle Systems Simulation and Testing	Industry Awards	Zero Emission Heavy Duty Drayage Truck Demonstration	Zero Emission Heavy Duty Drayage Truck Demonstration	VSS115	South Coast Air Quality Management District	Brian Choe	Industry	$ -	$ 4,169,000	$ -	$ -	
DOE	Vehicle Systems Simulation and Testing	Industry Awards	Houston Zero Emission Delivery Vehicle Deployment Project	Houston Zero Emission Delivery Vehicle Deployment Project	VSS116	Houston-Galveston Area Council	Christine Smith	Industry	$ -	$ 920,000	$ -	$ -	
DOE	Vehicle Systems Simulation and Testing	Industry Awards	Hydrogen Fuel-Cell Electric Hybrid Truck Demonstration	Hydrogen Fuel-Cell Electric Hybrid Truck Demonstration	VSS117	Houston-Galveston Area Council	Christine Smith	Industry	$ -	$ 1,075,000	$ -	$ -	
DOE	Vehicle Systems Simulation and Testing	Interagency Agreements	EPA IAA -Hydraulic Range Extender for MD BEV	EPA IAA -Hydraulic Range Extender for MD BEV	N/A	U.S. Environmental Protection Agency	N/A	Other	$ -	$ 1,000,000	$ -	$ -	Funding repurposed for support of EPA testing procedures by ORNL. Work is to develop test procedures for power pack testing in support of Phase 2 of the EPA GHG rulemaking
DOE	Vehicle Systems Simulation and Testing	Lab & Field Vehicle Evaluations	Medium and Heavy Duty In-Use Fleet Field Evaluations	Medium and Heavy-Duty Vehicle Field Evaluations	VSS001	National Renewable Energy Laboratory (NREL)	Kevin Walkowicz	Lab	$ -	$ 900,000	$ 1,150,000	$ 300,000	
DOE	Vehicle Systems Simulation and Testing	Lab & Field Vehicle Evaluations	Medium Truck Duty Cycle (MTDC) Part 2	Truck Duty Cycle and Performance Data Collection and Analysis Program	VSS002	Oak Ridge National Laboratory (ORNL)	Tim LaClair	Lab	$ -	$ 350,000	$ -	$ -	Project end in FY 2012
DOE	Vehicle Systems Simulation and Testing	Lab & Field Vehicle Evaluations	Vehicle Systems Integration Laboratory	Vehicle Systems Integration (VSI) Research Laboratory at ORNL	VSS035	Oak Ridge National Laboratory (ORNL)	David Smith	Lab	$ -	$ 1,100,000	$ 750,000	$ -	Project end in FY 2013
DOE	Vehicle Systems Simulation and Testing	Lab & Field Vehicle Evaluations	MD PHEV/EV Data Collection and Reporting	Real-World PHEV Fuel Economy Prediction	VSS047	National Renewable Energy Laboratory (NREL)	Jeffrey Gonder	Lab	$ -	$ 250,000	$ -	$ -	Project end in FY 2012
DOE	Vehicle Systems Simulation and Testing	Lab & Field Vehicle Evaluations	Fleet DNA: Common drive cycle database	Fleet DNA	VSS119	National Renewable Energy Laboratory (NREL)	Kevin Walkowicz	Lab	$ -	$ 400,000	$ 500,000	$ 325,000	

21CTP Project Inventory

Agency	Technology Area	Sub-grouping	Internal Project Title	Public Review Project Title	Project ID	Recipient	Principal Investigator	Lab/Industry/Univ Funding	ARRA Funding	2012 Funding	2013 Funding	2014 Funding	Notes
DOE	Vehicle Systems Simulation and Testing	Recovery Act	Interstate Grid Electrification Project (Cascade Sierra Solutions)	Interstate Grid Electrification Project (Cascade Sierra Solutions)	ARRAVT070	Cascade Sierra Solutions	Jon Gustafson	Industry	$ 22,200,000	$ -	$ -	$ -	
DOE	Vehicle Systems Simulation and Testing	Recovery Act	Smith Electric Vehicles	Smith Electric Vehicles: Advanced Vehicle Electrification + Transportation Sector Electrification	ARRAVT072	Smith Electric	Robin Mackie	Industry	$ 32,000,000	$ -	$ -	$ -	
DOE	Vehicle Systems Simulation and Testing	Recovery Act	Daimler SuperTruck	Recovery Act – Class 8 Truck Freight Efficiency Improvement Project	ARRAVT080	Daimler Trucks	Derek Rotz	Industry	$ 47,486,735	$ -	$ -	$ -	
DOE	Vehicle Systems Simulation and Testing	Recovery Act	Cummins/Peterbilt SuperTruck	Technology and System Level Demonstration of Highly Efficient and Clean, Diesel Powered Class 8 Trucks	ARRAVT081	Cummins Inc.	Ken Damon	Industry	$ 38,800,000	$ -	$ -	$ -	
DOE	Vehicle Systems Simulation and Testing	Recovery Act	Plug-in Hybrid Electric Medium Duty Commercial Fleet Demonstration & Evaluation	Plug-in Hybrid Electric Medium Duty Commercial Fleet Demonstration & Evaluation	ARRAVT083	South Coast Air Quality Management District	Matt Miyasato	State/Local	$ 45,443,325	$ -	$ -	$ -	Listed as ARRAVT068 in 2011 AMR
DOE	Vehicle Systems Simulation and Testing	Vehicle Simulation & Modeling	CoolCab Truck Thermal Load and Idle Reduction	CoolCab Test and Evaluation	VSS037	National Renewable Energy Laboratory (NREL)	Jason Lustbader	Lab	$ -	$ 350,000	$ 500,000	$ 200,000	
DOE	Vehicle Systems Simulation and Testing	Vehicle Simulation & Modeling	MD and HD Electric Drive Vehicle Simulation and Analysis	Medium- and Heavy-Duty Electric Drive Vehicle Simulation and Analysis	VSS043	National Renewable Energy Laboratory (NREL)	Robb Barnitt	Lab	$ -	$ 300,000	$ -	$ -	Project end in FY 2012
DOE	Vehicle Systems Simulation and Testing	Vehicle Simulation & Modeling	Hydraulic HEV Fuel Consumption Potential	Hydraulic Hybrid Vehicle Fuel Consumption Potential	VSS071	Argonne National Laboratory (ANL)	Namwook Kim, Aymeric Rousseau	Lab	$ -	$ 300,000	$ -	$ -	Project end in FY 2012
DOE	Vehicle Systems Simulation and Testing	Vehicle Simulation & Modeling	CoolCalc HVAC Tool	CoolCab Test and Evaluation & CoolCalc HVAC Tool Development	VSS075	National Renewable Energy Laboratory (NREL)	Jason A. Lustbader	Lab	$ -	$ 225,000	$ 700,000	$ 300,000	
DOE	Vehicle Systems Simulation and Testing	Vehicle Simulation & Modeling	Advanced HD Engine Systems and Emissions Control Modeling and Analysis	Advanced Heavy-Duty Engine Systems and Emissions Control Modeling and Analysis	VSS089	Oak Ridge National Laboratory (ORNL)	C. Stuart Daw	Lab	$ -	$ 325,000	$ 390,000	$ 250,000	
DOE	Vehicle Systems Simulation and Testing	Vehicle Simulation & Modeling	Army National Guard Vehicle Selection for Hybridization	N/A	N/A	Oak Ridge National Laboratory (ORNL)	Perry Jones	Lab	$ -	$ 100,000	$ -	$ -	Project end in FY 2012
DOE	Vehicle Systems Simulation and Testing	Vehicle Systems Optimization	DOE/DOD Parasitic Energy Loss Collaboration	DOE/DOD Parasitic Energy Loss Collaboration	VSS005	Argonne National Laboratory (ANL)	George Fenske	Lab	$ -	$ 250,000	$ 200,000	$ 170,000	
DOE	Vehicle Systems Simulation and Testing	Vehicle Systems Optimization	DOE's Effort to Reduce Truck Aerodynamic Drag for Significant Impact on Fuel Economy through Joint Experiments and Computations	DOE's Effort to Reduce Truck Aerodynamic Drag through Joint Experiments and Computations	VSS006	Lawrence Livermore National Laboratory (LLNL)	Kambiz Salari	Lab	$ -	$ 650,000	$ 900,000	$ 600,000	
DOE	Vehicle Systems Simulation and Testing	Vehicle Systems Optimization	Development of High Power density Driveline for Vehicle Efficiency Improvement	Development of High Power Density Driveline for Vehicles	VSS058	Argonne National Laboratory (ANL)	Oyelayo Ajayi	Lab	$ -	$ 350,000	$ 300,000	$ 350,000	
DOE	Vehicle Systems Simulation and Testing	Vehicle Systems Optimization	CRADA with PACCAR – Experimental Investigation in Cooling Boiling in a half-heated Circular Tube	CRADA with PACCAR Experimental Investigation in Coolant Boiling in a Half-Heated Circular Tube	VSS079	Argonne National Laboratory (ANL)	Wen Yu	Lab	$ -	$ 300,000	$ 200,000	$ -	Project end in FY 2013
DOE	Vehicle Systems Simulation and Testing	Vehicle Systems Optimization	ANL-Cummins CRADA: "Integrated External Aerodynamic and Underhood Thermal Analysis for Heavy Vehicles"	ANL-Cummins CRADA: "Integrated External Aerodynamic and Underhood Thermal Analysis for Heavy Vehicles"	VSS080	Argonne National Laboratory (ANL)	Tanju Sofu	Lab	$ -	$ 100,000	$ -	$ 125,000	
DOE	Vehicle Systems Simulation and Testing	Vehicle Systems Optimization	Development of Nanofluids for Cooling Power Electronics for HEVs	Development of Nanofluids for Cooling Power Electronics for Hybrid Electric Vehicles	VSS112	Argonne National Laboratory (ANL)	Dileep Singh	Lab	$ -	$ 150,000	$ 75,000	$ 350,000	
DOE	Vehicle Systems Simulation and Testing	Vehicle Systems Optimization	Thermal Control Thru Air Side Evaporative Heat Removal	N/A	N/A	Argonne National Laboratory (ANL)	Dileep Singh	Lab	$ -	$ 150,000	$ -	$ 200,000	Documented in FY 2013 Annual Progress Report
DOE	Vehicle Systems Simulation and Testing	Vehicle Systems Optimization	Autonomous Intelligent Hybrid Propulsion Systems	Autonomous Intelligent Hybrid Propulsion Systems	VSS107	Oak Ridge National Laboratory (ORNL)	Andreas Malikopoulos	Lab	$ -	$ 200,000	$ 200,000	$ 100,000	
DOE	Vehicle Systems Simulation and Testing	Vehicle Systems Optimization	Cummins MD & HD Accessory Hybridization CRADA	N/A	N/A	Oak Ridge National Laboratory (ORNL)	David Smith	Lab	$ -	$ -	$ 250,000	$ 43,000	
DOE	Vehicle Systems Simulation and Testing	Vehicle Systems Optimization	CRADA with PACCAR - Efficient Cooling of Heavy-Duty Truck Engines with Coolant Boiling at Higher Pressures	N/A	N/A	Argonne National Laboratory (ANL)	Wen Yu	Lab	$ -	$ -	$ -	$ 125,000	Follow-on to project VSS079
DOE	VMT Reduction and Legacy Fleet Improvement	Driver Feedback	Next Generation Environmentally Friendly Driving Feedback Systems Research and Development	Next Generation Environmentally Friendly Driving Feedback Systems Research and Development	VSS086	The Regents Of The University Of California -University Of California	Matthew Barth	University	$ -	$ 436,631	$ 376,301	$ 320,803	Project end in FY 2014
DOE	VMT Reduction and Legacy Fleet Improvement	Driver Feedback	Look-ahead Driver Feedback and Powertrain Management	Look-ahead Driver Feedback and Powertrain Management	VSS087	Eaton Corporation -Eaton Innovation Center	Rajeev Verma	Industry	$ -	$ 286,141	$ 341,000	$ 121,000	Project end in FY 2014
DOE	VMT Reduction and Legacy Fleet Improvement	Tire Technology	Materials Approach to Fuel Efficient Tires	Materials Approach to Fuel Efficient Tires	VSS084	P P G Industries Inc -Monroeville Technical Center	Peter Votruba-Drzal	Industry	$ -	$ 186,000	$ 675,000	$ 167,000	Project end in FY 2014

21CTP Project Inventory

Agency	Technology Area	Sub-grouping	Internal Project Title	Public Review Project Title	Project ID	Recipient	Principal Investigator	Lab/Industry/Univ Funding	ARRA Funding	2012 Funding	2013 Funding	2014 Funding	Notes
DOE	VMT Reduction and Legacy Fleet Improvement	Tire Technology	System for Automatically Maintaining Pressure in a Commercial Truck Tire	System for Automatically Maintaining Pressure in a Commercial Truck Tire	VSS085	The Goodyear Tire & Rubber Company -Goodyear	Robert Benedict	Industry	$ -	$ 571,189	$ 713,810	$ 161,535	Project end in FY 2015
DOT	NHTSA	Safety Systems	Safety Systems	N/A	N/A	Various	N/A		$ -	$ 500,000	$ 600,000	$ 500,000	
DOT	NHTSA	Heavy Vehicles	Heavy Vehicles	N/A	N/A	Various	N/A		$ -	$ 2,100,000	$ 2,400,000	$ 2,100,000	
DOT	NHTSA	Crash Avoidance	Crash Avoidance	N/A	N/A	Various	N/A		$ -	$ 700,000	$ 1,100,000	$ 1,000,000	
DOT	NHTSA	ITS	ITS	N/A	N/A	Various	N/A		$ -	$ 2,500,000	$ 2,000,000	$ 3,700,000	
DOT	FMCSA	Safety	Update Federal Motor Carrier Safety Regulations to address electric-drive commercial vehicles	N/A	N/A	Various	N/A		$ -	$ -	$ -	$ 150,000	FY 2014 funds, period of performance 9/2013 to 9/2014
DOT	FMCSA	Efficiency	Smartphone app for monitoring truck fuel economy	N/A	N/A	Various	N/A		$ -	$ 750,000	$ -	$ -	FY 2012 funds, period of performance 6/13 to 8/15
DOT	FMCSA	Idle Reduction	Idle reduction technology	N/A	N/A	Various	N/A		$ -	$ -	$ -	$ 287,400	FY 2014 funds, task expected to be awarded by 12/14
DOT	FMCSA	Efficiency	SmartPark real-time parking availability information	N/A	N/A	Various	N/A		$ -	$ 2,706,750	$ -	$ -	Funds are from FY 2007-2012, period of performance 9/08 to 9/15: http://www.fmcsa.dot.gov/research-and-analysis/technology/smartpark-real-time-parking-availability
DOT	FMCSA	Safety	Web-based course to train commercial vehicle inspectors on how to detect leaks from natural gas & propane trucks & buses	N/A	N/A	Various	N/A		$ -	$ -	$ -	$ 150,000	FY 2014 funds, period of performance 9/2013 to 9/2015
DOT	FMCSA	Safety	Web-based course to familiarize commercial vehicle inspectors with safety aspects of electric drive commercial vehicles	N/A	N/A	Various	N/A		$ -	$ -	$ -	$ 150,000	FY 15 funds (contract not yet awarded
DOT	FMCSA	Safety	Wireless Roadside Inspection	N/A	N/A	Various	N/A		$ -	$ 3,000,000	$ -	$ -	Funding year not specified (http://www.fmcsa.dot.gov/research-and-analysis/technology/wireless-roadside-inspection-field-operations-test)
EPA	SmartWay		SmartWay Transport Partnership	N/A	N/A	Various	N/A		$ -	$ 2,700,000	$ 2,700,000	$ 2,700,000	Estimated funding for SmartWay from EPA OIG report - 12-P-0747 (report gives 2012 funding, assume flat for 2013 and 2014) - SmartWay is part of EPA Climate Protection Program (FY 2015 budget justification page 203, FY 14 budget page 214)
EPA	Testing		Federal Vehicle and Fuels Standards and Certification	N/A	N/A	Various	N/A		$ -	$ 8,810,230	$ 8,685,800	$ 9,650,000	Estimated funding for HD portion of testing work (split 90/10 with light-duty) - from FY 2015 budget justification page 92, FY 14 budget justification page 88
	165								$ 185,930,060	$ 150,175,756	$ 133,926,369	$ 116,775,459	

Summary of 21CTP Goals

Goal Number	Goal Text
E-1	Develop and demonstrate an emissions compliant engine system for Class 7-8 highway trucks that achieves 50% brake thermal efficiency in an over-the-road cruise condition, improving the engine system fuel efficiency by about 20% (from approximately 42% thermal efficiency today). (2015)
E-2	Research and develop technologies which achieve a stretch thermal efficiency goal of 55% in prototype engine systems in the lab. (This efficiency gain would be equivalent to an additional 10% gain in over-the-road fuel economy when prototype concepts are fully developed for the market.) (2015)
E-3	Through experiments and models with FACE fuels and other projects, determine the most essential fuel properties, including renewables, needed to achieve 55% engine brake efficiency. (2014)
E-4	Identify alternatives to fossil petroleum based fuels and technology pathways (vehicle, fuels, and infrastructure) to a sustainable, long-term fuel supply.
HE-1	Ability to attain fuel consumption reductions (compared to today's conventional, non-hybridized heavy-duty vehicles) in a commercially viable manner.
HE-2	Develop a hybrid system with a design life of 15 years.
HE-3	Achieve cost targets for energy storage ($45 per kW and/or $500/ kW- hour for an energy battery by 2017; $40 per kW and/or $300/ kW- hour for a power battery by 2020; and cost of overall battery pack should not exceed cost of the cells themselves by more than 20% by 2016) and for e-machines ($23/kilowatt by 2016).
HH-1	N/A - EPA program cancelled
HH-2	N/A - EPA program cancelled
HH-3	N/A - EPA program cancelled
HH-4	N/A - EPA program cancelled
HH-5	N/A - EPA program cancelled
PD-1	Develop and demonstrate advanced technology concepts that reduce the aerodynamic drag of a Class 8 highway tractor-trailer combination by 20%. Evaluate a stretch goal of 30% reduction in aerodynamic drag. (2021
PD-2	Develop and demonstrate low rolling resistance tires that can reduce vehicle rolling resistance and wheel weight for a Class 8 tractor-trailer. Demonstrate 35% reduction in rolling resistance. (2021)
PD-3	Develop and demonstrate technologies that reduce essential auxiliary loads by 50% for Class 8 tractor-trailers. (2021)
PD-4	Develop and demonstrate engine, transmission, and driveline systems that enhance engine cycle operating efficiency and reduce friction losses. (2021)
PD-5	Develop and demonstrate lightweight material and manufacturing processes that lead to a 10% reduction in tare weight for a tractor/trailer combination. Establish a long-term stretch goal of reducing combined vehicle weight by 20%. (2021)
PD-6	Increase heat-load rejected by thermal management systems by 20% without increasing radiator size. Develop and demonstrate technologies that reduce powertrain and driveline losses by 50%. (2021)
IR-1	Promote the incorporation of idle reduction (IR) equipment on new trucks as fuel saving devices as they are identified through the DOE SuperTruck Initiative.
IR-2	Establish a nationwide multi-mode IR education program.
IR-3	Work with OEM truck manufacturers to obtain data on the number of new trucks being ordered with IR options.

Summary of 21CTP Goals

IR-4	Conduct a fleet survey to gather data on the amount of in-use idling hours that are accumulated by type of heavy-duty vehicle.
IR-5	Analyze data from the EPA SmartWay Transport Partnership to measure fuel savings and emissions reductions associated with the various type of IR equipment available.
IR-6	Develop improved IR systems to minimize fuel required, cost, and weight to meet hotel functions in sleeper cabs.
S-1	The 21CTP will work collaboratively with DOT to enhance safety primarily through a variety of crash avoidance strategies that include on-board vehicle technologies as well as operationally-focused programs designed to reduce crash risk. The overall goals of this collaboration are to 1) ensure that advancements in truck design and technology to improve fuel efficiency do not have any negative impacts on safety; and (2) conversely, to ensure that efforts to improve safety to not reduce efficiency—and, where possible actually contribute to improvements in overall motor carrier industry system efficiency.
EO-1	Develop and demonstrate technologies that minimize the impact of driver behavior for optimal acceleration efficiency by automatically controlling vehicle accelerations at a level for which the engine operates in its most efficient operational state for the current environment. Driver feedback information devices can also be implemented as a retrofit option for existing vehicles.
EO-2	Develop simple tools for the trucking industry that will provide estimates of the fuel savings potential of advanced efficiency technologies and technology combinations depending on specific usage information of a particular fleet (measured drive cycle data). The tools will provide cost and benefit analyses for the selection of technologies on a case-by-case basis when representative drive cycles for an individual fleet or owner-operator are available (and recommendations to the fleet for obtaining the drive cycles can be provided).
EO-3	Conduct a study to identify proposed ITS/connected vehicle technologies that offer significant fuel savings and quantify the reduction in fuel consumption for technologies that offer the greatest benefits. Select one technology, evaluate the benefits for fuel consumption as a function of market penetration and identify the infrastructure needs and costs for deployment of the technology to a level at which the benefits of vehicle-to-vehicle (V2V) and vehicle-to-infrastructure (V2I) networking are realized.
EO-4	Establish a real-world test corridor for commercial vehicles focused on improving commercial vehicle operations, including fuel efficiency. The test corridor should include DSRC and WiMax technologies in the infrastructure, would involve one or more fleets enabled for DSRC/WiMax capability and outfitted with various applications designed for improved efficiency of commercial vehicles.
EO-5	Explore regulatory changes to permit the replacement of body-mounted mirrors with a camera-based system and quantify the fuel saving benefits associated with such a change.
EO-6	Demonstrate the fuel savings benefits and develop policy guidelines for extending the use of long combination vehicles (LCVs), particularly triple trailer units.
EO-7	Promote improved supply chain management strategies in the commercial freight industry with an objective to increase the loads carried per truck and reduce vehicle miles traveled.

Appendix E

Acronyms

21CTP 21st Century Truck Partnership

ABT averaging, banking, and trading
ACC American Chemistry Council; also, adaptive cruise control
ACE advanced combustion engine
ACES Advanced Collaborative Emissions Study
AFCI alternative fuel compression ignition
AFV alternative fuel vehicles
AHSS advanced high-strength steel
AMOX ammonia oxidation
AMR Annual Merit Review (DOE)
AMT automated manual transmission
ANL Argonne National Laboratory
APEEM advanced power electronics and electric motors
API American Petroleum Institute
APS Advanced Photon Source
APU auxiliary power unit
ARRA American Recovery and Reinvestment Act of 2009
ASTM American Society for Testing and Materials (now ASTM International)
ATA American Trucking Association
ATP-LD Advanced Technology Powertrains for Light-Duty Vehicles
ATRI American Transportation Research Institute
AVPTA Advanced Vehicle Power Technology Alliance

bhp-hr brake horsepower-hour
BMEP brake mean effective pressure
BSM blind spot monitoring
BTE brake thermal efficiency

C_d coefficient of drag
C_{rr} coefficient of rolling resistance
CAFE corporate average fuel economy
CAGR compound annual growth rate
CAMP Crash Advoidance Metrics Partnership
CARB California Air Resources Board
CEC California Energy Commission
CFD computational fluid dynamics
CGVW combined gross vehicle weight
CH_4 methane
CI compression ignition
CLEERS Cross-cut Lean Exhaust Emissions Reduction Simulation
CMB collision mitigation braking
CNG compressed natural gas
CO carbon monoxide
CO_2 carbon dioxide
CPF catalyzed particulate filter
CR compression ratio
CRADA cooperative research and development agreement
CRC Coordinating Research Council
CVT continuously variable transmission

DCN derived cetane number
DCT dual clutch transmission
DEER Directions in Engine-Efficiency and Emissions Research (conference); also, Diesel Engine-Efficiency and Emissions Research (conference)
DEF diesel exhaust fuel
DME dimethyl ether
DOC diesel oxidation catalyst; also, U.S. Department of Commerce
DOD U.S. Department of Defense
DOE U.S. Department of Energy
DOT U.S. Department of Transportation
DPF diesel particulate filter
DSRC dedicated short-range communications

EATS	exhaust aftertreatment system
EBS	electronically controlled braking system
ECM	electronic control module
EDT	electric drive technologies
EERE	Energy Efficiency and Renewable Energy (DOE Office of)
EHN	ethylhexyl nitrate
EGR	exhaust gas recirculation
EIA	Energy Information Administration
EISA	Energy Independence and Security Act of 2007
EMA	Truck and Engine Manufacturers Association
EPA	U.S. Environmental Protection Agency
EPAct	Energy Policy Act of 2005 (P.L. 109-58)
ES	energy storage
ESC	electronic stability control
FACE	fuels for advanced combustion engines
FAME	fatty acid methyl ester
FC	fuel consumption
F-CAM	Forward Collision Avoidance and Mitigation
FCVT	FreedomCAR and Vehicle Technologies
FCW	forward collision warning
FE	fuel economy
FHWA	Federal Highway Administration
FMCSA	Federal Motor Carrier Safety Administration (DOT)
FMEP	friction mean effective pressure
FMVSS	Federal Motor Vehicle Safety Standards
FSP	friction stir processing
F-T	Fischer-Tropsch
FTA	Federal Transit Administration
FTP	federal test procedure
FY	fiscal year
g/bhp-hr	grams per brake horsepower-hour
GCW	gross combined weight
GDI	gasoline direct injection
GEM	GHG emissions model
GHG	greenhouse gas
g/mi	grams per mile
GPS	global positioning system
GPU	Graphical Processor Unit
GTL	gas-to-liquid
GVW	gross vehicle weight
HC	hydrocarbon
HCCI	homogeneous-charge compression ignition
HD	heavy duty
HEV	hybrid electric vehicle
HHDDT	heavy heavy-duty diesel truck
HMI	human–machine interface
hp	horsepower
HPCR	high-pressure common rail
HPL	high-pressure loop
HTHS	high-temperature, high-shear
HTML	High-Temperature Materials Laboratory
HTUF	High-Efficiency Truck Users Forum
HVAC	heating, ventilation, and air conditioning
HVIP	Hybrid and Zero-Emission Truck and Bus Voucher Incentive Program
ICE	internal combustion engine
ICME	integrated computational materials engineering
IPM	Intelligent Powertrain Management
IQT	ignition quality test
IR	idle reduction
ISG	Integrated Starter Generator
ITS	Intelligent Transportation Systems
KOH	potassium hydroxide
kWh	kilowatt-hour
LCFS	Low Carbon Fuel Standard
LCM	lane change/merge
LCV	long combination vehicle
LDV	light-duty vehicle
LDW	lane departure warning
LED	light-emitting diode
LES	large eddy simulation
LLFC	leaner lifted flame combustion
LLNL	Lawrence Livermore National Laboratory
LNG	liquefied natural gas
LOC	loss of control
LSFC	load-specific fuel consumption
LTC	low-temperature combustion
LTCC	Large Truck Crash Causation
LTGC	low-temperature gasoline combustion
MD	medium-duty
MHDVs	medium- and heavy-duty vehicles
MIT	Massachusetts Institute of Technology
mpg	miles per gallon
mph	miles per hour
MY	model year
NAAQS	National Ambient Air Quality Standards
NACFE	North American Council for Freight Efficiency
NETL	National Energy Technology Laboratory
NGV	natural gas vehicle
NGWBS	next-generation wide base single (tire)
NHTSA	National Highway Traffic Safety Administration
N_2O	nitrous oxides
NO_x	oxides of nitrogen
NPBF	non-petroleum–based fuels

NRC	National Research Council
NREL	National Renewable Energy Laboratory
NVH	noise, vibration, and harshness
OBD	on-board diagnostic
OECD	Organisation for Economic Co-operation and Development
OEM	original equipment manufacturer
OMB	Office of Management and Budget
ORC	organic Rankine cycle
ORNL	Oak Ridge National Laboratory
PACCAR	Pacific Car and Foundry Company
PCCI	premixed charge compression ignition
PCI	passenger compartment intrusion
PCP	peak cylinder pressure
PDF	probability distribution function
PM	particulate matter
PMC	Project Management Center
PN	particle number
PNGV	Partnership for a New Generation of Vehicles
PNNL	Pacific Northwest National Laboratory
PPCI	partially premixed charge compression ignition
R&D	research and development
RCCI	reactivity-controlled compression ignition
RD&D	research, development, and demonstration
RF	radio frequency
RFS	Renewable Fuels Standard
rpm	revolutions per minute
RSC	roll stability control
RSP	renewable super premium
SAE	Society of Automotive Engineers
SC	single cylinder
SCAQMD	South Coast Air Quality Management District
SCFI	Stronach Center for Innovation
SCR	selective catalytic reduction
SET	Supplemental Emission Test
SI	spark-ignition
SMC	sheet molding compound
SOFC	solid oxide fuel cell
STEP	ShorePower Truck Electrification Project (DOE)
SUT	single unit truck
SUV	sport utility vehicle
TARDEC	Tank-Automotive Research, Development and Engineering Center
TCFB	Texas Clean Fleet Program
TEM	transmission electron microscopy
TPMS	tire pressure monitoring system
TRU	transportation refrigeration unit
UMTRI	University of Michigan Transportation Research Institute
U.S. DRIVE	U.S. Driving Research and Innovation for Vehicle Efficiency and Energy Sustainability
V2I	vehicle-to-infrastructure
V2V	vehicle-to-vehicle
VAM	Vehicle Acceleration Management
VERIFI	Virtual Engine Research Institute and Fuels Initiative
VG	variable geometry
VIUS	Vehicle Inventory and Use Survey
VMT	vehicle miles traveled
VSI	Vehicle Systems Integration (ORNL)
VSST	vehicle systems simulation and testing
VTO	Vehicles Technology Office (DOE)
WAAS	weighted aerodynamic-average speed
WHR	waste heat recovery
WIM	weigh-in-motion
WTW	well to wheel